Europe – On Air

Europe – On Air

Interwar Projects for Radio Broadcasting

Suzanne Lommers

Amsterdam 2012

The research has been made possible by: Netherlands Organisation for Scientific Research (NWO). This publication is made possible by: Eindhoven University of Technology, Foundation for the History of Technology and the Unger – Van Brero Fund.

ISBN 978 90 8964 435 0
e-ISBN 978 90 4851 665 0
e-ISBN 978 90 4851 666 7 (e-pub)

Design and typesetting: Ellen Bouma, Alkmaar, the Netherlands

Cover image: In 1936 European broadcasting experts working for the International Broadcasting Union (IBU) travelled to the United States to visit their American colleagues from the National Broadcasting Company (NBC). In the light of NBC's tenth anniversary, the IBU experts visited the station and shared their knowledge. As part of the visit, the European broadcasters participated in a live NBC broadcast. Source: Box 58 Photographies, IBU. Used by the courtesy of the European Broadcasting Union

Amsterdam University Press, Herengracht 221, NL-1016 BG Amsterdam
www.aup.nl

Acknowledgements

Sometimes things just fall into place. This is particularly true for the study in front of you. When I was young, I sometimes accompanied my father, who is an innovator, on his visits to clients. These clients were often working on technical projects in industrial areas that greatly resemble the campus at Eindhoven University of Technology in the Netherlands. Walking across the campus to attend my job interview, I immediately experienced a sense of déjà vu. In the following years I examined the connection between broadcasting activities and projects that aimed to unite Europe. Often I had to think of a postcard that I once bought at the age of sixteen and still have in my possession today. The card quotes Albert Einstein and reads: *Peace cannot be kept by force. It can only be achieved by understanding.* Who would have guessed that both the man and his thought would ever have a place in my dissertation research more than fifteen years later? That quote keeps on inspiring me. I hope this study, which combines these childhood memories with my personal interests in people, culture, music, and identity formation, will inspire you too.

My years at Eindhoven University of Technology have been an eye-opener in more ways than one. I came to Eindhoven with a background in economics and history. The combined practice of history of technology, science and technology studies, and philosophy has broadened my view on research and on life. I would like to thank the Netherlands Organization for Scientific Research who funded the Transnational Infrastructures and the Rise of Contemporary Europe project (TIE) of which my research is a part. I also thank Eindhoven University of Technology, the Unger-Van Brero Fund, and the Foundation for the History of Technology for their generous support that enabled me to publish this book.

I am particularly grateful to TIE project leader and supervisor Johan Schot. His guidance, his dedication, his enthusiasm and his faith in me have been inestimably valuable. Johan is an exceptional academic. Yet to me, he is above all an exceptional person. I also cherish the discussions I had with my co-supervisor Erik van der Vleuten. His expertise in writing and his precision when it comes to concepts regularly drove me crazy, but were invaluable.

It has been my pleasure to be part of the TIE project. Collaborating with people like Alexander Badenoch, Vincent Lagendijk, Frank Schipper, Irene Anastasiadou and our fantastic secretaries Lidwien Hollanders-Kuipers and Iris Custers-Houx

gives an impetus to scientific debate in a way that could never happen in an individual research environment. Ruth Oldenziel's enthusiastic support and advice have been invaluable for rounding out the manuscript for publication. I am grateful to Jan Korsten for his help with the publication process. I thank Vincent Lagendijk for editing the figures in this book, and I owe Alison Britton and Val Kidd for editing my texts. My special thanks go to Alexander Badenoch with whom I have shared numerous good times in archives and cafes in Geneva. What is more, without his advice on broadcasting-related matters, this book would not be what it is now. Alec is my unofficial second co-supervisor.

Historical research on international events requires extensive archive research. My many visits to Geneva have created lasting memories. I especially thank the European Broadcasting Union for allowing me to use their archives, for welcoming me with open arms, and for sharing with me their expertise. I am indebted to former Secretary-General Jean Réveillon, Director of the Radio Department Raina Konstantinova, Head of Communications Avril Mahon-Roberts, former Director of the Legal Department Werner Rumphorst as well as Armi Heikkinen, Eric Piraux, Georges Béry, Frans de Jong, Mathias Coinchon, David Wood, Pierre Duret, Pierre-Yves Tribolet, and all the girls in the communications department. Apart from the EBU, I have used the archives and library of the International Telecommunication Union. My thanks go to Heather Heywood and Kristine Clara who have done everything in their power to help me, even at a distance. For my visits to the League of Nations Archives, I thank Bernhardine Pejovic and Jacques Oberson. Finally, I could not have done without the archives of the Netherlands Institute for Sound and Vision. My gratitude goes to Erik Kappetijn for his enthusiasm and his endless trips from the basement to the study corners with cart upon cart full of program guides.

My participation in the Tensions of Europe network has further enhanced my thinking over the years. I found it particularly stimulating to be involved in the establishment of the Transmitting and Receiving Europe group. I thank Andreas Fickers for our discussions over the years and his suggestions on contextualization. I have enjoyed the numerous occasions where Nina Wormbs shared with me her insights on frequency allocation. I also thank Jennifer Spohrer for our correspondence on Radio Luxembourg and Karin Bijsterfeld for our discussions on the interwar canon for music. Last but not least, I am indebted to Christian Henrich-Franke, Alexander Badenoch, Christoph Classen, and Andy O'Dwyer for their enthusiasm, advice, and debate.

For me, the Tensions network is about excellent research, wonderful friendships, great parties, but above all about the *Tensations*. I thank my fellow singers

Judith Schueler and Alexander Badenoch, our drummer Andreas Fickers, our washtub base player Nil Disco, our sax player Donna Mehos, our keyboard player Frank Schipper, and our base player Paul Edwards for sharing with me the best months of my PhD research!

Nothing is more important than the love and support I receive from my friends and family every day. I have the honor to be surrounded by beautiful and caring people. Dear Jo-Anne, Brechje, Renske, Inge, Ton, Kirsten, Bram, Sybren, Marjolijn, Martijn, Yolanda, Miranda, Sónia and Marieke, I am so happy to have you in my life. I would never have been where I am today without my parents Rob and Nadine and two of the strongest and most sensitive women I have ever known: oma Jowie and oma Miepel. The strength and sensibility that characterized my grandmothers made me who I am. Yet, my parents are the firm pillars in my life. What their love, support and trust mean to me cannot be expressed in words. I dedicate this book to them.

Utrecht, December 6, 2011

Table of Contents

List of Illustrations

Tables

Figures

Abbreviations

ABC	Australian Broadcasting Company
AVRO	Algemeene Vereeniging Radio Omroep
BBC	British Broadcasting Company / British Broadcasting Corporation
BCIR	Bureau Catholique International de la Radiodiffusion
CBS	Columbia Broadcasting System
CCIF	International Telephone Consultative Committee
CCIR	International Radio-electricity Consultative Committee
CCIT	International Telegraph Consultative Committee
CIR	Comité International de la T.S.F, also known as Comité International Radioélectricité
CIRR	Centre International de Radiophonie Rurale
CLR	Compagnie luxembourgeoise de radiodiffusion, 1929-1953
CSF	Compagnie générale de télégraphie sans filistres
EBU	European Broadcasting Union
FRC	Federal Radio Committee
IARU	International Amateur Radio Union
IBC	International Broadcasting Company
IBU	International Broadcasting Union
ICIC	International Committee on Intellectual Co-Operation
ICW	International Council of Women
IECI	International Educational Cinematograph Institute
IIIC	International Institute on Intellectual Co-Operation
IRU	International Radiotelegraph Union
ITU	International Telegraph Union / International Telecommunication Union
KRO	Katholieke Radio Omroep
LoN	League of Nations
NAB	U.S. National Association of Broadcasters
NBC	National Broadcasting Company
NCRV	Nederlandsche Christelijke Radio-Vereeniging
NKPiT	Commissariat of Posts and Telegraphs of the USSR
NSF	Nederlandse Seintoestellen Fabriek
OCI	Intellectual Co-Operation Organization
OCT	Organization for Communications and Transit
OIRT	Organisation Internationale de Radio et Télévision / International Organization for Radio and Television
RCA	Radio Corporation of America
RRG	Reichs Rundfunk Gesellschaft m.b.H.
SLER	Société luxembourgeoise des études radioélectriques / Luxembourgeois Radio-Electric Exploratory Company
SSR	Société Suisse de Radiodiffusion
TSF	Télégraphie sans fils
UIR	Union Internationale de Radiophonie / Union Internationale de Radiodiffusion
UN	United Nations
UNESCO	United Nations Educational, Scientific and Cultural Organization
VARA	Vereeniging van Arbeiders Radio Amateurs
VPRO	Vrijzinnig Protestantse Radio Omroep

Chapter 1 Introduction
European Broadcasting Visions

In 1944 the British, Belgian and Polish engineers Peter Pendleton Eckersley, Auguste Hubert and B. Tenenbaum together sought to re-design the existing but by then paralyzed organization of broadcasting in Europe.[1] The world was at war and the Allied and Axis powers had been fighting one another with the help of modern technologies like radio broadcasting, wireless telephony, and telegraphy. Eckersley, Hubert, and Tenenbaum proposed a ready-made plan for a new start for broadcasting directly after the end of the war. Their "European Broadcasting Alliance" plan described in detail a broadcasting infrastructure that assured free expression to the various rival groups in Europe.[2] The three men chose the name of their plan with care. It expressed their belief in a diversity of opinions, cultures, and peoples, while pushing for an alliance between Allied and Axis powers that were using broadcasting to attack each other. The combination of diversity and alliance formed the core of the European Broadcasting Alliance plan.

The plan claimed that diversity is served and alliance is built "...if every European could be made conscious of his status as a European – a member of a nation within a wider group of nations."[3] This suggests that Eckersley, Hubert, and Tenenbaum considered it important that radio listeners in Europe were already aware, or would become aware, of their participation and inclusion in European society. In their eyes, broadcasting had an important and powerful role to play in creating this awareness. They therefore designed an internationally owned and operated broadcasting infrastructure that covered all corners of Europe.[4]

1 The designers of the plan had impressive international curricula vitae. Between 1919 and 1923 Eckersley was head of the experimental section of the design department at the Marconi Wireless Telegraph Company, and from 1923 to 1929 the first Chief Engineer of the BBC. Both at the BBC and internationally he widely influenced technological broadcasting developments. Hubert was a Belgian broadcasting pioneer and administrator of Radio-Belgique S.A. He was associated with SAIT (a Belgian Operating Company), was President of the International Radio Maritime Committee, and honorary member of the International Broadcasting Union. Tenenbaum was a Polish radio engineer, of whom little is known. S.n., "Mr. P. P. Eckersley: First Chief Engineer of the B.B.C.," *The Times*, March 15, 1963; Asa Briggs, "Peter Pendleton Eckersley," *Oxford Dictionary of National Biographies*, 2008, http://www.oxforddnb.com/view/article/37385; S.n., "Real International Broadcasting: Proposals for a European Alliance," *Wireless World* 50, no. 8 (1944): 234-235.

2 ITU, "Un réseau international de radiodiffusion en Europe," *Journal des Télécommunications* 12, no. 5 (1945): 62.

3 S.n., "Real International Broadcasting," 234.

4 ITU, "Un réseau international de radiodiffusion en Europe"; S.n., "Real International Broadcasting."

The European Broadcasting Alliance plan was for an international European broadcasting network *parallel* to and explicitly separate from the national broadcasting systems already in place by the 1940s (Figure 1.1). The proposed international network consisted of a web of interconnected landlines, with Bern, Brussels, and Vienna as the main so-called International Radio Centers. Via these international landlines, broadcasters could easily exchange programs that at their destination would be broadcast to listeners by means of a coupling with the wireless national networks. Such a network enabled listeners everywhere in Europe to follow international events live via their radio sets. An immense short wave connection at Tangiers in Africa would in turn link the European network, which included Soviet Russia and Turkey, with the rest of the world, assuring the global diffusion of selected European programs. In the eyes of Eckersley, Hubert, and Tenenbaum, the international service should have precedence at all times over the national wireless services. They thus consciously foregrounded the international part of their plan, and their ideas of a European consciousness, above the national part.

The European Broadcasting Alliance plan circulated among governments and the international telecommunication community, but eventually failed. When the British government officially turned down the plan, British manufacturers had already proposed another plan.[5] In this – "Post War European Broadcasting" – the manufacturers emphasized the inaccuracy of the pre-war broadcasting system that in their eyes had been heavily influenced by political considerations.[6] The allocation of frequencies on technical grounds should make it possible to avoid quarrelling over "national" channels. Frequencies should be allocated to stations on the basis of their fading radius, irrespective of nation-state boundaries. The power of the existing stations and the way their signals faded away as a result of geophysical circumstances, would then define the borders of Europe's nations.[7] In such a way, technology rather than political, ideological, or cultural visions defined the boundaries of European countries.

5 It remains unclear why the 1944 plan failed, and was turned down by the British government. There is reason to believe that Eckersley's reputation had become or had already been dubious for a while. In 1931 he and his new wife Frances Dorothy joined Oswald's Mosley's party. From 1935 onwards they spent their holidays in Germany "where they admired the progress that Germany was making under national socialism." His wife stayed in Germany after the outbreak of war and would be put on trial in Britain after the war for her international radio activities for Germany. ITU, "Un réseau international de radiodiffusion en Europe"; Asa Briggs, *The History of Broadcasting in the United Kingdom, Volume III: The War of Words* (Oxford: Oxford University Press, 1995), 127n. Quoted from Briggs, "Peter Pendleton Eckersley."

6 S.n., "Monthly Commentary: The Industry's Plan for Broadcasting," *Wireless World* 51, no. 9 (1945): 257.

7 S.n., "Plan for Europe: Broadcasting Allocations on Engineering Principles," *Wireless World* 51, no. 9 (1945): 258.

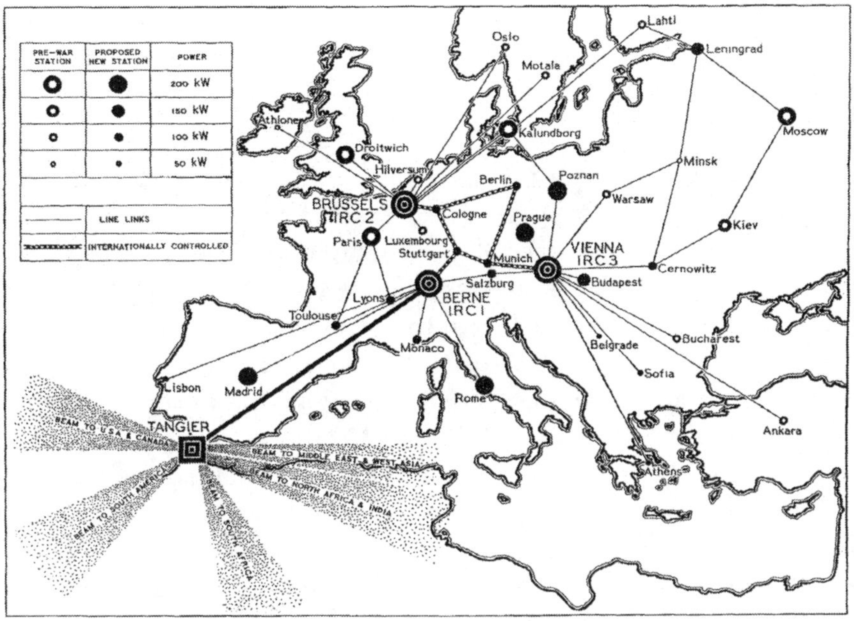

Figure 1.1 – The 1944 European Broadcasting Alliance proposal
Source: "Un réseau international de radiodiffusion en Europe," Journal des Télécommunications 12, no. 5 (1945), 63.

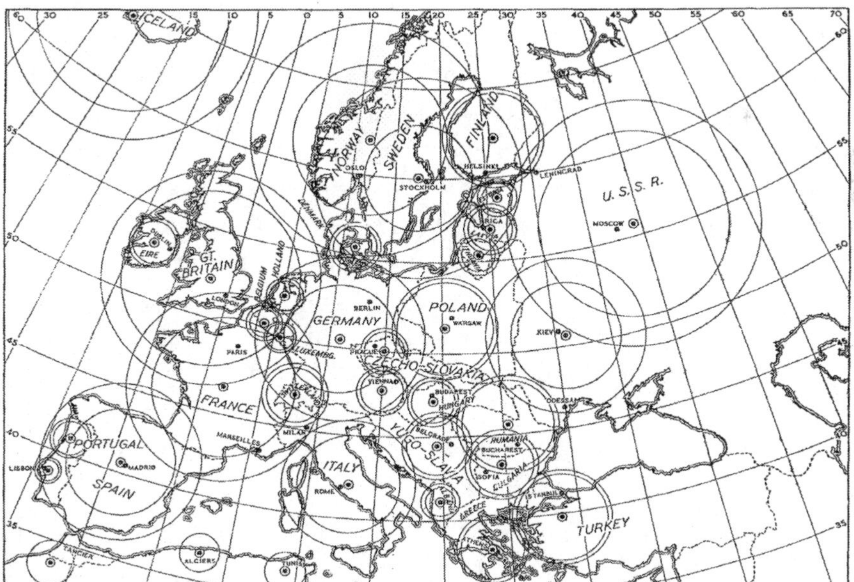

Figure 1.2 – Post War European Broadcasting Proposal of 1945
Source: S.n., "Plan for Europe: Broadcasting Allocations on Engineering Principles," Wireless World 51, 9 (1945), 261.

The manufacturers' approach to broadcasting was technical, bearing in mind their economic motivation. The war was having a considerable impact on their activities. They not only experienced a shortage of labor and materials, but also "a lack of detailed knowledge on conditions of use for which their productions must be designed."[8] A sub-committee of the British Radio Equipment Manufacturers' Association designed a plan to solve these war-driven problems. The Radio Industry Council, who finally issued the plan, was trying to avoid large discrepancies between post-war developments and pre-war technology. It wanted to guarantee the accuracy of the existing radio sets which had already been sold.[9] The plan encouraged re-allocating the infrastructure of wireless long and medium wave broadcasting which had been disturbed by the war. This infrastructure was already in place in Europe, but would be modified via a different approach to that used in the interwar years (Figure 1.2). The plan tried to resolve in the best possible way the economic slump, warranting the production and market share of the British radio industry.

The 1945 plan took no consideration of European unification. It favored the national above the European despite focusing on all of Europe geographically. Every country would have to be provided with "two 'national' programmes, as well as a system of localised transmissions to suit the special needs of any important regions or language groups in each country."[10] The manufacturers thus stressed the need to concentrate on the national as well as the regional and local communities within nation-states. Furthermore, a focus on the technical principles of broadcasting would enhance the quality of reception, provide a solution to interference, and would assure reception of foreign stations for each listener. Therefore listeners could only receive national programs from those foreign stations whose signals they managed to receive. The plan thus did not stipulate the need for reception of a station from each country in Europe, which also included Soviet Russia, Turkey, and all countries bordering the Mediterranean. The British manufacturers' plan made the national and local significantly more important than international reception or a European broadcasting service. The 1944 and 1945 plans thus proposed a European agenda of a very different nature, making choices that related the creation of broadcasting to European unification efforts in a different way.

8 S.n., "Monthly Commentary," 257.

9 Briggs shows how the Radio Industry Council in the area of television informed the Television Advisory Committee directly after the war "that it did not favour even minor changes in frequencies if they would mean that tuning changes had to be made to existing sets." Briggs, *The War of Words*, quote 181, 169, 173-174.

10 S.n,. "Plan for Europe," 258.

The 1944 and 1945 plans were the only two which spread to all corners of Europe with the aim of resolving the stalemate in broadcasting caused by the Second World War. Like the 1944 plan, for unknown reasons the 1945 plan failed as well. After the war, broadcasting organizations chose a third option, namely the establishment of two international non-governmental organizations who coordinated an international collaboration of national broadcasters. The Eastern European *Organisation Internationale de Radiodiffusion* (OIR, later to become OIRT when it started covering television as well) and the Western European Broadcasting Union continued the pre-war broadcasting of international collaboration coordinated by the International Broadcasting Union (IBU), albeit with some modifications.[11] After the war, the EBU and the OIRT and their international networks, Intervision and Eurovision, divided the European landscape into an eastern and a western European bloc.[12] On January 1, 1993, this bloc was dissolved when the OIRT merged into the EBU, maintaining the name of the EBU and the Eurovision network.[13]

Like the 1944 and 1945 plans for post-war broadcasting, the EBU initiative made specific choices. These were stressed on the occasion of its twenty-fifth anniversary. The EBU projected its work as a simple programming and engineering job taking place in an environment of pragmatism. According to EBU president Sir Charles Curran, it was "by the scrupulous avoidance of ideologies that we have managed to build a cohesive organization in which the common interest serves to promote mutual tolerance and differences."[14] His note should be seen within the context of the *War of the Black Heavens* that was the Cold War.[15] Curran suggests that the EBU tried to stay outside this war by using a non-political and non-ideological approach but his remark nevertheless was political and ideological. The EBU saw itself as an expert organization that connected national differences to

11 For a discussion on the whereabouts of the development of the OIRT and EBU see: Andreas Fickers and Suzanne Lommers, "Eventing Europe: Broadcasting and the Mediated Performances of Europe," in *Materializing Europe: Transnational Infrastructures and the Project of Europe*, ed. Andreas Fickers and Alexander Badenoch (London: Palgrave, 2010), 236-237; Léo Wallenborn, "From IBU to EBU: The Great European Broadcasting Crisis (Part I)," *EBU-Review* 29 (January 1978): 25-34; Léo Wallenborn, "From IBU to EBU: The Great European Broadcasting Crisis (Part II)," *EBU-Review* 29 (March 1978): 22-30; Charles E. Sherman, "Turmoil and Transition in International Broadcasting Organizations, 1938-50," *Journal of Broadcasting* 15, no. 3 (1971): 265-273.

12 Rüdiger Zeller, *Die EBU – Union Européene de Radio-Télévision (UER) – European Broadcasting Union (EBU): Internationale Rundfunkkooperation im Wandel* (Baden-Baden: Nomos Verlagsgesellschaft, 1999), 38.

13 S.n., "Merger: 1 January 1993: A New Era for Eurovision," *EBU Dossiers: 50 Years of Eurovision*, no. 1 (2004): 30-33.

14 EBU, *25 Years European Broadcasting Union* (Geneva: EBU, 1974), 3.

15 Michael Nelson, *War of the Black Heavens: The Battles of Western Broadcasting in the Cold War* (London; Washington: Brassey's, 1997).

common interests. It fostered the idea of a Europe of peacefully co-existing nation-states with differing traditions, political systems, and economic structures.[16] The EBU combined the focus on building Europe through broadcasting as contained in the Broadcasting Alliance plan, emphasizing the technical aspects in the British manufacturers plan.

The intriguing question is why the two failed plans and the EBU related broadcasting initiative refer so explicitly to European unification and to the importance of a technical approach? The answer is that they were building on a rich interwar legacy in which both aspects were central. After all, the EBU took over the activities of the IBU created in 1925. This book can be read as a history of the IBU, but it is more than that. It focuses on initiatives to develop European broadcasting infrastructures. Infrastructure initiatives, not only for broadcasting but also for transport, energy and other communication technologies, are routinely ignored in economic, political and cultural histories of Europe.[17] In many of these studies, infrastructures are assumed to be there, reducing distance in time and space.[18] Historians often perceive infrastructures as background factors. Thomas Misa and Johan Schot have argued that they should be foregrounded, and studied as vectors for a process they call "the hidden integration of Europe." In this process, technical organizations such as the IBU explicitly try to contribute to the construction of Europe through the development of European standards and networks. In this process they follow a strategy of technification or depolitization.[19]

The central question in *Europe – On Air* is how and why those involved in transnational broadcasting tried to contribute to European unification in the interwar years. European unification can refer to any project or process, whether social, cultural, political, and/or economic, which actors articulated in European

---

16 EBU, *25 Years European Broadcasting Union*, 4.

17 Some examples of histories that focus on other areas are: Eric Bussière, *La France, la Belgique et l'organisation économique de l'Europe, 1918-1935* (Paris: Comité pour l'histoire économique et financière de la France, 1992); Patricia Clavin, *The Great Depression in Europe, 1929-1939* (New York, N.Y.: St. Martin's Press, 2000); Norman Davies, *Europe: A History* (London: Pimlico, 1997); Paul Dukes, ed., *Frontiers of European culture* (Lewiston: E. Mellen Press, 1996); David W. Ellwood, *Rebuilding Europe: Western Europe, America and Postwar Reconstruction* (London; New York: Longman, 1992); Mary Fulbrook, ed., *Europe since 1945* (Oxford; New York: Oxford University Press, 2001); Heikki Mikkeli, *Europe as an Idea and an Identity* (New York: St. Martin's Press, 1998); Aviel Roshwald and Richard Stites, eds., *European Culture in the Great War: The Arts, Entertainment, and Propaganda, 1914-1918* (Cambridge: Cambridge University Press, 1999).

18 A noteworthy early exception is Albert Carreras, Andrea Giuntini, and Michèle Merger, eds., *European Networks, 19th – 20th Centuries: New Approaches to the Formation of a Transnational Transport and Communications System*, B8 Proceedings Eleventh International Economic History Congres Milan, September 1994 (Milan: Università Bocconi, 1994).

19 Thomas Misa and Johan Schot, "Inventing Europe: Technology and the Hidden Integration of Europe. Introduction," *History and Technology* 21, no. 1 (2005): 1-22.

terms. I will investigate the main question by focusing on five aspects: building new institutions such as the IBU; the creation of networks and standards; the rivalry amongst various broadcasting organizations in Europe; the attempt to save civilization by means of broadcasting; and finally the making and distribution of specific international programs. These five aspects represent the key areas where broadcasters and other interest groups made crucial choices that shaped the relationship between broadcasting and the process of European unification during the interwar years. The question remains, how do we research such a relationship?

Unraveling European Broadcasting

By following various broadcasting planners, this study examines broadcasting as an infrastructure. At first glance, people often relate broadcasting to entertainment and programs rather than to questions of infrastructure.[20] The Palgrave Dictionary of Transnational History however defines broadcasting as a "means to send out sound and/or (motion) pictures by means of radio waves through space for reception by the general public."[21] Broadcasting serves as a point-to-mass medium connecting one point to a large mass of people within a circular area like the wireless connection of stations to home sets via long, medium and short waves. On occasion, broadcasting also connects point A to point B via a network such as Eurovision. *Europe – On Air* therefore regards broadcasting as a spatial and connective technology which joins up various places in Europe and by doing so

20 For a historical evolution of the concept of infrastructures see: Dirk van Laak, "Infra-Strukturgeschichte," *Geschichte und Gesellschaft* 27 (2001): 367-393; Dirk van Laak, "Der Begriff „Infrastruktur" und was er vor seiner Erfindung besagte," *Archiv für Begriffsgeschichte* 41 (1999): 280-294; Dirk van Laak, "Technological Infrastructure, Concepts and Consequences," *ICON Journal of the International Committee for the History of Technology* 10 (2004): 53-64.

21 Andreas Fickers, "Broadcasting," in *The Palgrave Dictionary of Transnational History*, ed. Pierre-Yves Saunier and Akira Iriye (Basingstoke; New York: Palgrave Macmillan, 2009), 106. Initially radio referred to wireless point-to-point communication, like wireless telegraphy, also known as radiotelegraphy or radiotelephony. Later radio became a means of point-to-mass communication, connecting one point to a mass of people. For a historical development of the definition of "radio" see: Susan Douglas, *Inventing American Broadcasting, 1899-1922* (Baltimore: Johns Hopkins University Press, 1988), viii-xix. As the definition in the Palgrave dictionary shows broadcasting nowadays includes television as well. Television developed in a number of countries during the interwar years and provisions were made internationally to create standards that would facilitate cross-border experimentation. The main cross-border experiments with television would take place after the Second World War and thus falls outside the scope of this research. For the first multi-national cross-border experiments with television see: Andreas Fickers and Suzanne Lommers, "Eventing Europe: Broadcasting and the Mediated Performances of Europe," in *Materializing Europe: Transnational Infrastructures and the Project of Europe*, ed. Andreas Fickers and Alexander Badenoch (London: Palgrave, 2010), 225-251.

creates a specific European geography of connected and not-connected places.[22]

In order to unravel the relationship between infrastructures and European unification efforts further, *Europe – On Air* follows Thomas Hughes, and defines infrastructures as socio-technical systems.[23] Socio-technical systems are mixed systems. They consist among other things of technological and institutional, organizational, and legislative components.[24] Constructing a European broadcasting infrastructure implies, therefore, creating European institutions and regulations which match. All these components influence relationships between people. The actors involved in the project to construct a European broadcasting infrastructure were thus also engaged in unifying people in Europe, and they knew it. How the various components should fit, however, was not always clear from the start.

As the introductory plans show, infrastructure construction is an actor-driven process with competing designs.[25] Actors have to negotiate their designs and so the way they construct Europe. Hughes calls these actors system builders.[26] Organizations like the IBU and the EBU are such system builders. Individuals like the engineers Eckersley and Hubert or other interest groups like the British radio industry who designed the 1944 and 1945 plans are also such system builders. This study departs from the idea that organizations such as the International Broadcasting Union were Europe's system builders. They set up and formed *arenas* where various actors negotiated the future of transnational broadcasting and Europe simultaneously.[27] For this reason the IBU is an excellent site for answering my research question.

---

22 Michael G. Müller and Cornelius Torp, "Conceptualising Transnational Spaces in History," *European Review of History: Revue européenne d'histoire* 16, no. 5 (2009): 613-614.

23 Hughes examined electricity networks as socio-technical systems. His *Networks of Power* resulted in a new field of study that focused on Large Technical Systems (LTS) that not only concentrated on electricity, but on a wide variety of historical developments. For further reading see: Thomas Hughes, *Networks of Power: Electrification in Western Society, 1880-1930* (Baltimore: Johns Hopkins University Press, 1983); Renate Mayntz and Thomas Hughes, *The Development of Large Technical Systems* (Frankfurt am Main: Campus, 1988); Erik van der Vleuten, "Infrastructures and Societal Change: A View from the Large Technical System Field," *Technology Analysis & Strategic Management* 16, no. 3 (2004): 395-414; Erik van der Vleuten, "Understanding Network Societies: Two Decades of Large Technical System Studies," in *Networking Europe: Transnational Infrastructures and the Shaping of Europe, 1850-2000*, ed. Erik van der Vleuten and Arne Kaijser (Sagamore Beach: Science History Publications/USA, 2006), 279-314.

24 Thomas Hughes, "The Evolution of Large Technical Systems," in *The Social Construction of Technological Systems: New Directions in the Sociology and History of Technology*, ed. Wiebe Bijker, Thomas Hughes, and Trevor Pinch (Cambridge: MIT Press, 1987), 51-82.

25 Erik van der Vleuten and Arne Kaijser, "Prologue and Introduction: Transnational Networks and the Shaping of Contemporary Europe," in *Networking Europe: Transnational Infrastructures and the Shaping of Europe, 1850-2000*, ed. Erik van der Vleuten and Arne Kaijser (Sagamore Beach: Science History Publications/USA, 2006), 1-22; Van der Vleuten, "Understanding Network Societies," 308-309.

26 Hughes, "The Evolution of Large Technical Systems," 25-26.

27 Erik van der Vleuten et al., "Europe's System Builders: The Contested Shaping of Transnational Road, Electricity and Rail Networks," *Contemporary European History* 16, no. 3 (2007): 328.

It is important to emphasize one important characteristic of these system builders. They have not only been set up by experts, they also bring together experts and help to develop expert communities. Experts are professionals with a recognized expertise and competence in a particular domain and an authoritative claim to knowledge relevant for governments that want to regulate this domain.[28] They attend conferences and meetings of various governmental and non-governmental organizations that include social events. During these meetings they often developed strong informal and personal relationships that surpass national and institutional boundaries, and create what Christian Henrich Franke calls "a transnational culture" amongst experts.[29] The concept of expert communities therefore not only emphasizes the formal, but also the personal and informal aspect of these arenas.

Europe – On Air is a transnational history of broadcasting in a two-fold sense.[30] Firstly, it replaces state-centered history with one that includes state and non-state actors, such as "multinational corporations, religious organizations, regional communities, transnational private organizations, and even stateless persons."[31] Secondly the benefit of a transnational approach is that it does not privilege a specific spatial frame. It brings together in one setting the global, European, national, and local. This allows us to explore the relationships between these levels and their mutual entwinements. Such an approach implies that creating a European broadcasting space will be put in the context of the simultaneously constructed local, national and global spaces. Europe then becomes a constructive place for "a more precise interrogation of other, more essentialized, scales of the 'national' and the 'global.'"[32]

In summary: by examining the efforts of the actors involved in transnational broadcasting in their aim to achieve European unification, *Europe – On Air*

28 Peter M. Haas, "Introduction: Epistemic Communities and International Policy Coordination," *International Organization* 46, no. 1 (1992): 1-35. See also: Katrin Steffen and Martin Kohlrausch, "The Limits and Merits of Internationalism: Experts, the State and the International Community in Poland in the First Half of the Twentieth Century," *European Review of History: Revue européenne d'histoire* 16, no. 5 (2009): 716.

29 Christian Henrich-Franke, "Cookies for ITU: The Role of Social Networks in Standardization Processes," in *Bargaining Norms – Arguing Standards: Negotiating Technical Standards*, ed. Judith Schueler, Andreas Fickers, and Anique Hommels (The Hague: Stichting Toekomstbeeld der Techniek, 2008), 89.

30 Two articles that critically discuss and further explore the history of transnationalism are: Pierre-Yves Saunier, "Learning by Doing: Notes about the Making of the Palgrave Dictionary of Transnational History," *Journal of Modern Economic History* 6, no. 2 (2008): 159-179; Erik van der Vleuten, "Toward a Transnational History of Technology: Meanings, Promises, Pitfalls," *Technology and Culture* 49, no. 4 (2008): 974-994.

31 Akira Iriye, *Global Community: The Role of International Organizations in the Making of the Contemporary World* (Berkeley: University of California Press, 2002), vii-viii.

32 Alexander Badenoch and Andreas Fickers, "Introduction: Europe Materializing? Toward a Transnational History of European Infrastructures," in *Materializing Europe: Transnational Infrastructures and the Project of Europe*, ed. Alexander Badenoch and Andreas Fickers (London: Palgrave, 2010), 9.

studies transnational broadcasting as a socio-technical system with a specific negotiated spatial reach, and organizational structure. It perceives the International Broadcasting Union as one of the main international system builders of the time. The organization provided an arena and formed part of an expert community that negotiated spatial and functional relationships between broadcasting and various forms of European unification. This research strategy implies that alongside negotiations and relationships between actors, my book should also examine their motivations and visions. Ideas, visions and motivations were significant aspects of infrastructure design and unification efforts. Eckersley was aware of this aspect of his work. In his memoires he describes himself as "an inventor of mechanisms to serve ideas."[33] Both the mechanisms and ideas receive ample attention as building blocks for a "hidden" creation of European unification via broadcasting infrastructure development.

Transnational Infrastructures in Europe

This book is one of the products of the research program Transnational Infrastructures and the Rise of Contemporary Europe (TIE). The program began in 2002. It was adopted by the Tensions of Europe scholarly network as part of its research programs to examine the "hidden integration of Europe." The TIE project examined transnational infrastructures such as railroads, highways, air traffic, tunnels, electricity lines and communication networks as arenas that shaped, negotiated, and contested Europe in the twentieth century.[34] *Europe – On Air* squarely follows this tradition. In this section I summarize the findings that are relevant to my research. On top of that, I explore the history of broadcasting, albeit in a limited sense. I concentrate on studies which address the theme of my book directly.

TIE has approached infrastructure construction as non-linear and transnational processes with a special focus on human agency, on system builders.[35] The construction and use of infrastructures went hand in hand with the creation of a host of new international actors and institutions. In the course of the nineteenth

33 Peter P. Eckersley, *The Power Behind the Microphone* (London: J. Cape, 1941), 21.
34 The Transnational Infrastructures and the Rise of Contemporary Europe project has been funded by the Netherlands Organization for Scientific Research (NWO) in September 2002, Dossier number: 2777-53-001. For further information see: www.tie-project.nl where you can find a summary of the project: Johan Schot, "Transnational Infrastructures and the Rise of Contemporary Europe" (Working document no. 1, www.tie-project.nl, 2003). For details on Tensions of Europe see: www.tensionsofeurope.eu.
35 Van der Vleuten and Kaijser, "Prologue and Introduction"; Van der Vleuten, "Understanding Network Societies," 308-309.

and twentieth centuries, a wide variety of people, international governmental as well as non-governmental organizations have worked on infrastructure projects.[36] A wide number of international expert and often non-governmental international agencies were established to deal with infrastructure related problems that could not be resolved at the national level. In addition, TIE finds a broad involvement of international governmental organizations. In the course of the interwar years in particular, the technical committees of the League of Nations, like the Organization for Communications & Transit, focused on infrastructure development.[37] These international organizations were Europe's system builders, setting up arenas where infrastructures were negotiated in relation to Europe.[38]

The construction and use of these transnational infrastructures have created a specific philosophy on how to work together internationally in Europe. Johan Schot and Vincent Lagendijk label this philosophy technocratic internationalism.[39] At its basis lies first of all the idea that once they had established a connection, experts thought that their technical links would automatically create feelings of peace and understanding. They would serve public aims. In this tradition the League of Nations aimed to establish Radio Nations in the 1930s.[40] Experts believed in the myth of the network. Secondly, such network construction should preferably happen outside the sphere of nation-oriented state power politics. Only by "technifying" their discussions would experts be able to shape these peace-pursuing networks internationally in a way they considered suitable. Thirdly and finally, experts "worked in a range of competing and collaborating settings."[41] They

36 TIE has resulted in a number of detailed European transnational infrastructure histories: Irene Anastasiadou, *Constructing Iron Europe. Transnationalism and Railways in the Interbellum* (Amsterdam: Amsterdam University Press, 2012); Vincent Lagendijk, *Electrifying Europe: The Power of Europe in the Construction of Electricity Networks* (Amsterdam: Aksant, 2008); Frank Schipper, *Driving Europe: Building Europe on Roads in the Twentieth Century* (Amsterdam: Aksant, 2008); Judith Schueler, *Materialising Identity: The Co-Construction of the Gotthard Railway and Swiss National Identity* (Amsterdam: Aksant, 2008); Waqar Zaidi, "Technology and the Reconstruction of International Relations: Liberal Internationalist Proposals for the Internationalisaton of Aviation and the International Control of Atomic Energy in Britain, USA and France, 1920-1950" (Ph.D. diss., University of London (Imperial College), 2008).

37 Frank Schipper, Vincent Lagendijk, and Irene Anastasiadou, "New Connections for an Old Continent: Rail, Road and Electricity in the League of Nations Organisation for Communications and Transit," in *Materializing Europe: Transnational Infrastructures and the Project of Europe*, ed. Alexander Badenoch and Andreas Fickers (London: Palgrave, 2010), 113-143.

38 Van der Vleuten et al., "Europe's System Builders."

39 Johan Schot and Vincent Lagendijk, "Technocratic Internationalism in the Interwar Years: Building Europe on Motorways and Electricity Networks," *Journal of Modern European History* 6, no. 2 (2008): 196-217.

40 Antoine Fleury, "La Suisse et Radio Nations," in *The League of Nations in Retrospect: Proceedings of the Symposium* (Berlin; New York: Walter de Gruyter, 1983), 196-220.

41 Johan Schot and Frank Schipper, "Experts and European Transport Integration, 1945-1958," *Journal of European Public Policy* 18, no. 2 (March 1, 2011): 289; Anthony D. Smith, "National Identity and the Idea of European Unity," *International Affairs* 68, no. 1 (1992): 57.

could have national and international roles at one and the same time. Following their multiple roles, they could *fine-tune* their national and international activities and interests when working on transnational infrastructure projects.[42] Over the years the philosophy of technocratic internationalism inspired the emergence of specific kinds of European governance structures, European markets, and a host of European visions, symbols, and images. Jean Monnet's idea of a "gradualist and sector-specific approach towards general European unification" for instance, contributed to the establishment of the European Coal and Steel Community and is just one of many examples.[43]

The TIE results contradict the dominant viewpoint that infrastructures were built chiefly to serve the national, political, and economic agendas of the system builders. Armand Mattelart for instance finds that the great power of unifying peoples should not be attributed to communication and broadcasting technologies. They were used as instruments of national power during the two world wars.[44] The idea of technocratic internationalism, however, is that the motivations for national power politics and international peace efforts do not necessarily exclude one another. They could co-exist and perhaps even become interrelated in the material construction of the infrastructure.[45]

The philosophy of technocratic internationalism has contributed considerably to the kind of network created by societal cohesion. Over time, choices for network construction have influenced the linking, non-linking, as well as de-linking of Europe or parts of Europe.[46] TIE finds that international networks have mostly been constructed by linking national and local systems. In turn, these infrastructures also created an interrelation of identities. A local project like the Swiss Gotthard Tunnel, for instance, has been a transnational construct. Local, national, and international factors have been fine-tuned and intertwined, recreating Swiss identity in an international world.[47] These findings show that infrastructure

---

42 Schot and Lagendijk, "Technocratic Internationalism in the Interwar Years," 198-199; Johan Schot, "Transnational Infrastructures and the Origins of European Integration," in *Materializing Europe: Transnational Infrastructures and the Project of Europe*, ed. Alexander Badenoch and Andreas Fickers (London: Palgrave, 2010), 98-99.

43 Monnet argued that one should strive after integration in specific sectors. These integration efforts would then automatically spill over into other sectors. Ole Waever, "Europe Since 1945: Crisis to Renewal," in *The History of the Idea of Europe*, ed. Jan van der Dussen and Kevin Wilson (London: Routledge, 1993), 167.

44 Armand Mattelart, *Networking the World, 1794-2000* (Minneapolis: University of Minnesota Press, 2000); Nelson, *War of the Black Heavens*.

45 Van der Vleuten, "Infrastructures and Societal Change: A View from the Large Technical System Field," 395-398.

46 Van der Vleuten, "Toward a Transnational History of Technology: Meanings, Promises, Pitfalls."

47 Schueler, *Materialising Identity*.

construction can fine-tune and link a variety of spaces, ideas, goals, and identities simultaneously.

In the same way, we have to transcend some simplistic notions of Europe. Many European integration studies have a more restricted sense of space when it comes to Europe. They consider Europe to be a synonym for the European Union.[48] The TIE program concludes that engineers have often viewed Europe as a natural unit of infrastructure connection, even though at times they had the intention to focus on the global ultimately.[49] The examination of the shaping of Europe, implicitly as well as explicitly, always integrates other spaces such as the global, the national, and the local.[50] The task therefore is to inquire what Europe meant to different people at different points in time.

TIE shows that Europe has always been more than a geographical area. In line with histories of plans for Europe, it finds that Europe itself has also been an idea or a project.[51] Europe-oriented infrastructure plans similar to the 1944 and 1945 ones for post-war broadcasting often served as important guiding principles in the course of infrastructure projects. These maps even discursively communicate an image of Europe.[52] Ideas for global infrastructure projects have often been developed in relation to visions of European civilization and European culture.[53] According to Michael Wintle, such European cultural identities will never

---

48 For instance see: Chris Shore, *Building Europe: The Cultural Politics of European Integration* (London; New York: Routledge, 2000); Antje Wiener and Thomas Diez, eds., *European Integration Theory* (Oxford; New York: Oxford University Press, 2004).

49 Schot and Lagendijk, "Technocratic Internationalism in the Interwar Years," 199-200; Schot, "Transnational Infrastructures and the Origins of European Integration," 99.

50 Badenoch and Fickers, "Introduction," 9.

51 According to Davies, geographically one could speak of a "tidal Europe" with boundaries subject to change over time. Davies, *Europe: A History*, 9. A wide variety of literature discusses Europe as an idea or a project including Kevin Wilson and Jan van der Dussen, eds., *The History of the Idea of Europe* (London: Routledge, 1993); Anthony Pagden, ed., *The Idea of Europe: From Antiquity to the European Union* (Cambridge: Cambridge University Press, 2001); Carl Pegg, *Evolution of the European Idea, 1914-1932* (Chapel Hill: University of North Carolina Press, 1983); Gerard Delanty, *Inventing Europe: Idea, Identity, Reality* (New York: St. Martin's Press, 1995); Derek Heater, *The Idea of European Unity* (Leicester: Leicester University Press, 1992); Denis De Rougemont, *The Idea of Europe* (New York: Macmillan, 1966); Menno Spiering and Michael Wintle, eds., *Ideas of Europe since 1914: The legacy of the First World War* (New York: Palgrave, 2002).

52 Alexander Badenoch, "Touring between War and Peace: Imagining the 'Transcontinental Motorway,' 1930-1950," *Journal of Transport History* 28, no. 2 (2007); Alexander Badenoch, "Myths of the European Network: Constructions of Cohesion in Infrastructure Maps," in *Materializing Europe: Transnational Infrastructures and the Project of Europe*, ed. Alexander Badenoch and Andreas Fickers (London: Palgrave, 2010), 47-77; Schipper, *Driving Europe: Building Europe on Roads in the Twentieth Century*, 21.

53 For a fine study of European ideas in infrastructure development see: Lagendijk, *Electrifying Europe*; Vincent Lagendijk, "'To Consolidate Peace'? The International Electro-Technical Community and the Grid for a United States of Europe, 1929-1937," *Journal of Contemporary History* (forthcoming 2012). For histories of ideas of Europe amongst others see: Pim den Boer, "Europe to 1914: The Making of an Idea," in *The History of the Idea of Europe*, ed. Kevin Wilson and Jan van der Dussen (London: Routledge, 1993), 13-82;

be anything but a set of aspirations and images.[54] TIE also finds that Europe has always been contested and contradictory. It agrees with John Borneman and Nick Fowler's argument that Europe exists solely within its historical context.[55]

The media are important for building feelings of unity and would therefore be a crucial research site to examine European unification. Nowadays we consider the media to be the first means of simply diffusing visions, sentiments, information and the like among large groups of people. Broadcasting technology is easily available for a variety of social, cultural and political ideologies, and can easily influence feelings of unity or disunity. Broadcasting infrastructures are subject to the same dynamics as other infrastructures. Their overtly cultural content and the different ways this can be molded, nevertheless raise the question whether or not the creation of broadcasting infrastructures adheres to the philosophy of technocratic internationalism.

In the course of the twentieth century, various system builders constructed a number of broadcasting infrastructures with a number of purposes in mind. In Europe the Roman Catholic Church for instance established Vatican Radio to spread the faith; French investors established Radio Luxembourg for commercial purposes; and the IBU created an international network for public broadcasters. These different networks and different ideologies had to be negotiated and fine-tuned to make broadcasting in Europe effective. Moreover, system builders also negotiated these different purposes into programs and suitable musical constructs. The large differences in systems and ideologies initially suggest that broadcasting construction and the creation of European unity via broadcasting do not necessarily need to go hand in hand with technocratic internationalism.

Europe – On Air will follow the TIE tradition in shedding light on the relationship between transnational broadcasting infrastructure construction and Europe. Unlike most infrastructures, broadcasting has an artistic side to it which, as we shall see later, involved huge debates about what kinds of music served what purposes. The potential to create specific feelings of unity was an important element of these debates. *Europe – On Air* seeks to include this artistic side, which may add new or additional insights to the TIE program.

Denys Hay, *Europe: The Emergence of an Idea* (New York: Harper & Row, 1966).

54 Fickers and Lommers, "Eventing Europe," 236-237; EBU, *25 Years European Broadcasting Union*, 4; Léo Wallenborn, "From IBU to EBU: The Great European Broadcasting Crisis (Part I)," *EBU-Review* 29, no. 1 (January 1978): 25-34; Léo Wallenborn, "From IBU to EBU: The Great European Broadcasting Crisis (Part II)," *EBU-Review* 29, no. 2 (March 1978): 22-30; Charles E. Sherman, "The International Broadcasting Union: A Study in Practical Internationalism," *EBU-Review* 25, no. 3 (1974): 32-36.

55 John Borneman and Nick Fowler, "Europeanization," *Annual Review of Anthropology* 26 (1997): 49; Bo Stråth, "Introduction: Europe as a Discourse," in *Europe and the Other and Europe as the Other*, ed. Bo Stråth (Brussels; New York: P.I.E.-P. Lang, 2000), 10.

Broadcasting and Society-Building

Unlike the TIE literature, broadcasting histories give valuable information on national, international, and transnational broadcasting activities. Broadcasting histories above all tend to focus on national histories and national activities. Although *Europe – On Air* concentrates on transnational efforts, these national histories provide important points for reflection.

First of all, in line with broadcasting histories, this study considers programming efforts to be an important part of transnational broadcasting construction. One of the most marked differences between the two plans for post-war broadcasting discussed earlier in the chapter, lies in the way these plans deal with programming efforts. The 1944 plan openly underlines the importance of creating unity in Europe via programs. It suggests a program exchange between European broadcasters as well as live broadcasts from one point in Europe across the entire continent. In contrast, the British radio industry's 1945 plan largely ignores the programming side. This only states that a national broadcasting service should cover the entire nation by making space for two national channels to cover localities. According to Paddy Scannell and David Cardiff, the history of broadcasting was a history of *discovery* rather than a history of *a priori* knowledge of what broadcasting actually was.[56] It was one thing to discover you could send a signal. It was another thing to discover "what" programs to broadcast and how to realize that. Program construction efforts therefore are an important place to examine "hidden" European unification efforts.

Secondly, ever since the 1980s, these national broadcasting histories have related culture and societal cohesion to broadcasting development. Susan Douglas argues that the establishment of nation-wide broadcasting in the United States increased interconnections and a sense of belonging between U.S. inhabitants. It drew the countryside, the poor, the uneducated, and the housebound closer into the heart of U.S. society.[57] Additionally, international broadcasting stations like the first empire stations that linked colonies to their mother lands, as well as war-related stations like Radio Free Europe created a globally further interconnected world.[58] Broadcasting

---

56 Paddy Scannell and David Cardiff, "Serving the Nation: Public Service Broadcasting before the War," in *Popular Culture: Past and Present*, ed. Bernard Waites, Tony Bennett, and Graham Martin (London: Croom Helm London, 1982), 161-188.

57 Douglas, *Inventory American Broadcasting*, 306. For extensive national histories of broadcasting see: Scannell and Cardiff, "Serving the Nation"; Heinz Pohle, *Der Rundfunk als Instrument der Politik: Zur Geschichte des Deutschen Rundfunk von 1923/38.* (Hamburg: Verlag Hans Bredow-Institut, 1955); Alexander Badenoch, *Voices in Ruins: West German Radio Across the 1945 Divide* (Basingstoke: Palgrave Macmillan, 2008).

58 David M. Abshire, *International Broadcasting: A New Dimension of Western Diplomacy*, The Washington Papers (Beverly Hills: Sage Publications, 1976); Donald R. Browne, *International Radio Broadcasting*

connected people in different places and of different social classes to one another by addressing them in their homes.[59] People within the nation were bonded spatially.

Thirdly, as have histories of infrastructures in Europe, national broadcasting histories have found that these processes were not uncontested. Although several early histories assumed that broadcasting developed linearly, research during the 1990s proves otherwise.[60] Europe has a long tradition of state intervention in nation building, whereas the United States has a long tradition of commercial broadcasting. Both traditions were contested. In the United States commercial advertising sponsored broadcasting would win over public and education oriented broadcasting after no less than fifteen years of contestations. Developments in Europe were not uniform either. In the United Kingdom and the Netherlands, commercial broadcasting would not stand a chance against public broadcasting in the long run.[61] In France, however, public broadcasting activities would gradually suppress commercial broadcasting, although never entirely.[62] Different structures co-existed and changed in importance over time.[63] The choices made in the course of these processes often provoked heated debates among government officials, broadcasting and wireless radio telegraphy engineers, and societal interest groups.

While discussing these choices, broadcasting communities negotiated visions of their nation-state. Radio only "*appears*…to be basically empty and disinterested

---

(New York: Praeger, 1982); Nelson, *War of the Black Heavens*; Ana Paula Silva, "Shaping the Portuguese Empire in the Twentieth Century: The Telegraph and the Radio," *ICON Journal of the International Committee for the History of Technology* 7 (2001): 106-122; James Wood, *History of International Broadcasting* (London: Peregrinus, 1992).

59 Susan Douglas, *Listening in: Radio and the American Imagination* (Minneapolis, Minn.; London: University of Minnesota Press, 2004); Michele Hilmes, *Only Connect: A Cultural History of Broadcasting in the United States* (Belmont, CA: Wadsworth/Thomson Learning, 2002); Shaun Moores, "'The Box on the Dresser': Memories of Early Radio and Everyday Life," *Media, Culture and Society* 10 (1988): 23-40.

60 See for instance: Erik Barnouw, *A History of Broadcasting in the United States: Vol. I-III* (New York: Oxford University Press, 1970); Pohle, *Der Rundfunk als Instrument der Politik*.

61 For the contested U.S. negotiations see: Douglas, *Inventing American Broadcasting*; Robert W. McChesney, *Telecommunications, Mass Media, and Democracy: The Battle for the Control of U.S. Broadcasting, 1928-1935* (New York: Oxford University Press, 1993); Susan Smulyan, *Selling Radio: The Commercialization of American Broadcasting, 1920-1934* (Washington: Smithsonian Institution Press, 1994). For the contested process of British broadcasting see: Asa Briggs, *The History of Broadcasting in the United Kingdom, Volume I-V* (Oxford: Oxford University Press, 1961); Scannell and Cardiff, "Serving the Nation"; Seán Street, *Crossing the Ether: British Public Service Radio and Commercial Competition, 1922-1945* (Eastleigh: John Libbey Publishing, 2006). For the Netherlands see: Onno de Wit, "Radio tussen Verzuiling en Individualisering," in *Techniek in Nederland in de Twintigste Eeuw: Deel V Transport en Communicatie*, ed. Johan Schot and Onno de Wit (Eindhoven: Stichting Historie der Techniek, 2002), 208-211.

62 Christian Brochand, *Histoire générale de la radio et de la télévision en France* (Paris: La Documentation française, 1994); Cécile Méadel, *Histoire de la radio des années trente: Du sans-filiste à l'auditeur* (Paris: Anthropos; Diffusion Economica, 1994).

63 Richard McMahon, "Cultures of Communication: New Historical Perspectives," *European Review of History: Revue européenne d'histoire* 16, no. 1 (2009): 11.

[italics – s.l.]," Alexander Badenoch argues.[64] Radio is not a neutral medium and can be given any kind of meaning. According to these national histories, broadcasting focused on building national broadcasting and national societies. Apparently broadcasters not only specifically opted to design their broadcasts around the nation state, they also made explicit decisions about the form and shape these national societies would have in their programs. These choices impacted on that music and those groups that were united or disunited via programs. The acknowledgement that choices have formed the basis of broadcasting places national broadcasting history in a different perspective. Broadcasters actively tried to shape society by constructing programs, and chose to create national feelings of belonging.

Several national histories question the actual term "national." Some German histories conclude that local and regional broadcasting activities in Germany often rather identified their programming activities with regional instead of German identity.[65] Michele Hilmes not only addresses the mutual construction of U.S. and British broadcasting, but also examines transnational processes in national spheres.[66] She finds that in the United States, black bands often received no sponsorship, which hindered their performing in broadcast programs. As such, in the interwar years, broadcast programs expressed an American culture that made little or no reference to American black band music. Furthermore, where some groups took no part in such American broadcasts, others "were spoken for" without having their own voice, and could be marked as being different from the mainstream.[67] By building programs, broadcasters at the same time actively built a united as well as a disunited nation-state. By regarding national programs as transnational constructs, Hilmes thus finds that broadcasters created a kind of unity resulting from the negotiation of the local and the national.

From the 1970s onwards, the international dimension of radio and television broadcasting has increasingly caught the attention of broadcasting historians. Several institutional histories give an indication of those international system builders who were active in broadcasting. They describe the rise of international governmental organizations like the International Telecommunication Union (ITU), the representative of national PTT Administrations, as well as international

64 Alexander Badenoch, *Voices in Ruins*, 224.

65 For an example of such developments in German broadcasting see: Badenoch, *Voices in Ruins*; Brian Currid, *A National Acoustics: Music and Mass Publicity in Weimar and Nazi Germany* (Minneapolis: University of Minnesota Press, 2006); Winfried B. Lerg, *Rundfunkpolitik in der Weimarer Republik, Band 1* (München: Deutscher Taschenbuch Verlag, 1980).

66 Michele Hilmes, *Network Nations: A Transnational History of British and American Broadcasting* (New York: Routledge, 2011); Hilmes, *Only Connect*.

67 Hilmes, *Only Connect*, 78, 93, 97.

non-governmental organizations like the European Broadcasting Union (EBU), or the International Broadcasting Union (IBU), as the representatives of public broadcasters. These institutional histories discuss their organizational structure and their main activities. They find that such international organizations merely coordinate, fine-tune, and advise rather than construct and produce programs themselves.[68]

Other histories examine the television network and cross-border programming activities of these institutions from an international relations perspective. They find that in particular the EBU and the OIRT created exchange networks, which on occasion could be connected. The EBU and OIRT then applied these networks for the exchange of national programs mostly, reducing production costs considerably.[69] These histories nonetheless tend to examine the contested nature of the construction and interconnection processes only to a limited degree. These decisions and negotiations nonetheless are central to understanding the transnational relations between broadcasters, organizations, nations, and regions.

The late 1990s and 2000s marked a new era in broadcasting history. The approach towards these histories shifted in the direction of a European perspective. The European Television History Network established in 2004, for instance, promotes comparative historical research on television from a European perspective.[70] Over the years, the Network has re-evaluated television history in many European countries, identifying similarities, differences, "gaps and missing links in the research landscape" in Europe.[71] During these years, similar questions have been posed regarding other technologies like wireless telegraphy, telephony, and radio broadcasting.

Some of these studies address the negotiation of infrastructure developments and standards. They focus on aspects of network construction like the allocation of frequencies or television line standards, the way these were negotiated internationally, and why. These studies argue that those who negotiated such standards favored their nation building activities over the building of international relations.[72] National

68 Harold A. Fisher, *The EBU: Model for Regional Cooperation in Broadcasting* (Lexington, Ky.: Association for Education in Journalism, 1980), 5.

69 Burton Paulu, *Radio and Television Broadcasting in Eastern Europe* (Minneapolis: University of Minnesota Press, 1974), 244; Ernest Eugster, *Television Programming Across National Boundaries: The EBU and OIRT Experience* (Dedham, MA: Artech House, 1983).

70 The European Television History Network has been established to connect scholars, archivists and institutions that work on television history. For more information see: http://cms.let.uu.nl/ethn/

71 Andreas Fickers and Catherine Johnson, "Transnational Television History: A Comparative Approach," *Media History* 16, no. 1 (2010): 5.

72 Andreas Fickers, "National Barriers for an Imag(e)ined European Community: The Technopolitical Frames of Postwar Television Development in Europe," in *Northern Lights: Film and Media Studies Yearbook 2005*, vol. 4, 2006, 15-35; Christian Henrich-Franke, *Globale Regulierungsproblematiken in historisch-*

interests often impeded smooth fine-tuning of international television line-standards or international frequency plans for wireless and wired communication and broadcasting. Sometimes specific parties could even be excluded from standards deliberately, as was the case with Radio Luxembourg and frequency plans for wireless broadcasting.[73] International standardization attempts highlight the fragmentation of the international in favor of national interests.

Research focusing on the development of the Eurovision programming network after World War II draws similar conclusions.[74] These studies show that the Eurovision network, which nowadays organizes the Eurovision Song Contest, mostly transmitted national broadcasting programs across borders. In 1967 already, the coordinator of this network, the EBU, organized the worldwide live broadcasting of the television program *notre monde*, Our World. For one moment Our World materially connected the whole world, a true watershed in international television. Ever since, the network has relayed and exchanged national music, programs or news across the continent. These mostly highlighted national feelings of belonging.

Building on these insights, in 2007 the European Television History Network rethought television history by coming up with the idea of "going transnational." Andreas Fickers and Catherine Johnson created a special issue on transnational television history. The initiative questions the national positioning of television developments and unravels new interrelations and entwinements. The volume reveals the diversity in transnational flows and transfers "not only across institutions and through the movement of people and programmes, but also within the production and aesthetics of programmes themselves."[75] What is more, the approach creates a refinement of prior research, for instance with respect to the Eurovision program exchange. The majority of the programs traded via the EBU Screening Sessions or diffused via the Eurovision network in reality shared a kind of commonness because of their low degree of national and cultural specificity.[76] These transnational television histories thus provide valuable new insights on prior approaches.

---

er Perspektive: Der Fall des Funkfrequenzspektrums 1945-1988 (Baden-Baden: Nomos Verlagsgesellschaft, 2006); Wormbs, "Standardising Early Broadcasting in Europe."

73 Jennifer Spohrer, "Ruling the Airwaves: Radio Luxembourg and the Origins of European National Broadcasting, 1929-1950" (Ph.D. diss., Columbia University, 2008).

74 Wolfgang Degenhardt and Elisabeth Strautz, *Auf der Suche nach dem europäischen Programm: Die Eurovision 1954-1970* (Baden-Baden: Nomos Verlagsgesellschaft, 1999); Wolfgang Degenhardt, *Die Entstehung und Entwicklung der europäischen Partnerschaft im Fernsehbereich 1950-1970: Zur historischen Betrachtung eines komplexen Sensemaking-Prozesses* (Siegen: Universität Siegen, 2002); Christian Henrich-Franke, "Die «EBU Screening Sessions»: Wandlungen des europäischen Markts für Fernsehprogramme 1963-1985," *Rundfunk und Geschichte* 31, no. 1 (2005): 17-25.

75 Fickers and Johnson, "Transnational Television History," 5-6.

76 Christian Henrich-Franke, "Creating Transnationality through an International Organization?," *Media History* 16, no. 1 (2010): 67-81.

Broadcasting has captured the attention of researchers from within the Tensions of Europe network as well. Along with broadcasting historians they form the Transmitting and Receiving Europe project (TRANS). TRANS seeks to provide a thorough history of transnational broadcasting technology and programming within Europe.[77] The project focuses on a broad ensemble of radio and television technologies, which it places at the forefront of East-West relations during the Cold War era. The methodological approach consists of a combined focus on horizontal and vertical dynamics that together define a European communication space. Particularly valuable is the focus on both the transmission and reception side of the medium that together make up the horizontal dimension. Vertically, the processes are characterized by multi-level negotiations in the material, institutional, and symbolic dimension which interact with one another. Thanks to such a dynamic approach, TRANS fills an important gap in Cold War and European broadcasting infrastructure history. It promises a broad and encompassing approach.

Following the TRANS tradition, *Europe – On Air* mostly covers the transmission side of broadcasting. First of all, the book contributes to the material dimension by examining the interactive development of a multitude of broadcasting network technologies, such as the long waves, medium waves, and short waves, as well as the relaying of music via cable and wire. Secondly, the book's institutional dimension lays bare the formal and informal expert networks that developed within and beyond the institutional boundaries of international governmental and non-governmental organizations. Thirdly, the book provides a symbolic dimension by examining how stakeholders negotiated visions and constructed a European culture and civilization via their material and institutional standards as well as their cultural programming and music activities. *Europe – On Air* examines these dimensions for a period which is not the main focus of the volume currently being prepared by TRANS: the interwar years. TRANS, however, strongly suggests that these years are a key part of the story. Since recently, several TRANS researchers have been performing new studies which particularly address the interwar years, above all with a view to technical questions.[78]

77 Transmitting and Receiving Europe (EUROTRANS) is a project within the ESF Inventing Europe program. Currently, TRANS is completing a book project "Airy Curtains in the European Ether: Broadcasting and the Cold War" edited by Christian Henrich Franke, Alexander Badenoch, and Andreas Fickers that will be forthcoming with Nomos Verlaggesellschaft.

78 See the work of Nina Wormbs and Andreas Fickers. Nina Wormbs, "Standardising Early Broadcasting in Europe: A Form of Regulation," in *Bargaining Norms – Arguing Standards: Negotiating Technical Standards*, ed. Judith Schueler, Andreas Fickers, and Anique Hommels (The Hague: Stichting Toekomst-

Europe – On Air aims to give an integrated history of transnational broadcasting and European unification efforts for the interwar period. It will contribute to broadcasting histories by taking a transnational approach to broadcasting construction as well as the plans for European civilization and culture. The most important contribution to broadcasting history perhaps lies in examining the negotiation of international standards related to the programming side. What is more, it examines what a veritable international program similar to Our World looked like, how it came about and why. In particular the musical component of these programs has long been recognized as a universal language. The universal understanding of music could be employed for unification efforts beyond national boundaries. At the same time, it could also be employed for disunity, as the example of the rejection of American black band music shows. Unification attempts with their inherent elements of disunity define the relationship between broadcasting and Europe.

Methodology and Outline

To realize a transnational perspective on broadcasting and on Europe, the research for *Europe – On Air* can best be viewed as a journey through sources that have received little attention from historians. When travelling, one comes across great sights, possibilities, new insights, and rewarding hikes. Nonetheless, no journey is without pitfalls, drawbacks, or sites one would have loved to visit, but never managed to. These sites might have turned the journey into an all-time and entirely complete travel experience. Such has been the journey of *Europe – On Air*: adventurous, full of rewards, but not without its pitfalls, and unvisited sites.

Starting out from the concept of a relationship between transnational broadcasting and Europe, this study began by consulting contemporary journals. On the one hand, elites established "Europe" oriented journals such as the Pan-European Movement journal *Paneuropa* (1924-1938), *l'Européen* (1929-1940), and *L'Europe Nouvelle* (1918-1940) to discuss the future of Europe. With the exception of *L'Européen*, these journals hardly consider technology matters. They focus on economic and political questions instead. The financer and leader of *L'Européen*, Lyonnais parliamentarian and industrialist Étienne Fougère, had a personal interest in broadcasting. He regularly frequented and sponsored the IBU.[79] His *L'Européen*

beeld der Techniek, 2008), 112-120; Andreas Fickers, "In Search of the 'Greenwich of the Air': Techno-Political Diplomacy in European Frequency Allocation and Control (1925-1952)," *History and Technology* (forthcoming).

79 Etienne Deschamps, "L'Européen (1929-1940): A Cultural Review at the Heart of the Debate on European Identity," *European Review of History* 9, no. 1 (2002): 85, 89.

approaches Europe through a cultural lens, providing detailed program schedules from various stations in Europe, sometimes with additional discussions. In contrast, there are journals for broadcasting experts, like *Wireless World* (1913-1945). They discuss developments in all aspects of broadcasting with a preference for the technology and the organization of broadcasting itself. These journals however give little information about their broader international context. Both types of journals provide important information, but few detail on the nature of the relationship between broadcasting and Europe.

So where should one look? In the view of Jérôme Bourdon, the national central organization of archives and the organization of political science foster "la prison mentale de la nation."[80] In these archives, Europe often appears as a dimension of national identity rather than as a possibility for cross-border unification. Examining visions of Europe from the perspective of national archives alone would therefore cause a teleological domination of the national in the European. It would immediately push all local or foreign initiatives to the background. Many broadcasting and European histories base their research on national archives. A conversation on writing transnational history urges more and ample research in the archives of both intergovernmental and international non-governmental organizations.[81]

In recent years, research in archives of intergovernmental organizations like those of the League of Nations have developed fresh and rewarding new viewpoints on European history. Comparative research on Europe-related questions in general requires insurmountable sums of money, access to archives in all corners of Europe, as well as knowledge of northern, western, southern, and eastern European languages. In the archives of intergovernmental and international non-governmental organizations, every European country, their polities, arguments, and doubts come together in one place. Comparative and national histories mostly stress the pre-World War I European balance-of-power system, finding that especially the north and east of Europe formed the periphery areas.[82] Recent consultation of the archives of the League of Nations reveals that in reality, the smaller and often eastern European countries actively participated in the work and policy of

80 Jérôme Bourdon, "Comment écrire une histoire transnationale des médias? L'exemple de la télévision en Europe," *Les Temps des Médias. Dossier: Espaces européens et transferts culturels* 11, no. 2 (2008): 165.
81 Christopher Allan Bayly et al., "AHR Conversation: On Transnational History," *The American History Review* 111, no. 5 (2006): 1461.
82 Mikael af Malmborg and Bo Stråth, "Introduction: The National Meanings of Europe," in *The Meaning of Europe: Variety and Contention within and among Nations*, ed. Mikael af Malmborg and Bo Stråth (Oxford; New York: Berg, 2002), 13.

the League.[83] Archives of international organizations therefore form an excellent research site. They give new perspectives on the many national broadcasting histories and histories of Europe.

This study takes the largely unexplored archives of the non-governmental International Broadcasting Union, Geneva (IBU, 1925-1950), and parts of the archives of the European Broadcasting Union (EBU, 1950-present) as its points of departure. Additionally, the study explores the archives of the intergovernmental League of Nations, Geneva (LoN, 1919-1940), and International Telecommunication Union, Geneva (ITU, 1919-1940). Complementary to primary sources like meeting minutes and proceedings, these archives contain so-called "self-congratulatory" material. Via brochures, memorial booklets, and leaflets these organizations positively communicate their achievements, visions, and activities. Furthermore, their libraries contain series of their own journals, like the *Journal télégraphique* published by the ITU, the *Radiodiffusion* as well as the *Bulletin mensuel* issued by the IBU, and the EBU-Review issued by the EBU. These journals provide inside information on work-related topics. They are either written to inform members and other interested parties, or by the members themselves to give local, or bottom-up insights. These journals thus open up two-directional provision of information. Since little secondary literature exists dealing with transnational broadcasting in the interwar years, *Europe – On Air* relies mainly on these archives.

International archives have remained relatively untouched when it comes to broadcasting questions. They therefore provide valuable new perspectives in addition to the many national broadcasting histories. Experience with these international archives over the past years, nevertheless, reveals that the best transnational histories should be based on information retrieved from personal, company, national, as well as international state and non-state archives. Archivists at international organizations are often selective in the type of documents they decide to store. Working notes, minutes from sub-, sub- sub-committees, or personal notes from employees often do not make it into storage. The archives often show traces of communication and collaborative lines with other interest groups like the radio industry, but give no detail on the nature of and content of these contacts.[84]

83 Schipper, Lagendijk, and Anastasiadou, "New Connections for an Old Continent."

84 For an in-depth discussion of radio industry in Europe see: Pascal Griset, "Innovation and Radio Industry in Europe during the Interwar Period," in *Innovations in the European Economy between the Wars*, ed. François Caron, Paul Erker, and Wolfram Fischer (Berlin; New York: Walter de Gruyter, 1995), 37-63; Andreas Fickers, "Sichtbar hörbar! Radioapparat und Stadt: Knoten im vernetzten Kommunikationsraum," in *Zentralität und Raumgefüge der Grossstadte im 20. Jahrhundert*, ed. Clemens Zimmermann (Stuttgart: Franz Steiner Verlang Stuttgart, 2006), 83-103; Keith Geddes and Gordon Bussey, *The Setmakers: A History*

Moreover, many important discussions, especially those on ideas, frequently happened over lunch, dinner, or during coffee breaks, and can be traced in these international archives only to a limited degree.[85] At times, parts of practical discussions, exchanges of views, specific actors as well as reasons behind construction efforts remain unnoticed or unclear when examining these international archives.

Europe – On Air studies these international archives to examine new perspectives that may crop up on the relationship between broadcasting and European unification. The liabilities of this approach do not greatly affect the conclusions of this study. These conclusions stand, but might have been contextualized even more broadly if it had been possible to consult other archives as well. The presence of radio industry or radio amateurs might have been communicated more pronouncedly, for instance. Whereas initially the broadcasting organizations were established by and composed of members of the radio industry, in the course of the interwar years, these actors became inherently different. Direct links continued to exist and collaboration remained. Whereas the nature of the collaboration disappears from this transnational history as an effect of the archives, the results of their collaborations remain the same and so have their place in this history.

Interviews and informal talks conducted over the past four years with employees in all layers of the EBU have brought the archive documents to life. They give a vivid impression of the dynamic world of transnational broadcasting in Europe. There appears to be a considerable degree of continuity in company culture between the EBU and its predecessor the IBU. The interviews and talks give personal accounts of national and transnational experiences. They give insight into the relationship between organizations, between various parts of such organizations, and the way employees were involved in shaping a European and global society. Those people who work mainly on technical or practical questions, like the filling of time slots, laugh at the idea that the EBU is trying to create a particular kind of European society. Higher up in the organization or committees, more people consider that the discussion of Europe should be part of their daily tasks. The internal dynamics of transnational expert communities seem to matter and the interviews shed light on ways to interpret the IBU and EBU archives.

Europe – On Air embarked on its journey in the archives of an international organization that took on the activity of broadcasting in 1925: the International Broadcasting Union. IBU archives reveal other important international organizations and influential experts who negotiated broadcasting in the interwar years.

of the Radio and Television Industry (London: BREMA, 1991).

85 Frank Schipper encountered similar problems in his research on road transport in Europe. Schipper, *Driving Europe: Building Europe on Roads in the Twentieth Century*, 34.

Europe – On Air has also studied these organizations' archives and they shed light on the five main broadcasting problems identified by the IBU. These have been the institutionalization of broadcasting, the materialization of networks, the interactions with other socio-technical systems, politico-ideological problems related to war and peace, and international program construction. Like the trip that *Europe – On Air* has made through these sources, the construction of broadcasting and the shaping of Europe have been a travel experience for contemporaries. This study follows their journey by devoting a chapter to each one of these five main broadcasting problems as sites where experts constructed broadcasting and shaped Europe.

Chapter 2 begins by examining "the birth of an idea." It looks at the earliest interwar broadcasting developments in Europe in the direct aftermath of the First World War. We are introduced to the main players, their activities, and the problems they faced. This reveals how a new idea was born and negotiated by actors with different motivations and backgrounds. The resulting International Broadcasting Union would become the main arena for negotiating broadcasting problems in Europe. The chapter unravels how this new idea evolved within specific European intellectual and business contexts.

Chapters 3 and 4 focus on the material and organizational realization of broadcasting infrastructures in Europe. They above all examine how the relationship between broadcasting and Europe materialized into standards in frequency allocations as well as the construction of international relay networks. Chapter 3 concentrates on the efforts by the IBU that through time increasingly represented public broadcasting organizations. Chapter 4 reflects on the disputes the IBU encountered due to the growing diversity of broadcasting systems in Europe. We follow the route the IBU and systems like Radio Luxembourg, Radio Nations, Vatican Radio, and Radio Moscow took to negotiate the unity and disunity of "Europe," how they fine-tuned their activities, and henceforth their visions of Europe.

Programming issues are the subject of Chapters 5 and 6. Chapter 5 examines how the IBU and the League of Nations tried to resolve the problem of illicit propaganda broadcasting. Programs crossed borders and could be heard by neighboring countries. With newly defined boundaries in Europe after the war, these programs regularly disturbed international relations as well as efficient international broadcasting systems. Broadcasting communities therefore continuously reconsidered their programming plans in the light of war-peace related sentiments, which flourished on a continent with so many nation-states and cultures. In their eyes, the unity of Europe was at stake.

Finally, Chapter 6 focuses on how the IBU in collaboration with the League of Nations and the European music elites attempted to create international broadcasting programs. We examine especially the construction of international live programs and cross-border program exchange. These programs and exchanges were meant to increase understanding amongst listeners across as well as within borders. They would represent and diffuse a specific kind of European culture. The effort proved harder than it first appeared. There was no easy way to determine those characteristics that defined the ultimate European program with the best expression of "European" culture.

It remains rather difficult to determine the impact of all these efforts to build European unification via transnational broadcasting. Studying interwar listeners would be a different research project altogether. It would require labor-intensive research, ample knowledge of all European languages, and anthropological skills. Chapter 6 makes a very brief exploratory attempt. It gives an impression of the way and extent to which visions of Europe negotiated by the broadcasters trickled down into Dutch program guides which were used by Dutch listeners to determine their listening behavior. These guides contain domestic and international program schedules. Their lay-out and in-depth articles communicate visions on broadcasting and music to the listener.[86] Although these guides do not reveal the radio audiences' interpretation, they certainly show us how transnational activities and visions of Europe actually found their way into peoples' homes. The efforts that paved the way for these program guides began immediately after the First World War, when a number of inventors developed crucial technology for the transmission of sound. Marconi was one such man. Therefore the next chapter starts with him.

86 Since the broadcasting situation in the Netherlands during the interwar period formed quite an exception for the co-existence of five different public broadcasters, *Europe – On Air* explores two out of five program guides: *De Radiogids* of the socialist *Vereeniging van Arbeiders Radio Amateurs* (VARA), and *De Radiobode* of the *Algemeene Vereeniging Radio Omroep* (AVRO). Both the VARA and the AVRO concentrated on international collaborative affairs to a larger degree than the program guides of the other three organizations *De Omroepgids* of the *Nederlandsche Christelijke Radio-Vereeniging* (NCRV), *de Katholieke Radio Gids* of the Catholic *Katholieke Radio Omroep* (KRO), and *Vrije Geluiden* of the latitudinarian protestant *Vrijzinnig Protestantse Radio Omroep* (VPRO) suggest.

Chapter 2
Elites on the Barricades for Broadcasting

> [I believe] that broadcasting, properly handled, will make a material contribution towards greater understanding and amity between Nations, the cementing of home life and happiness of the individual.
>
> Guglielmo Marconi, London 1924[1]

Guglielmo Marconi (1874-1937) was a European and truly internationally oriented personality. He was born in Italy but lived in Great Britain for most of his adult life. Already during his lifetime the world celebrated his achievements, as an inventor and entrepreneur in long distance wireless telegraphy, telephony, and broadcasting.[2] Marconi is often quoted as being the first to transmit an actual signal – the letter "S" – over long distances across the ocean from Europe to the United States.[3] As a boy he was fascinated by the idea of transmitting signals through the air from point A to B. He created his own lab at his parents' home and began experimenting with wireless.

In the course of his career, he extended his engineering activities to business by establishing the globally targeted British Marconi Wireless Telegraph Company as well as a subsidiary in the United States. The American Marconi Company sold Marconi transmission and reception apparatus on a global scale.[4] His market

1 Arthur R. Burrows, *The History of Broadcasting* (London: Cassell, 1924), foreword.

2 For information on his life and work see: Bernard Louis Jacot and D.M.B. Collier, *Marconi – Master of Space: An Authorised Biography of the Marchese Marconi* (London: Hutchinson & Co, 1935); Giuseppe Pession, *Guglielmo Marconi con figure nel testo e 16 tavole in rotocalco*, I Grandi italiani: Collana di biografie diretta da Luigi Federzoni 15 (Torino: Unione tipografico-editrice torinese, 1941); Orrin E. Dunlap, *Marconi, the Man and his Wireless* (New York: Arno Press, 1971).

3 The invention of telegraphy, telephony, and broadcasting was not down to one single genius. It was "the result of numerous parallel inventions based both on a common ground of shared scientific and on individually acquired tacit knowledge." Marconi's work would not have been possible if a number of physicists like the British James Clerk Maxwell, the German Heinrich Hertz, the British Sir Oliver Lodge, and the Russian Alexander Popov had not also contributed to its development. Andreas Fickers and Pascal Griset, eds., *Eventing Europe: Electronic Information and Communication Spaces in Europe* (forthcoming with Palgrave Macmillan, 2013).

4 Daniel Headrick, *The Invisible Weapon: Telecommunications and International Politics, 1851-1945* (New York: Oxford University Press, 1991), 180-187.

expanded vastly to remote corners of the United States and Europe. The Marconi Company followed a developing European tradition of cartel capitalism, where means of production were privately owned and governments followed a *laissez-faire* policy.[5] Marconi collaborated in an international cartel with other leading wireless companies. Together with the French Compagnie Générale de Télégrafie sans filistres (CSF), the German Telefunken, and the American Radio Corporation of America (RCA), he agreed to supply goods and services only with uniform conditions of sale, prices, marketing, and production.[6] This European tradition of cartel capitalism made Marconi one of the most influential entrepreneurial and engineering wireless mentors of his age operating outside governments. His lifelong friend and RCA director David Sarnoff even became known as "Marconi's American apprentice."[7] When one remembers Marconi, one remembers his engineering and entrepreneurial achievements.

But there was more to the man. Marconi is a text book example of an engineer with what Johan Schot and Vincent Lagendijk call "a technocratic internationalist ideology."[8] Marconi believed that his invention would lead to peace, mutual understanding, and to the creation of societies. Such societies could encompass the nation, but also the home life, and the happiness of the individual listener. He did not see a dichotomy between the national and the international, adopting what one nowadays would call a transnational approach. According to Schot and Lagendijk, theories on society flourished widely in transnational engineering circles. Moreover, Craig Murphy finds that similar theories underlined the development of a European tradition of international organizations during the second half of the nineteenth century.[9] He suggests that this tradition built on

---

5 The United States employed a system of competitive capitalism. Its antitrust legislation outlawed the formation of cartels on U.S. territory. Europe on the contrary used cooperative capitalism, a system of cartel capitalism. While such a system is usually known as "international cartel capitalism" it is an example of a very European rather than American way of implementing capitalism. However, American companies could join such cartels on European territory. Harm Schröter, *Americanization of the European Economy: A Compact Survey of American Economic Influence in Europe since the 1880s* (Boston, MA: Springer, 2005), 32.

6 Pascal Griset, "Innovation and Radio Industry in Europe during the Interwar Period," in *Innovations in the European Economy between the Wars*, ed. François Caron, Paul Erker, and Wolfram Fischer (Berlin; New York: Walter de Gruyter, 1995), 37-63.

7 Kenneth Bilby, *The General: David Sarnoff and the Rise of the Communications Industry*, 1st ed. (New York: Harper & Row, 1986), 6, 21, 25.

8 Johan Schot and Vincent Lagendijk, "Technocratic Internationalism in the Interwar Years: Building Europe on Motorways and Electricity Networks," *Journal of Modern European History* 6, no. 2 (2008): 196-217.

9 Craig Murphy, *International Organization and Industrial Change: Global Governance since 1850* (New York: Oxford University Press, 1994), 48-49. For an example see the opening speech of the in 1865 established International Telegraph Union. Armand Mattelart, *Networking the World, 1794-2000* (Minneapolis: University of Minnesota Press, 2000), 20; Harold R.G. Greaves, *The League Committees and World Order: A Study of the Permanent Expert Committees of the League of Nations as an Instrument of International*

Enlightenment values, values based on reason, because the position and wealth of the aristocrats who developed the activities of these organizations were a direct continuation of the European *ancien régime*. Discussions on his life and work often overlook Marconi's connection with these European elites and his concepts of European society.

Marconi reached adolescence in an elite milieu where ideas celebrating Europe's supremacy and the grandeur of its civilization and culture circulated widely in a variety of forms.[10] He grew up as the son of a well-known Italian businessman and an Irish noblewoman, the granddaughter of whisky distiller John Jameson, and descendant of the Scottish Earl Haig.[11] Consequently, he lived his life as much among wireless experts as British and Italian elites. Marconi felt closely involved with international peace and collaboration. The Italian government appointed "senatore" Marconi to represent Italy during the Paris Peace Conference in 1919.[12] He was there when the Conference established the League of Nations, an organization designed to improve international understanding, peace, and welfare on a global scale. Marconi believed that the unifying activities of the League might well be the last hope for the salvation of Europe's civilization, one that had fallen into disarray during the Great War. He hoped his invention would improve understanding as he believed that "Communications between peoples widely separated in space and thought is undoubtedly the greatest weapon against the evils of misunderstanding and jealousy..."[13]

During the conference, Marconi did not spend all his time advocating international peace and understanding based on thoughts about Europe's civilization. He also used the opportunity to promote the medium of broadcasting in order to pursue these ends. Throughout his life, Marconi sought to achieve international peace wearing "multiple hats." He seems to have become a mentor for many broadcasters trying to relate their broadcasting projects to projects for Europe.

Marconi is a fine example of someone who worked on the *hidden integration* of Europe. Historians of European ideas, as well as social historians like Bo Stråth, argue that such debate remained restricted to the political and intellectual

Government (London: Oxford University Press; H. Milford, 1931).

10 Pim den Boer, "Europe to 1914: The Making of an Idea," in *The History of the Idea of Europe*, eds. Kevin Wilson and Jan van der Dussen (London: Routledge, 1993), 62-78.

11 Jacot and Collier, *Marconi – Master of Space*, 18-19.

12 The Italian government had given Marconi the honorary title for life "senatore" for his engineering and entrepreneurial achievements. With this title Marconi became a member of the Italian Upper Legislative Chamber. Dunlap, *Marconi, the Man and his Wireless*, 235-236.

13 Ibid., 251; Asa Briggs, *The History of Broadcasting in the United Kingdom, Volume I: The Birth of Broadcasting* (Oxford [u.a.]: Oxford University Press, 1961), 282. Originally the quote comes from *The Popular Wireless Weekly* (January 1924).

arenas. A contemporary professor of history, however, found that these elites consisted of politicians, intellectuals, economists, industrialists and bankers.[14] Marconi was such an industrialist. Nevertheless, he was also a respected member of the international European elites, a diplomat, and an engineer. His hybrid background facilitated collaboration with people and organizations working under the banner of international peace. You could say that Marconi was one of those people who worked on the unity of Europe. He connected cross-border technological projects to visions of transnational societies.

This chapter examines the early development of transnational radio broadcasting by focusing on the broadcasters. It begins by showing how during the turbulent post-war years people like Marconi designed, contested, and imagined early radio broadcasting developments. The fact that these activities took place in a tradition of European cartel capitalism outside the sphere of governments initially conflicted with the interests of European governments and international governmental organizations. We will show how broadcasting pioneers explored transnational institutional possibilities to fill the regulatory gap left by their governments and existing international organizations. They continued in a tradition of what Craig Murphy calls a European conference system.[15] The first pioneers established a transnational community of broadcasters and international non-governmental organizations from scratch. They departed from the European traditions of both cartel capitalism and international organizations.

Pioneering in the Private Sector

At first, broadcasting enthusiasts pioneered in the private sector. Post-war circumstances did not favor the development of broadcasting in Europe. While the radio industry in Europe and the United States focused on the development of wireless telephony, radio amateurs in the United States discovered its application for diffusing music from one point to a mass of people.[16] Marconi and others in companies similar to his, quickly adopted the idea of broadcasting, with the prospect of new markets. Enterprises like CSF, the Marconi Company and Telefunken immediately

14 Bo Stråth, "Introduction: Europe as a Discourse," in *Europe and the Other and Europe as the Other*, ed. Bo Stråth (Brussels; New York: P.I.E.-P. Lang, 2000), 20; Andrew W. Cordier, "European Union and the League of Nations," *Geneva Special Studies* 2, no. 6 (1931): 4.

15 Murphy, *International Organization and Industrial Change*, 48-49, 52, 56.

16 Radio broadcasting, or point-to-mass communication, is a social and cultural construct serving a different purpose than point-to-point communication, which is based on the same technology. Griset, "Innovation and Radio Industry in Europe during the Interwar Period," 40-41, 47.

experienced hindrance from already firmly established customs in wireless communications.

Most European governments had drawn up regulations for using the airwaves. They ruled the airwaves and indirectly maintained control over the development of wireless services. During the war, wireless radio-telephony had been used for espionage, and governments were eager to maintain this control. In their eyes, free experimentation with this means of communication was a liability for post-war peace negotiations, for national defense, and nation-building practices. They felt "...uneasy concerning the general political situation and the idea of private individuals 'listening in' to whatever might be radiated through the ether."[17] They believed that wireless technology was an instrument of national power politics and should remain out of the hands of public or private organizations. In particular, the technical logic of broadcasting made it a concern for governments. Broadcasting transmissions could interfere with other applications of wireless technology and thus were placed under the authority of national post and telegraph administrations.

The control of telecommunications had been contested since the establishment of the first telegraph networks in Europe.[18] Although the control of telegraphy developed unequally in different countries, telegraphy started off as a private affair. Railway companies developed the first lines, opening up communications to a select group of private parties like the police and the press. Over the years, parliamentary decisions put an end to such practices. In Europe, a predominant system developed whereby governments came to own the telegraph network that would be open to private users. The telephone developed similarly. Initially private companies had room to maneuver but by the early 1920s, governments directly or indirectly controlled the telephone networks in Europe. Private companies wanting to develop and exploit wireless radio generally depended on the goodwill of the national PTTs.

When radio industry and amateurs began experimenting with broadcasting, they needed a concession from the PTTs to use certain frequencies. The PTTs acted reluctantly and blocked a large number of private activities. Initially, governments in Europe even forbade enthusiasts from carrying out private experiments with wireless radio communications.[19] The international climate was too fragile to

17 Arthur R. Burrows, "Broadcasting outside the United States," *Annals of the American Academy of Political and Social Science* 177 (1935): 30.
18 For an elaborative examination of the negotiation of private versus public control of developing telecommunications ever since the start of postal services in Europe since the late twelfth century see the first chapter of: Fickers and Griset, *Eventing Europe*.
19 Burrows, *The History of Broadcasting*, 46; IBU, *Twenty Years of Activity of the International Broadcasting Union* (Geneva: IBU, 1945), 7.

be threatened by experiments with wireless technology that might hinder government employment of wireless telegraphy and telephony for national defense.[20] The war had damaged international relations considerably, and now government officials and international elites wanted to improve the political and economic climate. Another such war simply had to be avoided.

Although intended to create stability, the renegotiation of power relations during and after the Peace Conference in Paris in 1919 led instead to an unstable political situation.[21] Britain's influence had diminished slightly, France wished to bend international politics along French lines, the Allies forced full reparations payments on Germany, and the creation of a belt of new states between Russia and Germany, the so-called *cordon sanitaire* or Shatter Zone, led to several upheavals in the eastern parts of Europe.[22] Furthermore, international commercial activities between nations only developed slowly. At this time "to say that Europe was at peace...was an exaggeration."[23]

Governments therefore ordered their PTTs to keep a close watch on activities by third parties which might encourage war. The PTTs granted the mushrooming and highly controlled, licensed radio amateur clubs certain liberties to expand. Moreover, the well-established international cartel of Marconi, Telefunken, RCA and CSF could also experiment with wireless within certain limits.[24] Other groups were shut out. Governments remained reluctant and with the exception of Germany and Russia, regularly blocked experiments. In 1920 the British government for instance temporarily banned Marconi from broadcasting experiments in favor of telephonic interest and those of the armed forces.[25] In France, the French PTT only reluctantly granted a concession to CSF more than a year after successful experiments.[26] Some

20 Andrew Crisell, *An Introductory History of British Broadcasting*, 2nd ed. (London; New York: Routledge, 2002), 17.

21 Margaret MacMillan, *Peacemakers: The Paris Conference of 1919 and its Attempt to End War* (London: John Murray, 2001).

22 Estonia, Latvia and Lithuania gained independence in 1918, whereas Poland, Czechoslovakia, Yugoslavia and Hungary were created one year later. Norman Davies, *Europe: A History* (London: Pimlico, 1997), 921, 926; Peter Bugge, "'Shatter Zones': The Creation and Re-Creation of Europe's East," in *Ideas of Europe Since 1914: The Legacy of the First World War*, ed. Menno Spiering and Michael Wintle (New York: Palgrave, 2002), 47-68.

23 Davies, *Europe: A History*, 926.

24 For a discussion on the role of amateurs in Great Britain and Germany see: Briggs, *The Birth of Broadcasting*, 46-53; Heinz Pohle, *Der Rundfunk als Instrument der Politik: Zur Geschichte des deutschen Rundfunk von 1923/38* (Hamburg: Verlag Hans Bredow-Institut, 1955), 25. For a discussion on the role of radio industry see: Griset, "Innovation and Radio Industry in Europe during the Interwar Period," 37-63; Christian Brochand, *Histoire générale de la radio et de la télévision en France* (Paris: La Documentation française, 1994), 621; Léonard Laborie, "La France, L'Europe et l'ordre international des communications (1865-1959)" (Ph.D. diss., Université Paris IV – Sorbonne, 2006), 568.

25 Crisell, *An Introductory History of British Broadcasting*, 17.

26 For a discussion of the diversity in developments in Europe see: chapter 1 of Fickers and Griset,

countries had no legislation to regulate the airwaves. The Luxembourg government for instance considered the air something it could not possess and the Polish government had never made legislation about granting concessions for wireless broadcasting.[27] European governments consequently hindered the opening up of a national broadcasting space, in some instances leaving a national legislative vacuum.

Radio amateur clubs and industry lobbied along with their PTTs and performed numerous experiments for four years. Only in 1922 did the PTTs grant their first broadcasting concessions, with varying degrees of provision. In Western Europe, the radio industry was responsible for the first broadcasting organizations. The British Broadcasting Company (BBC) was a consortium of manufacturers which had to comply with a License and General Agreement between the Corporation and the Postmaster General. In practice, the Postmaster seldom used his power.[28] The BBC worked with Marconi apparatus and employed Marconi employees like Peter Eckersley and Arthur Burrows. It aimed to educate its audiences via high culture programs.[29] In France, CSF established the private station Radio Paris while smaller radio set manufacturers created commercial stations like *Poste Parisien*, *Radio Côte d'Azur*, and *Radio Lyon* to increase the sale of their sets.[30] Only the German government was quick to take control of broadcasting. Though *Reichsrundfunkgesellschaft mbH* maintained an unofficial link with Telefunken from the start, its government prevented private parties from creating broadcasting stations.[31] Thus in Western Europe the international radio cartels

---

Eventing Europe.

27 Jennifer Spohrer, "Ruling the Airwaves: Radio Luxembourg and the Origins of European National Broadcasting, 1929-1950" (Ph.D. diss., Columbia University, 2008), 27-28; Ryszard Miazek, *Przeminelo z radiem: opowiesc o Zygmuncie Chamcu – zalozycielu i pierwszym dyrektorze Polskiego Radia* (Rzeszów; Warszawa: Wydawnictwo Wyzszej Szkoly Informatyki i Zarzadzania; Polskie Radio, 2005), 66.

28 The License stated that the BBC had to comply with the Post Office's technical conditions for the use of wavelengths as well as the barring of broadcasting advertisement or any other sponsored material. The British government centralized broadcasting and turned the BBC into a government regulated organization in 1927. Paddy Scannell and David Cardiff, "Serving the Nation: Public Service Broadcasting before the War," in *Popular Culture: Past and Present*, ed. Bernard Waites, Tony Bennett, and Graham Martin (London: Croom Helm London, 1982), 162-163.

29 Griset, "Innovation and Radio Industry in Europe during the Interwar Period," 37.

30 Brochand, *Histoire générale de la radio et de la télévision en France*; Seán Street, *Crossing the Ether: British Public Service Radio and Commercial Competition, 1922-1945* (Eastleigh: John Libbey Publishing, 2006), 234-253.

31 Though broadcasting in Germany also originated via the efforts of the private GmbH, a station that informed and educated laborers, the German government very quickly controlled radio broadcasting nationally. It appointed the experienced and soon to be postmaster Hans Bredow, director of the German radio company Telefunken as Director of the new Reichs Rundfunk Gesellschaft. He collaborated closely with his managing directors Kurt Magnus and Heinrich Giesecke, the latter directly representing the German *Reichspostministerium* within the RRG. Pohle, *Der Rundfunk als Instrument der Politik*, 24; Winfried B. Lerg, *Rundfunkpolitik in der Weimarer Republik, Band 1* (München: Deutscher Taschenbuch Verlag,

were operators of stations as well as manufacturers of radio sets. A patchwork of different systems evolved, some related to governments and varying from stations that served commercial purposes only to ones that pursued nation-building via high culture programs.

Initially Eastern Europe appeared to lack expertise in broadcasting and set manufacturing. Its broadcasting industry, seemingly small or non-existent, mostly developed on the initiative of well-established individuals with personal and economic interests. In Czechoslovakia, for instance, Ladislav Sourek had long been a wireless enthusiast, held a degree both in law and finance, and was the director of *Radioslavia*, a company that dealt in the construction and sale of wireless telegraphy and telephony apparatus.[32] With two others, Sourek established the first Czech broadcasting organization *Radiojournal* soon after the establishment of the BBC in Great Britain.[33] Responsible for the organizational side, Sourek visited the BBC in 1922. After his return he decided to organize *Radiojournal* following the British example.[34] Furthermore, he collaborated with the Marconi Company, Telefunken, and CSF to facilitate a swift import into Czechoslovakia of receiver sets that would secure a nation-wide radio audience. In contrast to the west of Europe, the international cartel in the east acted as producers of receiver sets, whereas influential people apparently established the broadcasting organizations that acted as operators. The BBC model and the international radio industry cartel appeared to have considerable influence on the development of East European broadcasting.

The underlying personal networks became an important prerequisite for the diffusion of broadcasting across Europe. Frequent international exchange of knowledge strengthened several pioneering broadcasting efforts, as happened in Poland.[35] When Marconi visited the Paris Peace Conference in 1919, he met Zygmunt Chamiec, a member of the Polish delegation. Chamiec belonged to an important Polish noble family, was a nationally well-known banker, and had a diverse education ranging from medicine, treasury, and finance to musicology. He had lived in various European countries, and was fluent in French, German, Russian, and Italian. In addition to official duties, informally Chamiec and

1980), 154, 201, 253.

32 My thanks go to Jiří Janáč who provided and translated this entry for me. J. Tomeš a kolektiv, *Český biografický slovník XX. století*, vol. 3 (Praha: Litomyšl; Paseka; P. Meissner, 1999). Entry: Sourek, Ladislav.

33 Miroslav Krupicka, *Czech Radio History* (Radio Prague, 1996), http://www.radio.cz/en/article/49702. Accessed July 21, 2008.

34 Eckhard Jirgens, *Der deutsche Rundfunk der 1. Tschechoslowakischen Republik: Eine Bestandsaufnahme* (Frankfurt am Main: Lang, 2005), 35-36.

35 Miazek, *Przeminelo z radiem*, 59, 310. My warm thanks go to Dagmara Jajesniak-Quast who found this very well-written and documented biography, as well as to Pavel Bernat who translated the Polish.

Marconi discussed the latest wireless broadcasting developments. Marconi invited Chamiec to Britain to study his work.

After some months in the Marconi laboratories, Chamiec returned to Poland where he lobbied for a concession that would allow him to establish a broadcasting station. He encountered severe difficulties and called for the help of the international cartel of CSF, Telefunken, and Marconi to put pressure on the Polish PTT. Having been taught by Marconi, Chamiec's experience of broadcasting was not only the result of Marconi's personal expertise, but also became intertwined with Marconi's visions of community building via broadcasting that exceeded national boundaries. Several years after *Polskie Radjo* was established, in 1928 Chamiec even received the French Order of the Legion of Honor for his "remarkable achievements in promoting peaceful cooperation across the airwaves."[36]

Chamiec and Marconi believed that broadcasting could facilitate international peace and understanding. Both belonged to the European noble elites and tried to forge direct links between their activities and ideological debates on the future of an international society.[37] They combined their economic interests and nation-building ideals. But others broadcasters were part of the European noble elites as well. Men like Maurice Rambert, director of Swiss broadcasting, and Marquis Solari, vice-president of Italian broadcasting not only had an economic interest in broadcasting, but wanted to achieve peace and understanding via broadcasting as well.[38] Moreover, important men like Sourek in Czechoslovakia were pursuing similar routes. By 1925 several socio-technical systems for broadcasting co-existed in Europe. In addition to various private stations in France, the Netherlands, and Yugoslavia, which operated for commercial purposes only, that is for the sale of

36 Ibid., 311.

37 These men were not always driven by the same motives or visions. Nevertheless, because of their work and their connections with the European elites, they frequently travelled abroad for politics, meetings, education, etc. A selection of biographies: H.S. von Heister, "Heinrich Giesecke und Kurt Magnus," in *Funkköpfe: 46 Literarische Porträts*, ed. Karl Wilczynski (Berlin: Verl. Funk-Dienst, 1927), 8-10; Reinhard Schlögl, *Oskar Czeija: Radio- und Fernsehpionier, Unternehmer, Abenteurer* (Wien: Böhlau, 2005); IBU, *Maurice Rambert, 1866-1941* (Geneva: IBU, 1941); Dunlap, *Marconi, the Man and his Wireless*; Bilby, *The General*; Tomeš a kolektiv, *Český biografický slovník XX. století*, 3; Miazek, *Przeminelo z radiem*.

38 Sourek and Chamiec were not the only ones who met Marconi and became influenced by his work and thoughts on the role of broadcasting in society. Marconi also maintained long lasting personal connections with the Italian broadcasting organization. Its Vice President, the Italian nobleman Marquis Solari, worked with Marconi for over 26 years. When broadcasters started to collaborate internationally, Solari referred to his connection with Marconi that "gives me a certain right to express to you my views, hoping that the decisions you are going to take will have the best effects on the relations and the good understanding between different countries." "...me donnent un certain droit pour vous exprimer mes vues, en espérant que les décisions que vous aller prendre auront le meilleur effet sur les relations et la bonne entente entre différentes pays." "Conférence européenne d'ingénieurs en radiophonie, tenue au siège de la Société des Nations, Séance 6 juillet 1925," 3, P.V. et documents complémentaires Commission Technique, 1925-1930 [hereafter: IBU].

receiver sets, many private broadcasters had an interest in society building activities. Consequently, private broadcasting initiatives within Europe paved the way for what we nowadays call public broadcasting.

U.S. broadcasters created a different broadcasting system. They constructed a national broadcasting infrastructure that valued commerce above cultural education and unity. Although they shared the unifying prospects of broadcasting with their European colleagues, they made different choices in the process. Most U.S. broadcasters agreed that broadcasting reduced class-differences, allowing for the development of a stronger and better integrated population. Farmers, the poor, the housebound, and the uneducated were likely to profit the most as they were drawn into society. To supporters of commercial broadcasting, such unity emerged thanks to the simple material linking of listeners across the country tuning in to the same programs. Even though the educated community shared these beliefs, they approached the medium from a more high-culture perspective. In their eyes, broadcasting provided an opportunity "to exert social control through culture," by affirming and extending "cultural tastes and norms of the upper ties of the middle class."[39] Though these two groups agreed on the unifying character of broadcasting, they disagreed on how to achieve it and on the kind of unity they should create.

At first, private U.S. broadcasters developed both viewpoints freely. While industry focused on wireless telephony only, radio amateurs established the application of wireless for broadcasting. They discovered point-to-mass communication while experimenting with point-to-point communications. By the time industry anticipated these developments around 1920, U.S. wireless broadcasting infrastructures were already well established.[40] Department stores, newspapers, churches, universities as well as the major radio-telecommunication companies AT&T, Westinghouse, and General Electric all established radio stations. Soon, the congestion of the air forced these groups to share their wavelengths. The U.S. government took a *laissez-faire* approach, paving the way for creating a radio trust of the private corporations RCA, AT&T, Westinghouse, and General Electric. This trust aimed to control U.S. broadcasting where the government did not interfere.[41] It "had the technical, financial, and organizational resources to shape programming content, to influence public policy, and to determine how broadcasting

39 Susan Douglas, *Inventing American Broadcasting, 1899-1922* (Baltimore: Johns Hopkins University Press, 1988), 314.

40 Douglas, *Inventing American Broadcasting*, 299-300.

41 Robert W. McChesney, *Telecommunications, Mass Media, and Democracy: The Battle for the Control of U.S. Broadcasting, 1928-1935* (New York: Oxford University Press, 1993), 262-263.

would maximize profits," Susan Douglas shows.[42] Government took upon itself the right to only intervene in affairs on behalf of American citizens, assigning wavelengths if required.[43] Consequently this radio trust of private corporations quickly gained power in most areas of broadcasting.

The radio trust would influence the expansion of broadcasting in the years to come. Already in 1926, RCA established the first nation-wide network, the National Broadcasting Company (NBC), followed by a second one in 1927, the Columbia Broadcasting System (CBS). These networks connected many already existing local radio stations via telephone cable, allowing for the simultaneous broadcasting of one radio program across the continent. Furthermore, industry began regular sponsorship of programs, an idea heavily promoted by radio advertising boosters. These public relations advocates "actively marketed the idea to advertisers, advertising agents and broadcasters" and developed "a form of radio advertising that they argued listeners would tolerate and even enjoy."[44] Slowly the initially free expansion of broadcasting by ideologically different groups narrowed down to an industry-dominated and commercially-based radio broadcasting culture that ignored the ideas of the American educated and cultural elites.

These developments did not come about without resistance. As soon as people became aware of the contours of the new system "Americans, from a variety of backgrounds, reacted with utter disgust."[45] They developed public campaigns to make room for non-profit and noncommercial elements in broadcasting. This reform movement consisted of a colorful group of people from the American bourgeoisie, intellectuals, civic activists, to elements of the labor movement and of the press. Together they fought the 1927 Radio Act, generally considered to be the first provisional statute for the "regulation" of U.S. airwaves. The Radio Act left little ground for society related projects by the non-profit sector or the American elite community. When the newly established Federal Radio Committee (FRC) built on the Radio Act and reallocated the frequencies, it downplayed the number of stations operated by non-profit parties by nearly one hundred. The FRC stressed "that Congress had given no indication as how to determine the meaning of public interest, convenience, or necessity."[46] Consequently, it had decided in

42 Douglas, *Inventing American Broadcasting*, 316.

43 Ibid., 315.

44 The most extensive study on this topic is: Susan Smulyan, *Selling Radio: The Commercialization of American Broadcasting, 1920-1934* (Washington: Smithsonian Institution Press, 1994), 72-86; Quoted from Jennifer Spohrer, "Paradise or Perdition? European Commercial Broadcasters and American Radio, 1920-1960" (presented at the ESF Inventing Europe Conference: Appropriating America, Making Europe, Amsterdam, 2009), 5.

45 McChesney, *Telecommunications, Mass Media, and Democracy*, 252.

46 These decisions came to be known as General Order 40. It assigned 40 clear channels, 34 regional

favor of stations with the best possible technical equipment. Without doubt, its decision to hear commercial stations with few non-profit exceptions influenced their choice substantially. The FRC clearly had been either incapable or unwilling to make room for the non-profit sector. In the end, the reform movement only managed to delay the domination of commercial broadcasting. After this defeat, commercial corporations could determine the course of American broadcasting as they pleased.

In the United States and in Europe the course of broadcasting settled down in the 1930s. In spite of personal overseas networks between American and European broadcasters, different dominating systems emerged.[47] This mainly appeared to be due to the role of European governments and elites. Governments in Europe exerted more control over granting concessions. In several instances they issued only one concession, limiting the number of stations that could be created within their country.[48] Furthermore, the European elites involved in the establishment of many broadcasting organizations actually had a greater influence than in the United States. Their drive to create a sense of national belonging via educational and cultural programs was strong. This caused the development of the medium to take a different course in the United States and Europe in spite of occasional contacts.

Practical problems hampered efficient national broadcasting in Europe. An explosion of new broadcasting organizations across the continent was causing technical and ideological difficulties. The growing number of transmissions created a vast increase in signal interference. "Reception became torture; the listener suffering constant interference between stations of different countries was the pitiable victim of these profligate signals in the form of whistling, grinding, cracking, groaning, which he remembers with horror to this day!"[49] Signals crossed borders and therefore the interference problems could not be resolved at national level. They required international solutions.

---

channels, "with the remaining frequencies to be low-power local channels that would accommodate thirty broadcasters in each zone [for convenience sake FRC had divided the country into five geographical zones]." Ibid., 25.

47 Marconi and David Sarnoff, director of RCA, were personal friends, and several Marconi employees as well as internationally active European broadcasters had occasionally traveled to or studied in the United States as well. "L'U.I.R. rend visite à la N.B.C.," 1936, box 58 Photographies, IBU; Bilby, *The General*.

48 The British Post Office, pushed several set manufacturers to collaborate and establish one broadcasting station together.

49 "La réception devient un martyre, l'auditeur souffrant des interférences constantes entre les stations des différents pays, est la pitoyable victime de ces émissions prodigues en sifflements, grincements, craquements, gémissements, dont il conserve, aujourd'hui encore, un souvenir horrifié!" Arno Huth, *La radiodiffusion: Puissance mondiale* (Paris: Librairie Gallimard, 1937), 46.

Ideologically, the post-war renegotiations of power relations and the redefining of national boundaries were the main grounds for debate. These could be issues about the material network, the broadcasting of "harmful" internationally sensitive propaganda to neighboring countries, or suitable standards for domestic day-to-day program content. In a complex post-war international climate, constructing a broadcasting medium to promote national community proved a difficult task.

Such a complex climate forced governments to also focus actively on nation-building. New countries like Czechoslovakia or countries whose borders had been redefined like Germany and Poland prioritized creating domestic unity in order to adapt to the new situation. Redrawing their boundaries had torn nations apart, separating groups from their compatriots. Consequently, many East European states had developed a "plural" ethnic character. Whereas in Western Europe the concept of a nation coincided with the boundaries of the nation-state, those in the east had a historical sense of belonging to a nation, to maintaining cultural family ties. These ties did not necessarily correspond with state boundaries. Nations increasingly crossed boundaries, signifying that domestically a nation-state included more nations than before. In this new situation, state leaders not only tried to create domestic cohesion, but also struggled to maintain peace with their neighbors.[50]

According to the broadcasters, their medium could serve as an educative and cultural instrument to pursue such national unity. In Western Europe all the early broadcasting stations primarily tried to employ their medium as an instrument to educate their listeners. They aimed to uplift and revive Europe's civilization and culture which had been shattered by the destruction of the Great War. Broadcasters in Central and Eastern Europe like the Czech *Radiojournal* or *Polskie Radjo* tended to address questions of national unity more directly in light of their national political situation.[51] They considered it their personal duty to employ their medium to help develop unity and consensus, even though they initially held little or no formal ties to governments.[52]

The attempt to employ broadcasting for the creation of unity in Eastern Europe was a struggle and problematic.[53] In Poland, new communities now had to be welcomed. In the meantime, outside its borders lived large communities of Poles

50 Zara S. Steiner, *The Lights that Failed: European International History, 1919-1933* (Oxford; New York: Oxford University Press, 2005), 510-533; Anthony D. Smith, "National Identity and the Idea of European Unity," *International Affairs* 68, no. 1 (1992): 61-62.

51 Scannell and Cardiff, "Serving the Nation"; Jirgens, *Der deutsche Rundfunk der 1. Tschechoslowakischen Republik*; Miazek, *Przeminelo z radiem*.

52 Jirgens, *Der deutsche Rundfunk der 1. Tschechoslowakischen Republik*.

53 Burrows, "Broadcasting outside the United States," 39.

whom Polish broadcasters considered part of their nation as well. The latter in particular caused problems. Director of *Polskie Radjo*, Chamiec found that broadcasting could establish a daily link with fellow Poles now living abroad in the newly defined German areas. Broadcasting across borders by newly established frontier stations enabled these groups to stay in touch with their language by listening in to daily domestic Polish broadcasts.[54] Such programs could easily lead to international tension if the other country felt offended by "harmful" broadcasting content. Such disputes over unintentional versus intentional "harmful" propaganda could not be resolved at national level. Broadcasters felt the need to address these ideological problems internationally.

Broadcasters also experienced difficulties in finding appropriate standards for the society-building content of their national daily broadcasts. First of all, they realized that broadcasting required an entirely different way of communicating with audiences than any other form of cultural arts or technology. Previously, people had always gone out to listen to music together. They could go to the theater to hear live music or listen to gramophone records at home.[55] Listening to music was a social experience. Early radio sets did not enable such mutual listening activities, nor did they provide the opportunity of direct contact between audience and performers. Initially, people could listen individually via a headphone, or together alone when it became possible to connect multiple headphones to one receiving set. Such early listening behavior brought all conversation and social activities to a halt. Only in the early 1930s did the first receiver sets appear with loudspeakers, enabling audiences to experience broadcasting as a truly social small-scale event within the walls of their own homes.[56] Broadcasters sought suitable and effective ways of communicating their message of national unity to their audiences via broadcasting programs which complied in the best possible way with the new technology.

54 Miazek, *Przeminelo z radiem*, 122. Note that broadcasting organizations did not necessarily have to construct stations near their frontiers to facilitate broadcasting to compatriots living near their borders. Country capitals or important centers for cultural development also lay in such border regions. Limiting the power of these frontier stations or not placing a station in such regions "would be to curtail the development of national broadcasting, which would no longer be an adequate service." LoN, *Broadcasting and Peace: Studies and Projects in the Matter of International Agreements* (Paris: International Institute of Intellectual Co-Operation, 1933), 216.

55 Heinz-Gerhard Haupt and Philipp Ther, *Europe and Beyond: Transfers, Networks and Markets for Musical Theatre in Modern Europe, 1740-1960*, Project proposal (Florence: European University Institute, 2008); Sophie Maisonneuve, "Le disque et la musique classique en Europe, 1877-1949: L'invention d'un médium musical, entre mutations d'écoute et formation d'un patrimoine" (Ph.D. diss., European University Institute, Department of History & Civilization, 2002).

56 Shaun Moores, "'The Box on the Dresser': Memories of Early Radio and Everyday Life," *Media, Culture and Society* 10 (1988): 23-40.

Figure 2.1 – Preparations for the Belgian radio news
Source: IBU, "La salle de rédaction du journal-parlé de l'INR, 1 feb 1935," box 58 Photographies, IBU. With special thanks for the European Broadcasting Union, Geneva, who allowed me the usage of their picture archive.

The rapidly growing and changing radio audience did not make matters any easier. With the diffusion of affordable receiver sets as well as the introduction of the loudspeaker, audiences started to develop their own program preferences.[57] Before regular broadcasting services began, radio amateurs had enjoyed "listening in the ether," enjoying the search for signals from far-away places. Consequently, the first radio station in the Netherlands oriented its "programs" towards a relatively technical listening audience.[58] By the time organizations began regular broadcasting, listeners increasingly cared about the content of the signals they received, giving broadcasting a place in their daily lives. They demanded more programs that reflected their personal preferences rather than those ordained by their

57 Brian Currid, *A National Acoustics: Music and Mass Publicity in Weimar and Nazi Germany* (Minneapolis: University of Minnesota Press, 2006); Jirgens, *Der deutsche Rundfunk der 1. Tschechoslowakischen Republik*; Scannell and Cardiff, "Serving the Nation."
58 Onno de Wit, "Radio tussen Verzuiling en Individualisering," in *Techniek in Nederland in de Twintigste Eeuw: Deel V Transport en Communicatie*, ed. Johan Schot and Onno de Wit (Eindhoven: Stichting Historie der Techniek, 2002), 208-211.

stations' high culture and educative cultural policies.[59] Audiences complained and pleaded for more low key and popular entertainment.

Though broadcasting employees also argued for a diversification of broadcast contents, the European broadcasting stations did not relent.[60] John Reith, director of the BBC in 1924 remarked when people complained: "we are apparently setting out to give the public what we think they need, and not what they want, but few know what they want, and very few know what they need."[61] The Czech *Radiojournal* compared so-called *Unterhaltungsmusik* with U.S. entertainment music. It would be "...tantamount to desertion."[62] Though reluctant to give in to their audiences, broadcasters still thought of suitable standards for program content that upheld their idea of high culture nation-building without losing their audiences. They felt that live broadcasts alone could achieve such results.[63] However, the costs of such educative and live programs were high. It was not easy to create a radio schedule that contained a considerable percentage of live programs.

Broadcasters thought that creating international standards could solve practical, technological, ideological, and programming issues. Negotiating such standards would enhance collaboration and resolve these problems to everyone's satisfaction. In the United States, broadcasters could organize their networks in relative uniformity. Europe's distinctive systems of administration and control prevented an easy resolution of such international questions.[64] European broadcasters agreed that only a coupling and adaptation of their systems via international agreements and coordination could settle these issues.

Lack of International Regulation

International broadcasting regulation was non-existent. International organizations dealing with communications or infrastructure projects in Europe had

59 Jirgens, *Der deutsche Rundfunk der 1. Tschechoslowakischen Republik*, 40.

60 BBC chief engineer Eckersley, who himself had featured in BBC programs for years, reflects on program policies in Europe in his memoires. He argues that improved reception of a plurality of waves would encourage specialization of genres on air with special jazz or classical music channels for instance, that would increase and diversify the supply of music. Peter P. Eckersley, *The Power Behind the Microphone* (London: J. Cape, 1941), 21.

61 Scannell and Cardiff, "Serving the Nation," 163. Quoted from John Reith, *Broadcast over Britain* (London: Hodder and Stoughton Ltd., 1924), 37.

62 "...einer Desertation gleich." Jirgens, *Der deutsche Rundfunk der 1. Tschechoslowakischen Republik*, 40-41.

63 Michele Hilmes, *Only Connect: A Cultural History of Broadcasting in the United States* (Belmont, CA: Wadsworth/Thomson Learning, 2002), 43.

64 IBU, *The Problems of Broadcasting* (Geneva: IBU, 1930), 105.

decided not to take on the emerging activity of broadcasting after the war. The International Telegraph Union (ITU) focused on wireless communications and technical committees of the new League of Nations dealt with infrastructure development. Neither wanted to develop or take the lead in the international development of broadcasting.

The ITU had been dealing with problems in point-to-point communications for over half a century. Since its establishment in 1865, it had sought to reach uniformity in the international telegraph systems. It coordinated the operation of telegraph networks and services and advanced the development of telegraph technologies. To these activities it added telephony in 1903 and radiotelephony in 1906.[65] The inclusion of radio-telephony went hand in hand with the unofficial establishment of the International Radiotelegraph Union (IRU) in 1906. Supervised by the ITU Bureau, the IRU focused on wireless radiotelegraph related questions.[66] Consequently, by 1922 the ITU possessed substantial knowledge on wireless communications and the accompanying international bottlenecks.

Nonetheless, ITU did not take up the new development of radio broadcasting. As international governmental organization, ITU represented national post and telegraph administrations all over the world. For its day-to-day activities it depended on the expertise of radio-electricity engineers in government employment. Though their transnational activities at ITU gave these engineers a certain amount of freedom to pursue their personal interests, government decrees restricted their ability to maneuver. Consequently, when European governments decided to restrict and hamper broadcasting experiments, these radio telegraphy and telephony engineers could not but ignore early international broadcasting problems. ITU had organized the previous World Conference of Telecommunications in Washington in 1920. This dealt with all aspects of wireless, but had no consequences for broadcasting.[67] Moreover, the one country that did pursue broadcasting freely, the United States, did not become an ITU member until 1932. Conversely, broadcasting in Europe was still in an experimental phase and not expected to proliferate for another three to four years. There seemed to be no sense of urgency in dealing with broadcasting related problems.

---

65 George A. Codding, *The International Telecommunication Union: An Experiment in International Cooperation* (Leiden: Brill, 1952), 32.

66 Though the ITU never established an actual Radiotelegraph Union, the term generally referred to "the group of countries signatory, or adherent, to the various Radiotelegraph Conventions." Codding, *The International Telecommunication Union*, 81.

67 The Washington Conference held in November 1920 drew up a draft Convention and a set of Regulations regarding telegraphy, telephony, cable, and radio-communications. These concerned point-to-point communications rather than point-to-mass communication media like wireless broadcasting. Ibid., 110.

Although the League of Nations had no international experience with wireless communication policies, since its foundation it had been seeking ways to contribute to international developments. In 1922 the League established a variety of "technical committees" designed to assure its expertise in Health Organization, Economic and Financial Organization, and an Organization for Communications and Transit (OCT).[68] With the acceptance of the League, participants at the 1919 Paris Peace Conference had opted for "a universal body of dialogue, of consensus, and arbitration, capable of avoiding conflict that peoples feared."[69] The League would be a "new path in international co-operation," not so much to avert war, but by being "an instrument to achieve a 'constructive peace' through building a 'new world order'..."[70] The organization built on the prevailing idea among the loosely coupled international elites that dialogue, joint efforts, and interests would promote peace, understanding, and international security. The technical committees would pursue these goals in a practical manner. OCT focused on infrastructure and quickly established sub-committees in the areas of rail, electricity and roads. However, it never created a permanent committee for broadcasting, telegraphy or telephony.

During the mid-1920s, the League's OCT performed an exploratory study of wireless communications. It established the temporary Special Committee on Enquiry of Telegraphic Questions, and invited several ITU experts to sit on the Committee.[71] On the basis of these exploratory studies, the Special Committee

68 For some early studies on the League of Nations see: Greaves, *The League Committees and World Order*; League of Nations, *Ten Years of World Co-Operation* (London: Secretariat of the League of Nations, 1930). For two recent and excellent studies see: Patricia Clavin and Jens-Wilhelm Wessel, "Transnationalism and the League of Nations: Understanding the Work of its Economic and Financial Organisation," *Contemporary European History* 14 (2005): 465-492; Frank Schipper, Vincent Lagendijk, and Irene Anastasiadou, "New Connections for an Old Continent: Rail, Road and Electricity in the League of Nations Organisation for Communications and Transit," in *Materializing Europe: Transnational Infrastructures and the Project of Europe*, ed. Alexander Badenoch and Andreas Fickers (London: Palgrave, 2010), 113-143.

69 "l'organe universel de dialogue, de consensus et d'arbitrage capable d'éviter le conflit que redoutaient les populations." Franck Théry, *Construire l'Europe dans les années vingt: L'action de l'Union paneuropéenne sur la scène franco-allemande, 1924-1932*, Euryopa (Genève: Institut européen de l'Université de Genève, 1998), 8.

70 Sir Arthur Salter, Director of the League's Economic and Financial Section, during the Economic Conference of the International Federation of League of Nations Societies in 1928. Jo-Anne Pemberton, "New Worlds for Old: The League of Nations in the Age of Electricity," *Review of International Studies* 28, no. 2 (2002): 314.

71 League of Nations (hereafter: LoN), *Advisory and Technical Committee for Communications and Transit, Procès-Verbal of the Fourth Session, held at Geneva, April 23-30, 1923*, LoN document series, C.486.M.202.1923.VIII (Geneva: LoN, 1923); LoN, *Advisory and Technical Committee for Communications and Transit, Procès-Verbal of the Fifth Session, held at Geneva, Aug. 29-Sept. 1, 1923*, LoN doc. ser., C.667.M.267.1923.VIII (Geneva: LoN, 1923); LoN, *Advisory and Technical Committee for Communications and Transit, Procès-Verbal of the Sixth Session, held at Geneva, March 12-14,1924*, LoN doc. ser., C.196.M.61.1924.VIII (Geneva: LoN, 1924); LoN, *Advisory and Technical Committee for Communications and Transit, Procès-Verbal of the Seventh Session, held at Geneva, Nov. 26-29, 1924*, LoN doc. ser.,

proposed organizing a world radio-telegraphic conference in 1924. Such a conference would require continuous collaboration between the League OCT and ITU.[72] To the dismay of several Assembly delegates, the League decided not to follow the Committee's advice. The delegates considered such a conference "absolutely necessary," and the League of Nations "the body best qualified to carry this through."[73] Nonethless, the League considered such a conference too premature. By late 1923, several issues on telegraphy had still not been resolved. Moreover, organizing such a conference now would disturb other international actions outside the sphere of the League concerning the revision of the 1920 Washington Telegraphy regulations, activities in which ITU was heavily involved.

In the meantime, the League explored the options for using the new medium for its own public relations. In the first half of the 1920s, the League established close contact with several broadcasting pioneers, and agreed to join in various cross-border radio broadcasting experiments. As a result, already on December 13, 1920 "some hundreds waited intently for the first sounds which would come from the gigantic loud speaker which had been set up" in the hall of the League of Nations in Geneva. Marconi performed an experimental relay of speech all the way from London. Though reception was poor, in between the shrill shrieking, the listeners could clearly hear a voice saying "Hello, Geneva!"[74] People were amazed.

In the ensuing years, a considerable number of experiments followed that centralized the relaying of important League events across Europe. Several times the League collaborated in relaying a presidential address, followed by the historical speeches by Aristide Briand and Gustav Stresemann in 1926 on the occasion of Germany's admission to the League of Nations.[75] In 1926 European broadcasters even managed to realize a live overseas connection with the United States,

C.72.M.36.1925.VIII (Geneva: LoN, 1925); LoN, *Advisory and Technical Committee for Communications and Transit, Procès-Verbal of the Eighth Session, held at Geneva, July 24-30, 1925*, LoN doc. ser., C.621.M.203.1925.VIII (Geneva: LoN, 1925).

72 LoN doc. ser., C.196.M.61.1924.VIII, 3. Initially the CCIF had been set up as a *Comité préliminaire pour la téléphonie à grande distance en Europe* in 1924, but became a permanent advisory body of the ITU in 1925.

73 Briggs, *The Birth of Broadcasting*, 283.

74 Jacot and Collier, *Marconi – Master of Space*, 170.

75 Broadcasters relayed the historical speech by Briand and Stresemann via cable to French, Swiss, and German stations. The Danish Copenhagen and Soroe stations in turn captured the signal from the German station in Köningswusterhausen, and relayed the speech all across Denmark. These wireless relay experiments had been complemented by more regional efforts as well. In June 1926, for instance, the Scandinavian countries, under the auspices of the *Ligue des Femmes pour la Paix et la Liberté* relayed a program of discourse and music. Though the factual relay from Copenhagen suffered from distortions due to submarine cables crossing the Sund, the relay from Stockholm was an excellent result. "Relais Internationaux, Résume de la situation présenté par le Secrétaire Général," Série 563, 15 juin 1927, 2, box 85, file Relais Général I, 1927 et précédents, IBU.

enabling League Secretary-General Eric Drummond to address U.S. audiences directly. Drummond drew a direct link between radio and the League of Nations.

> Radio and the League of Nations are both in their youth. They were born at approximately the same time, and are growing and developing on somewhat the same lines. Assuredly their fruitful co-operation will make for the peace of the world. How rapidly civilization has travelled!...It is not easy, even for those who serve the cause of peace, to realize the enormous new force...distance and time almost cease to have importance. ...The full development of radio must render international co-operation, and therefore the task of the League, infinitely easier.[76]

The League showed an interest in broadcasting from the very first. As a result, it not only participated in international relay experiments, but also showed its belief that broadcasting could be used for diffusing its aims and ideology. The League was in touch with early European broadcasters and acquainted itself with their activities.

Nonetheless, the League OCT did not take any concrete steps towards creating an international policy for broadcasting. By 1926 its originally intended as temporary Special Committee of Enquiry on Telegraphic Questions functioned as a semi-permanent body. On the basis of its relations with ITU and established contacts with the European broadcasters, the Special Committee decided to step down. A permanent committee on radio telephony within OCT was unnecessary, the Committee concluded. ITU had expertise for a long time in wireless communications. New, potentially interesting developments in radio broadcasting deserved space to expand. OCT decided to stay informed of further developments and reserved the right to study communication related questions of any nature if desired.[77] The Special Committee, OCT and ITU decided not to create international broadcasting regulations as the technology was too "young" and changing too quickly. The decision left room for international efforts that were not driven by governments.

76 "Radio speech by Eric Drummond to U.S. Audiences," 20 January 1925, catalogued as 14.48542X, 28231 X, 1-2, box 14, R-1140, file 48542/28231, LoN.

77 LoN, *Commission Consultative et Technique des Communications et du Transit, Procès-Verbal de la 9ième session, tenue à Genève, 12 au 17 juillet 1926*, LoN doc. ser., C.595.M.231.1926.VIII (Geneva: LoN, 1926), 28-30.

Establishing International Broadcasting Union

European broadcasters took the initiative. At the same time that broadcasting was developing in their countries, they began discussing how to deal with the associated international problems. These discussions were in keeping with their own tradition of cartel capitalism and also with another European tradition, that of international organizations.

Craig Murphy argues that in the second half of the nineteenth century, many international organizations were established in a European tradition of conference systems. Conference attendees were a mix of what Murphy calls "experienced commoners" and members of the European *ancien régime*. They believed in encouraging peace and understanding via practical projects, and also organized these events and projects in an atmosphere that breathed nineteenth century European civilized values. Murphy finds that these conferences led to the establishment of international organizations such as the International Telegraph Union in a process of four steps: First of all, conference participants desiring to create an international organization would *propose a design*. They reached agreement on the design, in other words the nature of the collaboration. Secondly, a *powerful individual* had to take on the responsibility of the project. The project for a new international organization would have to obtain the *support* of powerful individuals and/or an individual state. A final conference would then make the new international organization reality by *creating a constituency*. In Murphy's view, this European tradition of conference systems created an international organization with a very precise ideological discourse, and long-established sets of civilized behavior.[78]

Though European broadcasters agreed on the need for international collaboration, initially they encountered disagreement on the nature of such cooperation. Should the private companies seek government regulation or establish their own international organization of broadcasters working outside governments? The BBC supported the first idea, arguing that "wavelengths are matters for government sanctions" only.[79] Other parties pleaded in favor of a non-political organization, and proposed arranging a conference to address the issue. For its stance in favor of a political route, the BBC argued that such a conference would then have an unofficial character and only be able to make recommendations.[80] Private broadcasters were the wrong public to deal with such issues, it argued. Consequently, the BBC refused to attend such a gathering. Thus the disagreements proved no simple

78 Murphy, *International Organization and Industrial Change*, 52, 71.
79 Briggs, *The Birth of Broadcasting*, 284.
80 Ibid.

matter to overcome, especially when the biggest and most influential broadcasting organization in Europe, a role model for many, refused to join in.

In 1924 a powerful Swiss individual by the name of Maurice Rambert came forward to overcome these design problems. He took steps to organize a Preliminary Conference for an International Agreement on Wireless Telephony in Geneva on April 22-23, 1924. Rambert was director of the *Société Suisse de Radiodiffusion* (SSR) and owner of *Radio-Electrique S.A.*, the first Swiss organization to build and sell radio-electric apparatus. He was from a noble Swiss family, a passionate violin player and music composer, with a background in banking and engineering. He had received the first concession in Switzerland allowing him to pursue public radiotelephonic emissions.[81] By bringing together state administrations, private associations, radio-clubs, radio industry and the wireless press, Rambert hoped the conference would find a solution to the issue of wavelengths, filling the regulatory gap left by national governments and existing international organizations.[82] Rambert managed to follow through with the conference, and as such, achieved the first steps towards establishing an organization outside the governmental sphere. Though such an organization could make recommendations rather than binding decisions, these would suffice for the time being, the conference participants argued. In time, they would be adopted by national governments and international organizations to become legally binding.

Rambert's idea for a non-governmental international organization obtained the *support* of the League of Nations, the third stage pointed out by Murphy. Not only did the League of Nations kindly host these broadcaster meetings, a member of the League's Transit Section Robert Haas also attended. Though the conference ultimately achieved few practical results, it did lay the basis for an International Union of Broadcasters. The League's OCT seems to have supported this idea. Haas even pledged that if "necessary at a later stage, the League might be able to help the Union to endeavour to obtain action through Governments."[83] In addition to Haas, other participants were important international players. Mr. Boisnier of the International Labor Organization attended along with Mr. Calame of the Swiss PTT administration, an engineer of the Bell Telephone Manufacturing Mr. Insbruck, Mr. Pitlik of the Czech Ministry of Trade, and Mr. Edwards of the British periodical "The Wireless World."[84] Consequently, the idea for an international non-governmental broadcasting organization gained the support of a wide

81 IBU, *Maurice Rambert, 1866-1941*, 6-8.

82 IBU, *L'Union Internationale de Radiodiffusion: Son histoire – son activité* (Geneva: IBU, 1944), 7-14.

83 "European Conference of Broadcasting Engineers, July 6, 1925," 1, P.V. Commission Technique, IBU.

84 IBU, *Twenty Years*, 9-10.

variety of influential members of society and state. Thus there was also a substantial *constituency* for an international organization for broadcasting, the fourth stage pointed out by Murphy.

As a result of these preparatory meetings, European broadcasters together managed to give the constituency a firm basis. Lengthy negotiations between the various private broadcasters, especially between Rambert and BBC director John Reith who managed to settle their differences, followed in London on March 18-19, 1925.[85] On April 3-4, 1925, ten European broadcasters established the non-governmental *Union Internationale de Radiophonie* (UIR). The Union, also known by its English name International Broadcasting Union (IBU), had its headquarters in Geneva.[86] The route that led to the establishment of IBU seems to have followed the four preconditions described by Murphy. Its emergence suggests that the non-governmental IBU was established in a European tradition of international organizations, where according to Murphy, the ideology of the European *ancien régime* persisted in an adapted form. This ideology combined late nineteenth century nationalism, "the glue that would hold society together," with the international behavior of the *noblesse oblige*.[87] Such orientation seemed to match the personal backgrounds and ideological beliefs of broadcasters like Marconi, Rambert, Chamiec, and Solari, who had also established the first broadcasting organizations on the European continent.

Figure 2.2 – International Broadcasting Union logos in the interwar years
Source: IBU, Radiodiffusion (Geneva: IBU, 1935-1939); IBU, L'activité de l'Union Internationale de Radiodiffusion de 1939 à 1946 (Genève: IBU, 1946). Used by the courtesy of the European Broadcasting Union.

85 For the complex and tense negotiations between BBC director John Reith and Rambert see: Briggs, *The Birth of Broadcasting*, 310-315.

86 The organizations that founded IBU were from Austria, Belgium, Czechoslovakia, France, Germany, Great Britain, the Netherlands, Norway, Spain, and Switzerland. IBU, *Twenty Years*, 12, 65-72.

87 Murphy, *International Organization and Industrial Change*, 52-53.

The IBU aimed to resolve the international problems of broadcasting in a practical manner. In 1926 it explained that broadcasting "is a form of applied science *which compels* international collaboration for its successful development."[88] It aimed to deal with the given that wireless waves knew no frontiers, that stations everywhere in Europe were growing in power and number, and that broadcasting programs were a matter of private property only. Such diverse and complex problems required an encompassing international organization that accounted for all sorts of broadcasting related issues. As a result, the IBU founders quickly formed a complex group of broadcasting experts with astoundingly hybrid and diverse backgrounds such as law, finance, engineering, journalism, and musicology. These experts prioritized the coordination of broadcasting developments in general, the centralization of the study of all issues from a technical, legal, and social point-of-view, and the protection of the interests of broadcasting in relation to other means of wireless media.[89]

These activities would not only resolve practical problems, but also facilitate the economic interests of these broadcasters. The international centralization of research and development in quality of production, transmission, program content, and reception would facilitate the rationalization of production substantially. Moreover, the experts argued that their efforts would improve the quality of broadcasting to such a degree that the medium would attract a wider radio audience. With a growing demand for radio sets in Europe, the radio industry would receive an impressive return on investment. Since these early broadcasters often represented or had direct connections to the radio industry, they would also benefit from such return on investment. Their line of reasoning proved correct. In 1920 only a few thousand homes in Europe had receiver sets, in 1926 this number had grown to 5.8 million and to 16 million sets in 1931.[90] Not surprisingly, their economic interests recurred on the agenda of the Administrative Council at regular intervals in spite of its Statutes that made clear the public non-commercial character of the organization.[91]

The practical and businesslike approach went hand in hand with the idea of creating a social community as these broadcasters saw fit. According to Akira Iriye, "culture was being 'produced,' not automatically but through the medium of forces that were eager to re-establish some sort of order in...society. That these forces...were fundamentally business-oriented, stressing themes like prosperity, productivity, and

88 Stress is the original. IBU, *International Union of Broadcasting Organisations – Union Internationale de Radiophonie* (Geneva: IBU, 1926), 11.

89 Ibid.

90 The IBU assumed 3.8 persons per home of seven years and older listening to the radio. The number could even be 4 if guests were also listening in. IBU, *The Importance of Broadcasting* (Geneva: IBU, 1932), 6. Found in registry file 9G, box R-2601, file 13698/2081, LoN.

91 UIR, "Statuts," Série 176, 23 mars 1926, 1, box 33 l'Histoire de l'UIR – statuts, IBU.

efficiency, is clear."[92] The broadcasters combined their European tradition of cartel capitalism and the European tradition of international organizations within one institutional structure.

European governments created turmoil when they nationalized broadcasting. For the second time, they hindered broadcasting development and went against the grain of the cartel capitalist tradition. Initially, they had forbidden granting concessions for broadcasting in favor of national defense right after the war. Now they increasingly nationalized broadcasting domestically based on the same argument. With nationalization, governments increased their influence over the medium, not only by obtaining an official stake in the organization, but also by centralizing control over the medium. Various governments merged the existing stations in their countries which were "operated by amateur clubs, newspapers, radio manufacturers, and the state, into a single national broadcasting monopoly."[93]

Organizations other than those nationalized could only continue to a limited degree. Amateurs were only permitted to use the radio waves if they were members of the highly controlled radio amateur clubs. These could join the International Amateur Radio Union (IARU) established in 1925. For reasons of transparency, lists of these members circulated widely between national PTTs, governments, broadcasting organizations, and the International Telegraph Union. Registered amateurs had the right to broadcast under certain conditions, though in some countries their governments forbade them to broadcast at all, making them focus on wireless telegraphy alone.[94] Broadcasting operations in Europe increasingly belonged to the fortunate few.

Not all governments nationalized the medium in the same manner. Some placed it in the hands of one or more government departments, whereas others turned it into a public utility, obtaining a seat on the public company board. With the latter construction, broadcasting organizations often maintained a high degree of freedom in their cultural and programming activities. Commercial broadcasting by means of private ownership decreased, and by the late 1930s only existed in Europe to a

---

92 Akira Iriye, *Cultural Internationalism and World Order* (Baltimore: Johns Hopkins University Press, 1997), 84.

93 Spohrer, "Ruling the Airwaves," 12. Extracted from: Radiofonica Zagreb, "Organisation de la radiodiffusion en Europe, document revise," June 13, 1935, box 53, Organisation de la radiodiffusion, IBU.

94 Box 94bis, Stations d'amateurs, privées et clandestines, 1924-1949, IBU. In the course of the interwar years, several amateur clubs requested admission to the IBU. It always refused these requests for legally it agreed to only represent official broadcasting organizations with nationwide coverage. For an example from Germany see: "Letter from the German Verband der RadioHörer to the International Broadcasting Union," May 17, 1946; "Letter from the International Broadcasting Union to the German Verband der RadioHörer," June 14, 1946, box 2, Adhésions – recrutements, Démissions – Radiations, IBU.

limited extent. By 1937, public and private broadcasting continued to co-exist only in France, Spain, Luxembourg, and Yugoslavia.[95]

The nationalization of most European broadcasting stations posed institutional tensions internationally. Immediately after its establishment, the IBU encountered the situation that many but not all of its members changed from being a private enterprise to a public organization. Institutionally, the IBU community tried to fine-tune its private origins with the newly acquired and increasing influence of the state. Whereas governments wanted to construct broadcasting for nation-building and national defense, IBU broadcasters also wanted to help resolve international problems and encourage increased collaboration. Consequently, the IBU sought an institutional construction that made room for the national without interfering with their international goals.

Over the years, the IBU's institutional structure began to reflect these stronger national ties, while broadly maintaining its private origins (Figure 2.3). The IBU's highest organ, the General Assembly, consisted of representatives from each IBU member organization. Once many member organizations centralized, these representations increasingly denoted the nation-state. By allowing each member the right to vote, the General Assembly approved all decisions taken by the Administrative Council. Accordingly, institutionally the nation-state always had the final say.

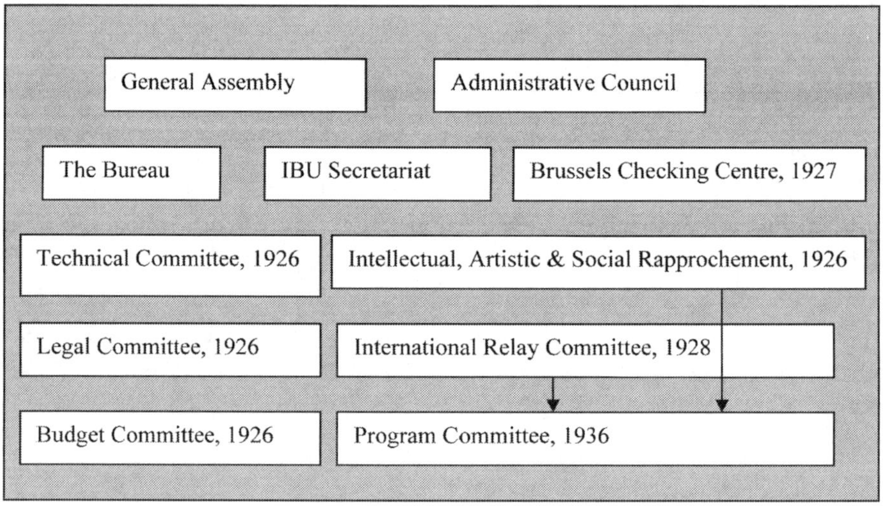

Figure 2.3 – Organizational structure of the International Broadcasting Union
Source: IBU, Twenty Years of Activity of the International Broadcasting Union (Geneva: IBU, 1945). Interpretation of the author.

95 Huth, *La radiodiffusion*, 37-38, 52-53.

All other IBU positions tended to reflect the private origins of the IBU. The Administrative Council, the IBU Bureau, and the executive committees reserved posts for individuals who represented all IBU members collectively. These people were supposed to be acting on their own behalf rather than as representatives of their organizations. Although we can question whether they placed their national interests first, their behavior probably differed with the occasion and per representative. The Administrative Council consisted of eleven people who represented all IBU members. Not surprisingly, the first Council was made up of many who had established a broadcasting station in their own country and co-established the IBU in the course of preparatory conference meetings. They were Vice-Admiral Carpendale from the United Kingdom, Heinrich Giesecke from Germany, Robert Tabouis from France, Rambert from Switzerland, Antoine Dubois from the Netherlands, Auguste Hubert from Belgium, Svoboda from Czechoslovakia, Skottun from Norway, and García from Spain.[96] They decided which studies the various committees were to perform. They defined the agenda of the IBU. The Council thus had a high degree of decision-making power, especially since they met twice a year whereas the Assembly met only once.

The IBU Secretariat or Bureau consisted of the Secretary General, the Presidency, and the *délégué-conseil* who not only maintained contact with other international organizations like the League of Nations, but also acted as "a permanent intermediary between the Council and the Director of the Office."[97] The day-to-day IBU staff was small. In 1925 it consisted of four people, and grew to a mere fourteen regular staff members in 1937. With the outbreak of World War II, the number of staff fell to four again.[98] Arthur Burrows, long-time colleague of Marconi and Eckersley as well as BBC program director, became the Union's Secretary General. He moved to Geneva to coordinate the IBU's daily activities. His colleague Vice-Admiral Charles Carpendale, BBC controller, became IBU President. Radio Paris representative Robert Tabouis and German RRG representative Heinrich Giesecke obtained the Vice-Presidency, a position they continued to hold until 1935. Maurice Rambert became the *délégué-conseil*. With the exception of Burrows, these men did not have to move to Geneva. They traveled back and forth. The Bureau performed the administration for both the Administrative Council and General Assembly, managing the Union's day-to-day activities and supporting IBU committees.

---

96 IBU, *Twenty Years*, 13-14.

97 "Statutes of the International Broadcasting Union, Adopted by an Extraordinary General Assembly of the International Broadcasting Union, Budapest in October 1930," 127, box 33 l'Histoire de l'UIR – statuts, IBU.

98 Ernest Eugster, *Television Programming Across National Boundaries: The EBU and OIRT Experience* (Dedham, MA: Artech House, 1983), 32.

The executive committees brought together experts in the various fields of broadcasting. The Technical Committee focused on coordinating cross-border infrastructures, electrical interference, or wave propagation issues. In 1927 the Director of the Technical Committee also became head of the new Brussels Checking Center. In Brussels, engineers performed practical tests with frequency allocations, surveying long waveband, as well as transmission stability. In March 1926 the Legal Committee took up all issues concerning illicit propaganda, copyright, and authors' rights. In July the Committee for Intellectual, Social, and Artistic Rapprochement began examining program exchange, for which it established a sub-committee to deal with international relaying over long distances. In principle it aimed to facilitate and develop the use of radio to encourage rapprochement between peoples. In 1928 its sub-committee became an independent body called the Committee on International Relays, but in 1936 merged into the Program Committee.[99] The Budget Committee took care of the Union's financial affairs.[100] These committees provided the IBU with specialized expertise in every area of broadcasting.

The executive committees were primarily responsible to the Administrative Council who determined most studies these committees were to carry out. In between Council meetings, the committees studied the issues then communicated their results to the Council. The committees met at regular intervals in Geneva or elsewhere depending on their workload. There seemed to be no exact rules for the frequency of meetings, which varied from five to twelve times a year. Having discussed and approved the studies, the Administrative Council would submit the results to the General Assembly for official approval and implementation. Though the nation-state always had the final word via the General Assembly, the major part of the IBU's work seems to have been conducted in accordance with the initial collaboration defined by the private broadcasters at the start of European broadcasting. The IBU thus created an institutional standard with a firm basis for national interests. The various broadcasting experts appear to have been able to negotiate amicably their national priorities alongside international needs. This followed their European tradition of cartel capitalism and international organizations.

The private traditions of early broadcasting organizations in Europe could continue because a very small group of broadcasters occupied the most important positions nationally and in the IBU. At first glance they remind us of the pre-war European balance-of-power situation. The BBC obtained IBU Presidency as well

99 IBU, *Twenty Years*, 16-17.

100 The Budget Committee had a minor role. It mostly seems to have managed the IBU membership fees as well as yearly expenditure.

as the position of Secretary-General, whereas France and Germany obtained Vice-Presidency. Several leading members were from countries other than the big three. The directors of the executive committees were from central and eastern Europe. Although it was a Frenchman, Raymond Braillard, who became director of the Technical Committee and Brussels Checking Center, the Czech Ladislav Sourek became director of the Legal Committee, the Austrian Oscar Czeija became director of the Rapprochement Committee, and Zygmunt Chamiec was Director of the International Relay Committee. Thus West and East European members occupied the key roles. In later years, North European members would also hold these posts in rotation. With the exception of southern Europe, over the years the important positions therefore reflected a relatively equal division of power between European members of the Union.

While creating institutional standards, these broadcasters fine-tuned the increasing national impact on their international activities, taking into account a European versus global sphere of influence. One might suggest that statutorily the IBU saw itself as an "international" organization representing the interests of European broadcasting organizations. In this sense "international" should equal "Europe." Nonetheless, in 1926 already, the IBU communicated that "As the majority of these problems, whether technical, legal or artistic, are of a worldwide character, it is felt that the time has come when the benefits arriving from this Union of broadcasters should be extended to broadcasters in all countries."[101] The IBU clearly defined broadcasting as a global issue without boundaries that required global collaboration. In 1927 the IBU finally set its institutional membership standards. These confirmed a very Eurocentric kind of globalism with a central position for European members and affairs. The so-called "active members" were from Europe, and "associate members" came from what the statutes called Extra-European countries. The former had the right to vote, the latter did not.[102] Institutionally Europe acquired a central position in relation to the rest of the world, not only as initial focus for activities, but also regarding membership.

Over the years the IBU became the main transnational hub for broadcasting related issues. Between March 1925 and 1938-1939 the IBU grew from ten founding European organizations to fifty-nine active and associate members.[103] Initially, the IBU had been "a pioneer in a new field," but soon encountered competition from newly established, more specialized international organizations.[104] For example the

101 IBU, *International Union of Broadcasting Organisations*, 11.
102 IBU, *Twenty Years*, 16.
103 Eugster, *Television Programming Across National Boundaries*, 32.
104 Ibid.

Comité International de la T.S.F, also known as *Comité International Radioélectricité* (CIR), established in 1922, focused on economic and legal issues. *Bureau Catholique International de la Radiodiffusion* (BCIR) was established in 1929 as the Catholic center for efforts relating to radio. BCIR had over thirty member nations. And finally in 1936, specialists in agriculture and rural radio initiated *Centre International de Radiophonie Rurale* (CIRR). CIRR aimed to facilitate the creation of rural broadcasting services in countries where no such service existed, using expertise already acquired in other countries.[105] The IBU met these other organizations at conferences and followed their activities with interest. Nevertheless, IBU's ambition to deal with the entire set of broadcasting issues continued to ensure its status as the premier transnational organization in broadcasting, a position it was keen to uphold.

Conclusion

Early European broadcasting took place locally, nationally as well as internationally. At first, broadcasting was a private activity typically outside the governmental sphere. National post and telegraph administrations granted the broadcasting concessions, and private broadcasting organizations then exploited and sometimes built the stations. This construction created mutual interdependencies between PTTs and broadcasting organizations that varied from country to country. Over the years, a wave of nationalization swept through Europe, which, with some exceptions, linked broadcasting more closely to national state policies. Broadcasting in Europe developed into different non-aligned systems.

The moment broadcasters began to shape broadcasting, they came across issues like cross-border electromagnetic interference, interactions with other systems, disturbed international relations as a result of illicit propaganda broadcasts, and defining the "right" international program. They recognized these issues could best be resolved internationally. National governments and international organizations like the League of Nations and the International Telegraph Union did not respond adequately to the need to regulate the new medium at international level. They left a legislative gap. For this reason, various broadcasting organizations established the International Broadcasting Union (IBU), which became the main transnational hub for broadcasting problems in Europe.

At the outset the organization was inherently European, with only European members. Over the years, it extended its services beyond European confines in

105 Huth, *La radiodiffusion*, 336-337.

line with its initial desire to be a global organization. The organization pursued a Eurocentric type of globalism which had two-fold forms. First, the IBU's policies suited a commercial European tradition of cartel capitalism based on collaboration and consensus seeking. Second, the broadcasting elites who established the IBU, believed in the European civilizing mission based on European Enlightenment values.

Both forms of Eurocentrism were deeply ingrained in the IBU's institutional structure and its activities. They continued to influence the practices of the IBU community of experts, despite substantial changes in broadcasting technology and organization. For example, the nationalization of broadcasting did not change the nature and commitment to seeking collaboration across borders. This was thanks to a shared tradition among experts and the fact that a very small group of broadcasting experts occupied the key positions nationally as well as internationally. They defined themselves as problem solving experts who did not see any discrepancy between their commercial and civilizing mission. Their position allowed them to negotiate decision-making procedures in the IBU with a minimum of nation-state interference in day-to-day affairs.

The first issue the IBU addressed after its formation was the urgent problem of electromagnetic interference beyond borders. This was a sensitive topic among the governments of various nation-states. Very soon the experts would find they could not resolve this on their own. They needed the collaboration of other international organizations.

Chapter 3
Europe in the Making

Europe is crowded with neighbours
In the world of broadcasting we are all neighbours.
Arthur Burrows[1]

With the above words Arthur Burrows ended his speech before the Committee for Moral Disarmament during the Disarmament Conference in 1933, organized by the League of Nations. Burrows was Secretary General of the International Broadcasting Union (IBU). He addressed a committee that considered the issue of moral disarmament its most important task. Above all, the committee aimed to change what it considered a dominant mindset amongst peoples across the world. This was based on nationalist thinking linked to warfare and the committee wanted to move towards a more "internationalist" mindset that would enhance peace and mutual understanding. Such a change would be a necessary precondition for any material disarmament leading to international peace. At committee meetings, broadcasting had come to the fore as an important instrument to achieve these aims.

Burrows had been invited by the International Students Association and he addressed these students directly at the end of his speech. In his opinion, their internationally oriented training qualified them to provide a critical note to the broadcasting organizations in their home countries. These critical notes would guarantee a just use of broadcasting to achieve international rapprochement. "You [international students]," Burrows stated, "in time may appear before the microphone in your own country and help your own people to realise the highest that patriotism extends beyond the State to which you are a loyal citizen, to the whole world to which you have been called to spend a life."[2] According to Burrows,

1 "Address (by Secretary-General A.R. Burrows) before the Conference on 'Moral Disarmament' arranged by the International Students' Associations, Geneva, April 1 1933," 8, box 74, file Propagande de l'Union où des Membres, Général II, 1932/1936, IBU.
2 "Address (by Secretary-General A.R. Burrows)," 8, box 74, file Propagande, Général II, IBU.

states were an integral part of the international world, but loyalty towards them also went beyond the borders of each state. "Europe is crowded with neighbours. In the world of broadcasting we are all neighbours," he concluded.

Burrows was conjuring up a Europe and a world of brotherhood between nation states. This was precisely the vision the IBU had in mind for its broadcasting infrastructure within and beyond Europe's borders. Although the IBU efforts to build such a system would dominate European broadcasting for most of the interwar years, other broadcasting organizations designed different systems. Stations like Radio Luxembourg and Radio Moscow were owned by a company and state respectively and specifically intended to transmit to audiences beyond their national boundaries. Chapters 3 and 4 examine how broadcasters negotiated the material and related organizational side of their infrastructures with their visions of Europe. We will explore how and why European broadcasting organizations fine-tuned their interests in an attempt to guarantee well-functioning broadcasting. While Chapter 3 focuses on the influential broadcasting system created by the IBU, Chapter 4 examines the battles that arose over Europe's borders. Other systems like Radio Moscow, Radio Nations, Vatican Radio, and Radio Luxembourg would challenge the standards developed by the IBU.

The present chapter focuses on the IBU efforts that initially imagine broadcasting in Europe relatively freely. The process turned out to be more difficult. Not only did IBU members have to agree on the kind of network to build, they also had to find their place as newcomers in a world of established organizations like the International Telegraph Union (ITU) and the League of Nations. This chapter concentrates on IBU's standardization efforts in wireless frequency allocation and its international relay activities. In the eyes of the IBU, these Europe-oriented efforts were two sides of the same coin. In the early 1930s, the development of short waves allowed for global coverage of broadcasts. The new opportunities challenged the European standards finding their way into the technology of broadcasting infrastructures in Europe. Technological developments made the IBU constantly rethink its broadcasting projects.

Invisible Interconnection

Wireless broadcasting activities interconnected Europe invisibly. Already during the early years of broadcasting, the number of radio stations had grown exponentially. This created considerable challenges for the slowly emerging European broadcasting community. In 1925 already, 87 European stations were operating in the available medium waveband for broadcasting, while no fewer than 37 new stations were poised to start in the short run.[3] As a result, broadcasters encountered increasing difficulties due to interference to their signals. Signals could be transmitted through the air via frequencies.[4] When two radio stations transmitted a signal over the same frequency, the signals clashed, and they interfered with each other. Audiences then heard "noise" rather than the actual broadcast. Moreover, if radio stations transmitted broadcasts over frequencies that lay (too) closely together, these broadcasts also caused interference, an electromagnetic disturbance of the other signal.[5] Consequently, only a limited number of signals could be transmitted through the airwaves at any one time. Frequencies were a scarce resource. The new IBU members considered it their primary task to find a way to regulate frequencies between broadcasting stations.

The way such frequencies were designed and allocated determined how the invisible side of broadcasting infrastructure projects would be achieved. Slightly simplified, the act of wireless broadcasting consisted of transmitting invisible electronic signals from radio stations possibly via relay stations, all the way to the receiver sets in people's homes. Decisions about the nature of this process could not be made without some understanding of the way one would like to link, de-link or non-link society. Visions of what society should look like thus found their way into agreements on frequency allocation processes.

To allocate frequencies effectively, broadcasters focused on a wide range of issues. Frequency standards in particular aimed to deal with interference by appointing specific frequencies to different stations with additional strict instructions. These additional standards could entail requirements about the kind of wave, the power of the station, the use of (non)directional antennas, the separation in Herzian kilocycles per second (kc/sec) between appointed frequencies, as well as the number of stations in operation. Such standards determined the overall geographical reach as well as the specifics of the network. They influenced the *radius* and *density* of the network. For example, as a rule, music and speech transmitted

3 The band between 550 and 270 meters. IBU, *Twenty Years of Activity of the International Broadcasting Union* (Geneva: IBU, 1945), 17.

4 A small explanatory note on frequencies: "Waves can be described in terms of wavelength (the distance from peak to peak) or frequency (how many peaks move through a particular point in a given amount of time; 1 Hertz (Hz) = 1 cycle per second). Wavelength was most commonly used in the earliest days of radio. After World War I, technical experts began using frequency rather than wavelength when discussing official matters, since frequency could be measured more accurately." Jennifer Spohrer, "Ruling the Airwaves: Radio Luxembourg and the Origins of European National Broadcasting, 1929-1950" (Ph.D. diss., Columbia University, 2008), 99. Though experts increasingly spoke of frequencies, they continued to refer to wavelength in terms of long, medium or short wave in addition to frequencies.

5 This means something else than "electrical interference", another issue the interwar broadcasting engineers discussed. Electrical interference was caused by electricity networks or railways etc.

over long waves could travel longer distances than over medium waves. As such, a long or medium wave reached listeners within a wider or narrower radius. Such factors enabled people deliberating frequency standards to design their broadcasting infrastructure projects as they pleased and to fulfill their vision of neighborhood in a variety of ways. Not surprisingly, such decisions could trigger heated debates on a continent with as many countries as Europe.

Over the years, the broadcasting expert community was continually re-negotiating broadcasting projects because of the exponential growth of stations in Europe and the rapid developments in technology. These encouraged the various parties to revise their projects at regular intervals. The IBU designed so-called "frequency plans" that were recommendations rather than legally binding agreements. In order to increase the effect of these plans and as newcomer in the international scene, the IBU sought the support of other international organizations like the League's Organization for Communications and Transit (OCT) and the ITU. The IBU hoped that collaboration with these governmental organizations would facilitate the implementation of IBU recommendations into international agreements and national law. Only then would its frequency standards and ideas about European broadcasting become legally binding for all states. This section of Chapter 3 focuses on the negotiation of standards for allocating frequencies. Discussions on technical standards like the power of stations, interference, or testing merely provide contextualization where necessary, for reasons of clarity.

The IBU prioritized creating a frequency allocation agreement. To deal with the problem of interference, the IBU organized a European Conference of Wireless Engineers in Geneva in July 1925. Radio engineers and broadcasting lawyers from around thirteen broadcasting organizations representing some 50 stations attended. These participants not only represented organizations that were already IBU members but also organizations that had been established shortly before, and would become IBU members in the future, like *Polskie Radjo*.[6] Resolving the problems proved no easy task for the participants. They had to reorganize 87 existing stations and include another 37 stations in the short term within the available wavebands consisting of so-called medium waves between 550 and 270 meters and long waves between 1,000 and 2,000 meters. They soon decided to focus on medium waves alone for the time being.

6 Ryszard Miazek, *Przeminelo z radiem: opowiesc o Zygmuncie Chamcu – zalozycielu i pierwszym dyrektorze Polskiego Radia* (Rzeszów; Warszawa: Wydawnictwo Wyzszej Szkoly Informatyki i Zarzadzania; Polskie Radio, 2005), 122.

To deal with the vast growth of stations, the IBU foresaw that it could only solve the problems by encouraging a spirit of cooperation among the conference participants. To achieve this, the opening session hosted by the League of Nations engendered an air of fellowship, stressing mutual European benefits above personal gain or national interests. IBU President Sir Charles Carpendale opened the conference. In his opening speech, he eloquently explained why there was a need for interactions and interrelationships between broadcasters in Europe, referring to the idyllic panoramic view from the League of Nations building.[7]

> An emission post which wishes to function in Europe as an isolated element will be in as dangerous a situation as a person without experience who attempts to pass through the eternal snow on the mountains that surround us. Such a post would risk complete failure, diverting millions of people from radiotelephony and the pleasures it brings.[8]

Everyone needed to understand that their interests did not stop at individual activities. BBC chief engineer Peter Eckersley who chaired the meeting, argued that such a plan would create "the greatest good for the greatest number of European listeners rather than for listeners of any particular country..."[9] It would create peace and understanding. These agreements in turn would feed in positively to the various personal interests.

The positive words during the opening session could not prevent tough negotiations over the drafted frequency plan. The engineers debated every aspect of the plan and did not agree on all points. Vice President of Italian broadcasting Marquis Solari for instance questioned why the draft plan favored large national broadcasting stations with a certain international standing over small broadcasting stations of national importance. He submitted an alternative plan.[10] While the

---

7 Vice-Admiral Sir Charles D. Carpendale (1874-1968) was a minister's son. He served in the British navy, becoming Rear Admiral. When he came in contact with BBC Director Sir John Reid after retiring from the navy in 1923, Reid offered him his deputy post of controller, a position which Carpendale filled until his retirement in 1938. In 1925 he became the IBU's first President, and thanks to his friendliness, genius and wisdom, was reelected annually until the position was finally passed on to the Swiss Maurice Rambert on 26 June 1935. http://www.ocotilloroad.com/geneal/carpendale1.html. Accessed January 20, 2009.

8 "Un poste émetteur qui voudrait fonctionner en Europe comme élément isolé serait dans une situation aussi dangereuse qu'une personne sans expérience qui tenterait de franchir les neiges éternelles des montagnes qui nous entourent. Tel poste risquerait d'échouer complètement et de détourner des milliers de personnes de la radiophonie et des plaisirs qu'elle procure." "Conférence Européenne d'ingénieurs en radiophonie, Séance 6 juillet 1925," 1, P.V. et documents complémentaires Commission Technique, 1925-1930, IBU.

9 Peter P. Eckersley, *The Power Behind the Microphone* (London: J. Cape, 1941), 87.

10 IBU archives contain no further information or discussion on this alternative plan. "International Union of Broadcasting Engineers: European Conference of Broadcasting Engineers, held at Geneva, First

Figure 3.1 – The first European Conference of Radiotelephony Engineers, July 1925
Source: "Première Conférence Européenne d'Ingénieurs en Radiophonie, juillet 1925," box 58 Photographies, IBU. Used by the courtesy of the European Broadcasting Union.

documents on the Geneva Plan focus to some extent on discussions between the engineers, these same documents pay barely any attention to the discussions between the engineers and participating lawyers. Fifteen years later, in his memoirs, Eckersley remembers the fierce discussions between engineers seeking practical international solutions and lawyers who had to account for their national laws. The PTT administrations which granted broadcasting concessions depended on national laws and often granted the concessions with additional strict conditions for use. Thus these national laws could not be ignored. Eckersley recalls regularly losing his patience, at which moments the French engineer Raymond Braillard proved to be a veritable diplomat.

Braillard's background was electric engineering and he worked on long distance wireless telegraphy at first in France, then in Belgium. For Belgium, he designed a wireless telegraph infrastructure for the Belgian Congo and created a national broadcasting service. In December 1925 he became head of the provisional IBU Technical Committee.[11] Braillard organized the establishment of the Brussels

Sitting," July 6, 1925, 3, P.V. et documents complémentaires Commission Technique, 1925-1930, IBU.
11 Asa Briggs, "Peter Pendleton Eckersley," *Oxford Dictionary of National Biographies*, 2008, http://www.oxforddnb.com/view/article/37385. Accessed November 4, 2008; Léonard Laborie, "La France,

Checking Center that performed day-to-day testing and control of broadcasting across the European continent. The center began in 1927, directed by Braillard himself. He continued in both these roles until the outbreak of war in 1940. According to Eckersley, Braillard managed to account for political implications underpinning technical proposals, translating "technical fact into the language of juridical compromise."[12] Braillard seemed to possess both the technical and diplomatic skills to point all concerned in the same direction, finding a balance between international benefits and national laws.

In between meetings in July, September and December, the draft plan was reworked into a "logical" formula, from which an "equal" allocation of frequencies could take place.[13] The formula allotted the common frequencies on the basis of the long existence of stations and by accounting for the population of a country, its geographical area as well as the necessity to duplicate broadcasts because of domestic language difficulties.[14] According to the plan, each country had the right to one exclusive wave and perhaps one or more commonly used waves. The plan further suggested that regrouping the stations with a separation of 20 kc/sec would reduce interference considerably, taking care of any accidental interference.[15] These decisions necessitated regular testing and evaluation of ideas. Consequently, each country should have a calibrated wave-meter and that country should transmit calibrated signals in order to "establish a standard measure for broadcasting."[16] Such technical, logical, and thus "honest" formulas in combination with these corresponding international standards for wireless broadcasting would enable an "equal" allocation of frequencies, something that the lawyers, as Eckersley recalled fifteen years later, could accept.

L'Europe et l'ordre international des communications (1865-1959)" (Ph.D. diss., Université Paris IV – Sorbonne, 2006), 572-573.

12 Eckersley, *The Power Behind the Microphone*, 88-89.

13 "Compte rendu de la réunion tenue à Bruxelles, le 16 décembre 1925," P.V. et documents complémentaires Commission Technique, 1925-1930, IBU.

14 For an extensive study on the 1926 formula see the work of Nina Wormbs: Nina Wormbs, "Standardizing Early Broadcasting in Europe: A Form of Regulation," in *Bargaining Norms – Arguing Standards: Negotiating Technical Standards*, ed. Judith Schueler, Andreas Fickers, and Anique Hommels (The Hague: Stichting Toekomstbeeld der Techniek, 2008), 112-120; Nina Wormbs, "'A Simple Arithmetic Task': Technopolitics in Early Frequency Planning for Broadcasting in Europe" (presented at the The NordMedia Conference, Karlstad, 2009); Nina Wormbs, "Invisible with Global Reach: Radio Spectrum for Broadcasting in Europe" (presented at the the Fourth Tensions of Europe Conference, Sofia, Bulgaria, 2010). "Compte rendu de la réunion, 16 décembre 1925, " 2, P.V. Commission Technique, 1925-1930, IBU.

15 IBU, *Twenty Years*, 17; Andreas Fickers, "Broadcasting as Critical Infrastructure: A Story of European Fine-Tuning and Techno-Political References" (presented at the Eurocrit International Workshop "Transnational Infrastructures: Coping with Scarcity and Vulnerability," Stockholm, 2008), 5-7.

16 Wormbs, "Standardising Early Broadcasting in Europe," 116.

Both in the initial proposal and its reworked logical formula, the engineers proposed to regulate the air by assigning wave lengths to European *countries* rather than *broadcasting organizations*. At first glance this choice might seem strange, because at the time, most IBU members were still private individuals with little or no ties to governments. Though the Geneva Plan documents give no indication, it is highly likely that such nation-state orientation was based on two factors. First of all, the concessions to broadcasting stations had been granted by their national PTT Administrations and were thus based on different national laws. Secondly, the formula simply seemed to reflect a dominant political idea of a Europe of nation-states that had been developing along with the rise of nationalism during the second half of the nineteenth century. People thought of Europe in terms of nation-states and national laws, with national administrations and forms of control. Thus a Geneva formula based on nation-states emerged, satisfying both political and technological contingencies.

The outcome of the formula, and thus the kind of neighborhood created via wireless, could be influenced considerably. One could, for instance, give extra weight to specific variables in the formula. Such weighting would affect the number of frequencies assigned to different countries. Ideally such an option would give those working on the formula room to maneuver, employing the formula as a guideline rather than as a given. Agreement was reached that the coefficient in the formula relating to the economic development of countries "calculated as the telegraphic and telephonic traffic of the country, as shown in international statistics," should for the moment have the extra weight, favoring the long existing stations as in the initial plan.[17] The decision to account for the weighting in favor of long existing stations enabled Eckersley and Braillard to use the formula to their advantage. The BBC received two wavelengths more than it deserved while Belgium received two rather than the one wavelength it was allowed to have.[18] The question remains to what extent the formula truly led to an unequal distribution of wavelengths.

Another discussion, on what Nina Wormbs calls "the drawing of Europe's borders" suggests a far more equal distribution of wavelengths. The conference participants debated heatedly whether frequencies should be reserved for countries that still had no stations in operation. Channels for these European countries could either be left blank for the moment, or allocated to countries with relatively well established broadcasting stations. Eckersley reflects: "Were we to deny ourselves the

17 Ibid., 116; "Conférence Européenne d'ingénieurs en radiophonie," 1925-1926, P.V. et documents complémentaires Commission Technique, 1925-1930, IBU.
18 Spohrer, "Ruling the Airwaves," 119.

use of precious wavelengths because they might be wanted in a year or ten years' time? Not-very-likely."[19] There is, however, reason to believe that the Geneva Plan did in fact make provision for countries with no operating broadcasting station or who were not yet members of the IBU. Albania, Bulgaria, Estonia, Greece, Ireland, Latvia, Lithuania, Poland, Portugal, Rumania, and Turkey were such countries.[20] Moreover, the Plan included the western part of Russia that would never enter the IBU. The final Plan made provision for 83 exclusive wavelengths and 16 common wavelengths between 200 and 580 meters, on the medium waveband.[21] The final plan therefore paid relatively equal attention to the various countries in Europe.

Not surprisingly, given the fact that the July conference had stressed the importance of allocating frequencies in *Europe*, the participants decided to design the Geneva Plan for what they called "the European zone." What such a zone entailed was far from clear in July. The initial plan accounted for the British Isles and the entire European continent west of Russia, even though many countries still had no broadcasting organization. Some months later, in December 1925, the engineers discussed the project for *L'Europe occidentale* (western Europe), wondering how they could account for interference from existing or projected stations in nearby countries or continents. The engineers suggested distinguishing between Europe, Asia, a northern Asia zone, a southern Asia zone, as well as an African zone, contending that each zone should be divided by meridians. The chairman of the meeting argued that he preferred to speak of continental zones defined by longitude and latitude rather than *zones mondiales*.[22] Discussions about the exact definition of Europe's boundaries became synonymous with a debate about meridians.

In the build-up to the December conference, participants and country representatives made proposals about Europe's boundaries. In November 1925, a Belgian representative proposed only integrating those European states situated between 10° West and 23° East of Greenwich. The European zone would then consist of 25 European countries excluding Russia.[23] At the start of the conference, various proposals were circulating on the definition of longitude and latitude. Another proposed a European zone with boundaries extending further to the west and east than the Belgian proposal. This definition suggested the meridian

19 Eckersley, *The Power Behind the Microphone*, 85.

20 Wormbs, "'A Simple Arithmetic Task': Technopolitics in Early Frequency Planning for Broadcasting in Europe."

21 IBU, *Twenty Years*, 18.

22 "Compte rendu de la réunion, 16 décembre 1925," 1, 3, P.V. Comité Technique, 1925-1930, IBU.

23 "Propositions de la délégation belge concernant la répartition des stations de radiodiffusion européennes opérant sur les longueurs d'ondes de 200 à 600 m," 9.11.25, 2, box 94 Stations à ondes longues – moyennes – intermédiaires, file Général, IBU. My thanks go to Nina Wormbs for making me aware of this document.

7.5° West and 37.5° East of Greenwich, with an intentional exclusion of European Turkey.[24] Such a definition would account for far more than 25 "European" countries. Whereas the proposals apparently disagreed on the western and eastern borders of the European zones, they agreed on Europe's northern and southern borders. Everyone clearly agreed these stopped at the glaciers in the north, and the countries bordering the Mediterranean like Italy, Spain or Greece in the south. Such boundary work contained decisive implications for the number of countries that would fall within the European zone.

The final Geneva Plan defined the European zone as that part of Europe lying between meridians 7°30 West of Greenwich and 32°30 East of Greenwich.[25] The definition cut off the most western parts of Ireland and Portugal, but integrated the most western parts of Russia as well as European Turkey. Russia formed part of the definition as far as its border regions were concerned. Interference could exist at Russia's borders between her transmissions and those elsewhere in Europe. The southern border is an interesting point to consider. Whereas the definitions agreed to accept a meridian continuing into northern Africa, the Geneva participants regarded North African territory as some kind of empty land. Spain and Italy formed the veritable southern borders of Europe in 1926. The Geneva Plan agreed on a somewhat technically defined boundary, while their actual southern boundary related more to those countries that states considered part of Europe. In contrast to the previous definitions of Europe, the final Geneva Plan definition included no fewer than 29 countries.

In order to reduce interference between stations in these 29 European countries, the drafters of the final plan approached the allocation procedure practically. Stations close in wavelength would have to lie further apart geographically. Stations with longer waves that traveled shorter distances and that were separated by some 10Khz would have to be situated at least 1000 meters apart. In contrast, the distance between stations with shorter wavelengths would have to be larger as such waves traveled further. In line with these rules, the

> actual plan was then made with a map, a thread and a box of pins. The thread was attached to the first pin, placed in a corner of Europe and with the appropriate length. It would then describe an arc on which the next

24 "Recommendations relatives au plan de répartition des stations européennes de radio-diffusion fonctionnant sur des longueurs d'ondes de 200 à 600 METRES (Ondes extrèmes exclues); présenté par la conférence d'ingénieurs, réunie à Bruxelles a la demande de l'Union Internationale de Radiophonie," Dec. 1925, 1, box 94 Stations à ondes longues – moyennes – intermédiaires, file Général, IBU.

25 "Note on the proposed plan for the allocation of wavelengths between the European Broadcasting Stations working on waves between 200m. and 600m," 12.12.25, 4, box 94 Stations à ondes longues – moyennes – intermédiaires, file Général, IBU.

Figure 3.2 – The European wireless network, 200-600m, March 1, 1926
Source: Box 94, Stations à ondes longues-moyennes-intermédiaires, IBU. Used by the courtesy of the European Broadcasting Union.

> wavelength could be placed. The procedure would then be repeated from there until all pins, representing radio stations, had been placed on the map.[26]

After submitting the Geneva Plan to the various PTT administrations, these then put it into force in November 1926. In spite of the nonbinding character of the plan, most broadcasting organizations and PTT administrations accepted it with its inherent projection of Europe.

Within months the IBU came up with the idea for an international broadcasting statute. Such statute would provide an international legislative basis for its recent work on frequency allocations. The League's OCT had followed closely the European conference of broadcasting engineers and attended several sessions. It

26 Nina Wormbs, "Technology-Dependent Commons: The Example of Frequency Spectrum for Broadcasting in Europe in the 1920s," *International Journal of the Commons* (submitted): 5.

became clear that the IBU and the League, as IBU President Carpendale put it, "are directly related by their mutual interests."[27] OCT member Robert Haas argued similarly that the main difference was that the IBU pursued the same ideas of cooperation and solidarity but mainly within the technical domain.[28] The League was delighted to see the work of the Union progressing and Haas pronounced that "in every way that we can, and without interfering with your strict and necessary independence, we will help you with all our means."[29] The IBU took the invitation quite literally. In the first half of 1926, the IBU officially approached the League of Nations on behalf of all its members to discuss an international statute for Radiotelephony. At the time, radiotelephony was the official technical term for the act of broadcasting. An international statute should free the way for creating a proper "international" space for wireless broadcasting.

The League's OCT quickly drew the IBU request for an international statute for Radiotelephony into its own sphere of interest.[30] The committee invited IBU Secretary General Arthur Burrows to expand upon the idea. In his address, Burrows reflected on the complexity and extraordinary pace at which broadcasting was diffusing across Europe. With such challenges ahead, the IBU would be incapable on its own of assuring efficiently working broadcasting. Governments should take the necessary legislative and administrative action. Though frequencies formed an important aspect of such wireless activities, an international statute needed to encompass other aspects of broadcasting, Burrows maintained. It should also include international regulation to prevent potential misuse of the medium by those who did not support the League's values of peace, understanding, and collaboration. An international statute with government support would resolve all broadcasting related problems and enable a more rigid control of broadcasting,

27 "La radiophonie et la Société des Nations sont étroitement liées par leurs besoins mutuels." "Conférence Européenne d'ingénieurs, 6 juillet 1925," 2, P.V. Commission Technique, 1925-1930, IBU.

28 "Conférence Européenne d'ingénieurs en radiophonie, 6 juillet 1925," 2-3, P.V. Commission Technique, 1925-1930, IBU. Asa Briggs, *The History of Broadcasting in the United Kingdom, Volume I: The Birth of Broadcasting* (Oxford [u.a.]: Oxford University Press, 1961), 282.

29 "...dans toute la mésure où nous le pourrons et sans toucher en rien à votre stricte indépendance nécessaire, nous vous aiderons de toutes nos forces." "Conférence Européenne d'ingénieurs en radiophonie, 6 juillet 1925," 3, P.V. Commission Technique, 1925-1930, IBU.

30 Initially the IBU proposed the League to discuss the idea with another League Committee, the International Committee on Intellectual Co-Operation (ICIC).The ICIC mainly dealt with cultural and scientific questions, something the IBU considered to fit well with the other broadcasting activities beyond frequency allocations. Whereas the ICIC showed little interest in broadcasting in 1926, the OCT argued that a statute to control radiotelephony fell within its own sphere of interest. The IBU could not but accept discussing such an international statute with OCT. "l'urgence qu'il y a à établir des statuts internationaux pour le contrôle de la radiophonie." "Annexe D. du rapport du Directeur, Extraits du procès verbal de la 9ième oct. 1926," Série 310, 1-2, box 92, file Société des Nations, Section des Communications et du Transit, Général, 1927-1930, IBU.

resulting in fewer incidents. Burrows clearly called for an international statute that was nothing more than a legislative extension of all the activities the IBU was working on already. The League was merely a means to an end.

Burrows' idea entailed more than an international agreement supported by national governments alone. He explicitly urged the deliberate exclusion and control of other groups of broadcasters. He was referring to those "broadcasters" who mostly operated locally outside the sphere of the IBU: the wireless amateurs. Burrows specified he did not oppose amateur research in itself, but argued that these activities should be integrated in the international statute to allow for close control of amateur broadcasting activities. Though from a technical viewpoint "private research might lead to great progress in broadcasting…it was more and more obvious that a great number of persons to whom licenses had been granted for the establishment of low-power transmitting stations were not carrying out serious research work."[31] Moreover, these people were unaware that their broadcasting activities "not only hindered the experiment of genuine research workers but also interfered with broadcasting from wireless stations."[32] Apparently amateur activities now interfered with the work of the IBU and its attempts to establish a Brussels Checking Center that in the near future would perform measurements in wireless interference. With that made clear, Burrows pleaded to control the amateurs at the very moment they were granted a concession. Broadcasting by this group hindered, or would even annul the good work of the League of Nations in the end, Burrows stressed dramatically.[33]

Burrows thus turned in a proposal to the League of Nations with a very particular vision of an international organization for broadcasting. He argued in favor of an international worldwide statute. This should not merely foster collaboration between nation-states, but rather be an agreement encompassing a wide diversity of actors, from governments, PTT administrations, and private as well as state owned broadcasting organizations, to amateur clubs and individual wireless amateurs. Burrows thus advocated an international agreement being concluded and supported by what this study calls a transnational community of broadcasting experts. Although it might not seem to feature at all in this proposal, Europe was in fact at the heart of this entire idea. Burrows felt that such a statute should be

31 LoN, Commission Consultative et Technique des Communications et du Transit, Procès-Verbal de la 9ième session, tenue à Genève, 12 au 17 juillet 1926, LoN doc. ser., C.595.M.231.1926.VIII (Geneva: LoN, 1926), 45, box 17, file Désarmement moral et propagande inadmissible, Général I, jusqu'à 1928, IBU.

32 LoN doc. ser., C.595.M.231.1926.VIII (Geneva: LoN, 1926), 45.

33 For an excellent account on the discovery of radio broadcasting see: Hugh G.J. Aitken, *Syntony and Spark: The Origins of Radio* (New York: John Wiley & Sons, 1976); Susan Douglas, *Inventing American Broadcasting, 1899-1922* (Baltimore: Johns Hopkins University Press, 1988).

nothing more than an official confirmation of the IBU's own recommendations. Since the IBU was merely a European organization, Burrows in reality imagined a European oriented statute for transnational broadcasting activities to be adopted by organizations outside the European zone.

After some discussion the OCT decided not to do anything about an international statute for Radiotelephony. The committee opposed in particular Burrows' comments about amateurs, since this particular group had been responsible for actually discovering broadcasting as well as a very simple wireless broadcasting transmission system. According to OCT member Mr. Stiévenard, "[i]t would be contrary to the objects of the Committee to support any action designed to restrict efforts which might promote a science still in its infancy, but calculated to produce enormous improvements in the means of communication open to mankind, the very *raison d'être* of the Committee."[34] Moreover, he thought that refusing concessions would only encourage the growth of clandestine stations, which should be prevented at all cost.[35] The OCT decided to adjourn the entire issue to be discussed at a later date. The committee recalled the longstanding expertise of the International Telegraph Union with respect to wireless questions, and the still young status of radiotelephony technology. The IBU clearly lacked national and international recognition as an expert organization in the field of broadcasting. The IBU idea was wiped off the table.

Other disappointments followed in Washington and Brussels in 1927 and 1928. Continuous changes in the status quo and complexity of broadcasting required a revision of the Geneva Plan within two years. Even though the Plan had proven relatively successful, with some 80% of European broadcasters deciding to comply, some 20% had chosen not to. The increasing complexity of the problem as well as the non-compliance by part of the IBU community continued to challenge the creation of an international frequency plan for broadcasters in Europe. The IBU Technical Committee stressed the urgency for its members to comply with and participate in these plans. As consolation, it stressed that these plans be seen as "an integrated set of amicable arrangements between the various radiotelephony organizations and it should be, in no case, a rigid and definitive plan satisfying all conditions of whatever era."[36] The urgency for such a new plan became even more

34 LoN doc. ser., C.595.M.231.1926.VIII (Geneva: LoN, 1926), 46.

35 Ibid., 45-46.

36 "Il convient d'insister sur ce que tout plan de complaisance répartition doit être considéré comme une intégrale d'arrangements amiables entre les différents organismes de radiophonie et ne saurait être, en aucun cas, un plan rigide et définitif satisfaisant à toutes les conditions de n'importe quelle époque." "Rapport de la Commission Technique au Conseil," Série 1268, 20 fév. 1929, 2, box 80ter, file Rapports

acute due to the outcome of the International Radio-communications conference in Washington organized on behalf of the ITU by the United States in 1927.

The Washington Conference became a decisive turning point for the way broadcasters could conceive their organization of broadcasting. This event in 1927 was the first time ITU integrated broadcasting in its activities. The previous conference in London in 1912 had featured telegraphy and early telephony standards such as maritime communications. The conference in Washington would have to take into account an immense development both in radiotelephony and radio broadcasting.[37] The IBU described the situation in Washington as one complicated "by a set of technical, political, and economic factors," in which it for the first time could negotiate frequency bands for broadcasting with the other means of telecommunications.[38] The ITU decided to establish the International Radio-electricity Consultative Committee (CCIR) that would consult the ITU on radio-electricity questions of any kind in the future.

Another important turning point concerned the allocation of wavebands between technologies rather than between countries. Conference participants negotiated the best options to allocate the available wavebands between the various means of communication. With this vast extension of wireless technologies, the United States proposed that wavebands and ranges of frequencies be allocated among wireless technologies. European participants on the other hand, favored an allocation of bands between countries. One country would then be assigned a waveband that it could arrange according to its own interests between the various wireless technologies it employed. The idea of organization by country evolved in line with Europe's late nineteenth century development of nationalism and state-centered organization of most wireless technologies. Nonetheless, the conference ultimately decided to follow the U.S. model.[39] Consequently, broadcasting was assigned several specific wave bands that still had to be allocated among broadcasting organizations in the various regions. Washington decided that regional agreements were acceptable for dealing with the allocation of frequencies.

The outcome of the Washington Conference was a disappointment for the European broadcasting society and for the IBU. On the first occasion when broadcasting was taken into account during a globally important conference on wireless, the new medium immediately had to succumb to other wireless applications. In

Généraux de la Commission Technique au Conseil à fin 1931, IBU.

37 "Commission Mixte (Conférence de Madrid) Rapport provisoire de la commission mixte concernant les amendements à apporter à La Convention de Washington (1927) et à règlement générale annexe," Série 2291, 5 janv. 1931, 3, registry file 9G, box R-2601, file 13698/2081, LoN.

38 "Commission Mixte," Série 2291, 3, registry file 9G, box R-2601, file 13698/2081, LoN.

39 Wormbs, "Invisible with Global Reach," 5-6.

compliance with the Geneva Plan, broadcasters could continue using the 545 to 200 meters (500 – 1,500 kc/sec to 550 – 1,500 kc/sec) band, with the exception of the 200 meter band. They were however severely limited in the band between 1,875 and 1,314 meters. The only positive outcome was that Washington finally set legal specifications for the long wavebands for broadcasting.[40] This reduction of available wavelengths together with the vastly increasing complexity of the European problem made the IBU organize the drafting of a new frequency plan by its engineers immediately. This new Brussels Plan went into effect in January 1928.

The new plan sought to improve the vastly deteriorating situation in the European airwaves. In 1927 the IBU had started preparing a follow-up meeting for engineers to work on a new plan. They designed a plan that as much as possible corrected for technological developments in transmission power, an enormous increase in the number of low power stations, and improved frequency measurement that enabled separation between frequencies of less than 20 kc/sec. In contrast with the Geneva Plan, the new plan reduced the space between the waves above 300 meters to some 9 kc/sec, while the majority of the common waves above 250 meters became exclusive waves.[41] Geographically, the Brussels Plan applied to a smaller European zone than the Geneva Plan, bordering Russia in the east.[42] Furthermore, the allocation of exclusive waves between stations did not change substantially. Changes in the density of the European wireless network mostly happened with commonly used waves, a development that had little effect on the shape of the overall network. With the exception of a smaller European zone, the difference between the Geneva and the Brussels Plan was relatively slight.

Despite the relative similarity of the two plans, the Brussels Plan failed to gain ground. The absence of government officials and PTT Administrations turned the Brussels meeting into a rather weak extract of the Geneva Plan. Director of the IBU Technical Committee Raymond Braillard at some point even reported to the Council:

> The Technical Committee does its best to heal the situation, but we can only act in the limits of our possibilities and call upon the goodwill of our members. We try to convince by the power of technical arguments, which reveals often successful. But what can a doctor do if the patient refuses to take the remedy that has been prescribed? Isn't it a fact that a few trouble

40 George A. Codding, *The International Telecommunication Union: An Experiment in International Cooperation* (Leiden: Brill, 1952), 122.
41 IBU, *Twenty Years*, 19-20.
42 "Rapport de la Commission Technique au Conseil," Série 1268, 3, box 80ter, file Rapports Généraux, IBU.

> makers can rain on someone's parade? And they of all people lament most strongly of being troubled.[43]

The problem with the Brussels Plan was that right after the Brussels meetings, other developments took over. At the Washington Conference the Czechoslovak PTT Administration promised to organize a regional conference in Prague to draft a plan for allocating the available wavebands for broadcasting in Europe.[44] The decision about this so-called Prague Conference was reached during the conference of the ITU, which had only recently decided to integrate frequency allocation into its sphere of interest. The decision was made without consulting the IBU. Both the disastrous outcome for wireless broadcasting in Washington and the new player in the field of frequency allocation in the form of the ITU, made the IBU increasingly aware of the need to gain international recognition for its activities. It had to firmly establish itself as the main European expert organization in broadcasting.

In 1928 the IBU made plans for a central broadcasting organization. After the failed proposal for an international statute for Radiotelephony, the IBU again approached the League of Nations for international support. Quite unexpectedly, the signing of the Kellogg-Briand Pact in 1928 opened up new prospects for international collaboration, providing a breakthrough for the IBU to collaborate with other international organizations. The Pact had eased international tensions considerably. International elites felt more confident to collaborate internationally, and the League's idea to further peace through increased interconnectedness and collaboration gained more widespread recognition.[45] To further this positive end that would ultimately lead to the famous Briand Memorandum of September 1929, the League's OCT decided to address the organization of frequency allocations in 1928.

When the IBU approached the League, it initially requested its OCT try and put the political misuse of broadcasting on the League Assembly's agenda. When the OCT discussed the request internally, it decided that such a political issue was not a topic for a technical committee like the OCT. Canadian OCT member, Mr. Dunning, nevertheless disagreed. In his view, the technical aspect of the issue was part of the OCT's remit. Dunning argued that a country like Canada experienced interference, not of a political but of a technical nature. People should be able

43 Quote translated by Fickers, "Broadcasting as Critical Infrastructure," 8. From "Raymond Braillard, Rapport de la réunion de la Commission Technique à Lausanne," 3, box 80ter, file Rapports Généraux de la Commission Technique au Conseil, à fin 1931, IBU.

44 Wormbs, "Invisible with Global Reach," 6; Codding, *The International Telecommunication Union*, 93-94.

45 Bo Stråth, "Introduction: Europe as a Discourse," in *Europe and the Other and Europe as the Other*, ed. Bo Stråth (Brussels; New York: P.I.E.-P. Lang, 2000), 15.

to communicate without interference of radio waves from neighboring countries. Here, he proposed, lay a task for OCT since such technical interference could affect international peace. Independent of the IBU request, Dunning proposed the League arrange some sort of international convention to resolve technical interference. OCT director Haas supported Dunning and referred to the good work of the IBU. Haas, however, felt it would be too premature to devise an international convention on frequencies, given the vast developments in broadcasting technology. He adjourned the meeting.

Dunning did not relinquish his idea. A few days later he argued that he considered it more and more apparent that an international convention was necessary if "national liberty in broadcasting was to be assured, particularly in countries adjacent to each other."[46] It was a matter of international arrangement in the interest of international goodwill and securing a national neighborhood. Such agreement would affect the sovereignty of states regarding their domestic broadcasting activities. Dunning's plea proved effective. The OCT decided to "discover some means of establishing international agreement to ensure an equitable distribution of wireless wave-lengths among the various countries, in order to diminish the probability of disturbance in wireless broadcasting."[47] An international convention would ensure the firm national roots of broadcasting within the broader international focus of the League. The OCT decided that an international convention portraying a global world of nation-states would be beneficial at national level too.

The League's idea for an international convention opened up debates on other ways of dealing with interference. OCT invited the director of the ITU Bureau, Dr. Raëber as well as IBU Secretary General Burrows to the first meeting of its newly established Committee for the Distribution of Wavelengths. The meeting hosted the discussion of a possible international agreement that assured an equal allocation of frequencies and a reduction of troubles with respect to radiotelephony.[48] Burrows immediately took the opportunity to express his concern about the Washington agreements and explained that the IBU had "recently produced a new European wave-plan by which it is hoped to make the best of what is an unfortunate state of affairs for the broadcasting services, which can do so much

---

46 LoN, "Second Committee, 1928 Assembly, Minutes of the Eighth meeting, held at Geneva, September 19 1928," LoN doc. ser., A.II/P.V.8, 3, box 92, file Société des Nations, Section des Communications et du Transit, Général, 1927-1930, IBU.

47 LoN, Organization for Communications and Transit, Report on the work of the organisation between the Eighth and Ninth ordinary Session of the Assembly, September 20 1928, LoN doc. ser., A.73.1928.VIII (Geneva: LoN, 1928), 3, box 92, file Société des Nations, Section des Communications et du Transit, Général, 1927-1930, IBU.

48 IBU, P.V. Conseil Administrative 1929, 190.

towards increasing mutual respect between nations."[49] With such an unfortunate state of affairs, Burrows argued, the time was right to create a Central Broadcasting Authority.

This Authority would be a newly established international institution with which the IBU could collaborate in order to make legally binding decisions relating to all nations. "It is not difficult to envisage," Burrows stated, "that in the near future special codes of international law will have to be created to regulate these problems."[50] He argued that all broadcasting related organizations and all nation-states would need to accept the decisions of such a global organization. Europe did not feature in this vision though it hovered in the background just like when the idea for an international statute for Radiotelephony was raised two years earlier.

The League, the IBU, and the ITU reached stalemate almost immediately when ITU director Raëber rejected the idea. The ITU had been organizing the allocation of frequencies to great satisfaction for years already and saw no need for League interference. The ITU could deal with the issue along similar lines to those it used for wireless telegraphy and telephony. In response, the League pointed out that the international problems of broadcasting concerned more than frequency questions alone and included copyright issues for instance. Thus the League supported some kind of broadcasting authority. Seelinger, the chair of the meeting pondered philosophically:

> Was the distribution of wave-lengths a purely technical problem? Did its effects reach far beyond the technical domain? Were the present organisations capable of settling the problem, once its scope had been defined? The greatest diversity of opinion on these subjects was quite legitimate at present. In any case, the speaker was sure that everyone agreed that the problem was both far-reaching and complex, and that the League could not dissociate itself from its various aspects.[51]

The idea of a Central Broadcasting Authority seems to have been too "big" and too complex to be decided at such short notice. The problems with technical interference were too pressing to await a solution.

The Prague Conference signified the breakthrough many had been waiting for.

49 LoN, Advisory and Technical Committee for Communications and Transit, Procès-Verbal of the Thirteenth Session held at Geneva, March 15th to 23rd, 1929, LoN doc. ser., C.216.M.81.1929.VIII (Geneva: LoN, 1929), 37, box 92, file Société des Nations, Section des Communications et du Transit, Général, 1927-1930, IBU.
50 LoN doc. ser., C.216.M.81.1929.VIII, 36.
51 Ibid., 39.

When the Czechoslovak PTT organized the intergovernmental European Radio-electric Conference in Prague in April 1929, it invited the IBU as expert organization to advise. This Conference officially brought together for the first time European PTT administrations and broadcasting organizations to discuss issues of frequency allocation for broadcasting. It had a more official character than the meeting organized by the IBU in Brussels. The new Prague Plan, unlike the 1927 Washington Convention, did not have the status of a legally binding treaty. As planned, the participants nevertheless managed to work the plan into a form of regional amendment to the Washington Convention. This Convention had stressed the option for regional allotments without specifying either the definition of a "region" or the form of agreement. Consequently, participants had a fair amount of freedom with the design.[52] In Prague the IBU and the ITU managed to resolve technical interference issues in a manner that could be realized in the short term.

The most important achievement of the Prague Conference was the way the IBU, the ITU, and the European PTT administrations arranged their collaboration concerning the allocation of frequencies. They decided to strengthen the official ties between the organizations, arranging a strict task division that secured effective frequency allocation plans. All parties accepted the IBU as the official international expert. As *primus inter parus,* the IBU would advise the ITU and PTT administrations on all questions related to broadcasting. Final decision power still remained with the ITU and PTT administrations, but in the meantime, they would be allowed to attend all IBU meetings as observers. Those PTT administrations that operated a radio station could now become ordinary IBU members as well. These decisions strengthened the role of PTT administrations in an organization like the IBU and the official character of frequency allocation discussions of any kind.[53] Shortly afterwards, the IBU amended its statutes and changed its official name from *Union International de Radiophonie* into *Union International de Radiodiffusion*, to clarify the task division between the IBU and the ITU. These decisions would influence profoundly any future options to conceive projects for European broadcasting.

These new ties could not prevent difficult negotiations over a new frequency plan in Prague. Within three years IBU membership had more than doubled and the Prague Plan would have to anticipate this large growth. The participation of many new members challenged the position of long-term well-established organizations like the BBC, the German RRG and French broadcasters. New broadcasting organizations could now actively defend their own interests, having a

52 Spohrer, "Ruling the Airwaves," 129-132.

53 IBU, *Twenty Years*, 23-24.

right to vote on the final plan.[54] In particular, small members questioned the old Geneva formula that in their view overvalued industry size. Finland, for instance, encouraged emphasis on geographical size, whereas multilingual countries like Czechoslovakia urged emphasis on the number of languages and political parties.[55] The relative shift in the balance of power might have been smaller than it seemed. Several small and relatively new broadcasting stations had already been involved in the 1926 Geneva negotiations.[56] Czechoslovakia, for instance, had already participated actively in the founding meetings of the IBU, whereas the Polish delegate had joined in the Geneva frequency discussions before Poland even had its own broadcasting station on air.[57] New rights to vote changed negotiations somewhat. They contributed to the complexity of the negotiations between more varied groups of stakeholders.

The Prague Conference might actually be regarded as a success for the IBU. At the opening meeting, most PTT administrations agreed to take the IBU Brussels Plan as the point of departure for discussion. The final plan was merely an adaptation of the previous Brussels Plan.[58] The Prague Plan extended the separation as defined in the Brussels Plan by 9 kc/sec to the bands below 300 meters, making the European broadcasting space more efficient, and allocated both medium and long waves. For the first time the plan included all East European broadcasting stations, moving the Eastern meridian as discussed in Geneva in 1926, from 32°30 to 40°30 East of Greenwich.[59] Almost all European governments accepted the Prague Plan. It reflected a European zone extending further eastwards and a vision of neighborhood on a more equal footing between smaller and larger broadcasting organizations, from west to east and north to south.

Shortly after the realization of the Prague Plan, the vast growth of broadcasting in Europe immediately challenged not only the plan itself but the IBU organization as a whole. Soon each European country possessed its own, and often nationalized, broadcasting service. IBU membership grew with 24 new broadcasting organizations to a total of 66 members between 1929 and 1934. Compared to 123 transmission stations in 1926 with an aggregate power of 116kw, by 1939 there

54 Ibid., 16.

55 Spohrer, "Ruling the Airwaves," 122.

56 Ibid.

57 Miazek, *Przeminelo z radiem*, 122; IBU, "Conférence européenne d'ingenieurs en radiophonie (Salle E.) Palais des Nations, Genève, 6&7 juillet 1925, Disposition des places," Série 16, registry file 14, box R-1139, file 43391/28231, jacket 1, LoN.

58 Spohrer, "Ruling the Airwaves," 128.

59 BUIT, *Documents de la Conférence Radioélectrique européenne de Prague 1929* (Berne: BUIT, 1929), 142; IBU, *Twenty Years*, 22.

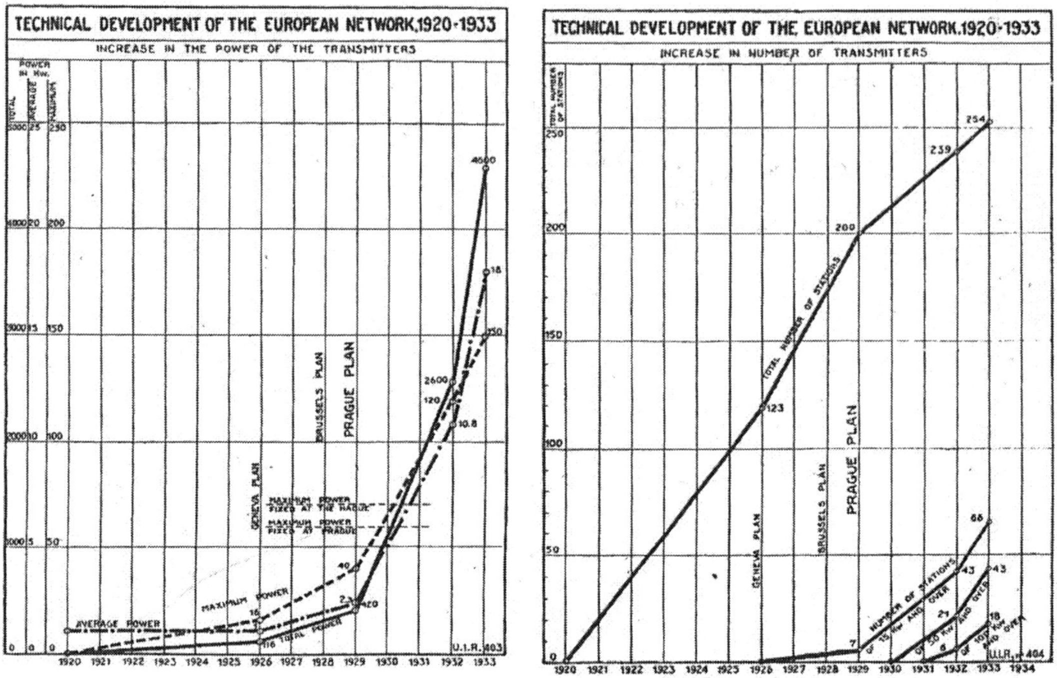

Figure 3.3 –Technical development of the European wireless network, 1920-1933
Source: IBU, Memorandum on the present situation of European broadcasting from the point of view of interference, and the best means by which this may be remedied (Brussels 1932), 15.

were 463 transmission stations with an aggregate power of no less than 11,750kw.[60] Compared to the early years, by the late 1920s growth was more and more the result of an increase in transmission power rather than an increase in the number of stations (Figure 3.3).[61] As a result of these changes, the IBU had to cope with a large growth in "meetings, conferences, and newly founded sub-commissions of both the juridical and programme committee."[62] The IBU extended its Brussels Checking Center to ensure the Prague Plan could be carried out. Collaborating with the ITU, the Checking Center distributed its control measurements as a kind of "soft power" to all broadcasting stations and European PTT administrations. Thus everybody could see who adhered to the plan and who did not. All these provisional measures proved to be an inadequate response to the vast changes. The Prague Plan would have to be adapted. What is more, the medium of broadcasting was also in dire need of a larger frequency band.

60 IBU, *Twenty Years*, 23, 49.
61 Arno Huth, *La radiodiffusion: Puissance mondiale* (Paris: Librairie Gallimard, 1937), 40.
62 Fickers, "Broadcasting as Critical Infrastructure," 10.

Under the Spanish sun, broadcasters reached political consent. Gathering in Madrid, with the Washington Convention disaster fresh in their minds, the European broadcasting community meticulously prepared to defend the extension of the wavebands assigned to broadcasting at the Fourth International Radio Telegraphic Conference in Madrid in September 1932. The IBU created a special "Mixed Committee" to prepare an official report containing recommendations for the Madrid government representatives and national plenipotentiaries.

The Mixed Committee was an inter-organizational collaboration of IBU members from the Council, Technical, and Legal Committees along with representatives from ITU and national PTT administrations. Together they produced a report proposing specific technical standards to resolve problems of stability and continuity of transmission, defining "interference," and the reduction of transmission power as well as the geographical distribution of stations. These standards would increase an efficient employment of wavebands for broadcasting.[63] The report detailed the grounds for a considerable expansion of the available wavebands for broadcasting. The wavebands assigned by Washington fell short considering the growth of broadcasting in recent years. Rumania, Czechoslovakia, Spain, Greece, Italy, Austria, and Portugal needed a long wave frequency to create a nation-wide broadcasting service.

Besides emphasizing the facilitation of national broadcasting, the Mixed Committee urged an accurate definition of Europe. They said Europe would also have to include "the non-European territories touching the Mediterranean and the Black Seas, as well as the African Atlantic North of the 30° parallel."[64] Though these territories had come under the 1926 Geneva and 1929 Prague definitions of Europe, the Mixed Committee meetings were the first time technical experts in the field of broadcasting specified a southern border reaching beyond countries bordering the Mediterranean like Italy and Spain. The committee argued that this definition would reduce rising inter-continental interference between broadcasting services worldwide and between broadcasting transmissions and those of other telecommunication means operating in the same wavebands in other parts of the world. The inter-organizational committee argued that integrating these recommendations in the Madrid Convention would improve broadcasting on a global scale.

63 "Commission Mixte," Série 2291, 2, registry file 9G, box R-2601, file 13698/2081, LoN.

64 "les territoires non européens touchant à la Mer Méditerranée, à la Mer Noire, ainsi qu'à la côte Atlantique de l'Afrique située au Nord du 30ème parallèle." Ibid., 9.

Over a period of 14 weeks, some 80 governments and 85 broadcasting organizations together with a League of Nations observer negotiated their broadcasting interests with a special role for the Mixed Committee Report. The final Madrid Convention was an improvement compared to the failure of the 1927 Washington Convention. Although the conference did not validate all the Mixed Committee recommendations, it did decide to enlarge the waveband for broadcasting by 41 kc/sec. This allowed for an increasing number of frequencies and thus an increasing number of nation-wide broadcasting services. The conference officially appointed the IBU as international broadcasting expert to modify the existing Prague Plan on the basis of this new waveband. The conference modified the statutes of the ITU, integrating a name change to "International Telecommunication Union" that symbolized ITU's terrain expanding beyond telegraphy alone. The Madrid Conference clearly attached value to the medium of broadcasting.

Above all, the Madrid Conference officially defined "Europe." To reduce global interference, the 1932 Convention defined the European Region "in the north and in the west by the natural boundaries of Europe, in the east by the meridian 40° East of Greenwich and in the south by the parallel 30° North, so as to include the western part of the U.S.S.R., and the territories bordering the Mediterranean, except parts of Arabia and Hejaz which are included in this sector."[65] Nowadays Arabia and Hejaz are part of Saudi Arabia. The definition was mostly a one-by-one adoption of the European one accepted at the Prague Conference in 1929.

The conference accepted the Mixed Committee proposal to change the description of the southern borders of Europe.[66] The new phrase in the definition indicates a change in the European broadcasting engineers' mentality. Although the Geneva Plan had already defined the southern boundary as the parallel 30° North continuing into the northern territories of Africa, in reality they were working with the southern political boundaries of Europe that stopped at the southern borders of Spain and Italy. The Madrid definition explicitly integrated *all* countries bordering the Mediterranean, allowing transmission signals to fade away into the Sahara in the south. This led to wireless experts increasingly making a technical concept of European boundaries more important than a political one. Overall, the Madrid Convention legalized the achievements of the Prague Conference and integrated the IBU's technical advice.

The 1933 Luzern Conference further solidified frequency allocation. The IBU organized this European Conference of Wireless Broadcasters to re-arrange the Prague

65 Codding, *The International Telecommunication Union*, 157, footnote 113.

66 "Commission mixte," Série 2291, 9, registry file 9G, box R-2601, file 13698/2081, LoN.

Plan into a new plan including the decisions made in Madrid. In preparation, broadcasting organizations from 26 countries together with 25 European telegraph administrations had drafted the allocation plan based on the IBU Prague Plan, which facilitated discussion considerably. The conference also statutorily laid down even closer bonds between (inter)national telecommunications and the IBU. The IBU agreed that delegates from PTT administrations as well as the ITU Bureau could participate actively in all IBU meetings. The IBU even went one step further and agreed that when administrations instructed the IBU to organize proposals for future conferences, the representatives of the administrations had the right to vote. Frequency allocation activities became more and more formal. Unlike the Prague Plan, the participation of Government plenipotentiaries gave the Luzern Plan a legally binding character for all parties. The plan went into force on January 15, 1934.[67]

Though the overall organization of the IBU continued to be non-governmental, Luzern increased government influence, and thus national influence, over the material infrastructure activities of the Union. An increasing number of broadcasting organizations become government controlled, and the more official ties between the administrations, the ITU, and the IBU, influenced government control over IBU frequency allocation activities. In the course of eight years, the initial European broadcasting society had grown in number and diversity. The fine-tuning activities had resulted in a European wireless network in which the imagined nation-wide wireless broadcasting infrastructures formed a firm standard within legally binding international conventions.

Though these early frequency standard negotiations focused on wireless broadcasting in Europe, "Europe" remained surprisingly absent, apart from the occasional quibbling about the meridians defining its boundaries. Whereas most discussions focused on "nationality" and "internationality", Europe's boundaries changed from a political vision in 1926 to a more technically inspired vision by 1932. The negotiations on frequency standards show a firm dominating role of the nation-state in these transnational Europe-centered efforts. Though the vast nationalization of broadcasting from 1927 onwards explains nation-wide broadcasting, this does not explain the privately owned broadcasting pioneers' deliberate choice in 1925 to concentrate on frequency allocations between countries rather than between organizations. Their choice suggests a historical path-dependency with late nineteenth century developments in wireless telegraphy and a little later with wireless telephony, something which should be examined in relation to rising nationalism during the second half of the nineteenth century.

67 IBU, *Twenty Years*, 25-26.

The national focus of the European wireless network formed part of a grander and explicitly "internationalist" scale.[68] Nation-wide wireless services should prevent interference and be complemented by another part of the infrastructure, an international relay network. The IBU wanted to build a relay network that would allow broadcasters to exchange high quality music internationally across borders. So the IBU chose to organize the European wireless network in such a way that the national was an inherent component of the international. These activities appear to be independent. In reality they were two sides of the same coin. Together they formed a European broadcasting network based on the idea of neighborhood and collaboration.

Wiring a Continent

Broadcasters wired the European continent in the interwar period. Already in 1925 the emerging European broadcasting society imagined broadcasting far beyond the level of the nation-state. They argued that national wireless networks should be complemented by a system allowing for the international exchange of local and national programs as well as joint international programs that furthered mutual understanding and rapprochement between peoples. To their dismay, in 1925 the status quo of broadcasting technology turned out to be far from effective and could not enable their projected high quality transmissions of music and speech on a Europe-wide scale. Programs should be able to include "high culture" and high quality music like symphony concerts. This genre, however, required very broad Herzian frequency ranges compared to other types of music, especially in respect of ordinary speech. Whereas the European wireless network lacked the power to allow people to listen in over long distances without annoying background noise, the technology of short wave transmissions was still in its infancy in the mid-1920s.[69] The IBU had to find other solutions to achieve high quality broadcasts across the European continent.

The IBU found its solution in the developing long-distance telephone networks that relayed music on a regular basis in Europe, the United States, and Canada. Broadcasting organizations like the American CBS and NBC employed telephone

68 For extensive studies on the allocation of frequencies during the interwar years see: Fickers, "Broadcasting as Critical Infrastructure"; Spohrer, "Ruling the Airwaves"; Wormbs, "Standardising Early Broadcasting in Europe."

69 Huth, *La radiodiffusion*, 39-41; Trevor I. Williams, *A History of Technology: Vol VII The Twentieth Century c. 1900 to c. 1950 Part II* (Oxford: Clarendon Press, 1978), 1255.

cables on a daily basis to diffuse programs between their radio stations.[70] CBS and NBC created the programs in their national studios. They relayed them via dedicated telephone lines to the many local stations that had joined their networks. These local stations in turn could broadcast the programs over their own local frequencies, assuring a nation-wide reception of the program.[71] British, German, and Swedish IBU members had also adopted the practice of relaying music over their long distance telephone networks across their countries. They could relay programs from one station to another over a maximum of 700 kilometers at night without the technical defects that occurred by broadcasting via wireless over similar distances.[72] The BBC for instance used the telephone network to broadcast the same program simultaneously from different stations, a system they called simultaneous broadcasting.[73] All in all, a relay network was meant to complement the wireless networks, relaying music that could then be broadcast over the stations' home channels. European broadcasters were well aware of such technical relaying developments. In time, an international telephone network capable of relaying music over long distances would enable a regular supply of high quality music to all European audiences.

A relay network required collaboration between different technical experts. The idea was not simple nor a short-term solution to the lack of high-quality Europe-wide broadcasting. The IBU had no connection to or influence over the construction of telephone networks in general. The European long distance telephone network constructed from the early 1920s onwards, which in some cases enabled international telephone conversations, was not able to transmit the broad band of frequencies accompanying musical performances. For instance: the correct transmission of speech required between 200 and 2500 kc/sec while music required no less than 50 or 100 up to 10,000 kc/sec. "The highest note of a piccolo has

70 "Relais Internationaux, Résume de la situation présenté par le Secrétaire Général," Série 563, 15 juin 1927, 2, box 85, file Relais, Général I, 1927 et précédents, IBU; IBU, *The Problems of Broadcasting* (Geneva: IBU, 1930), 105.

71 Jennifer Spohrer, "Paradise or Perdition? European Commercial Broadcasters and American Radio, 1920-1960" (presented at the ESF Inventing Europe Conference: Appropriating America, Making Europe, Amsterdam, 2009), 6-7.

72 "Relais Internationaux," Série 563, 2, box 85, file Relais, Général I, IBU; IBU, *The Problems of Broadcasting*, 105.

73 Simultaneous broadcasting was the act of transmitting one program over several radio stations within one country at the same time. This increased the efficiency of the network and reduced interference considerably. The British relay network allowed the same program to be supplied to all stations. Asa Briggs, *The History of Broadcasting in the United Kingdom, Volume II: The Golden Age of Wireless* (Oxford [u.a.]: Oxford University Press, 1995), 27; Paddy Scannell and David Cardiff, "Serving the Nation: Public Service Broadcasting before the War," in *Popular Culture: Past and Present*, ed. Bernard Waites, Tony Bennett, and Graham Martin (London: Croom Helm London, 1982), 165.

a frequency of 4750, but even more numerous frequencies are produced by orchestra music."[74] Having explored various options, in 1925 the IBU turned to the International Telephone Consultative Committee (CCIF), the advisory body of the ITU dealing with long-distance telephony.

The IBU approached the CCIF "for a systematic examination…of the possibility of creating an international network for telephonic circuits specially suited to the needs of broadcasting."[75] When the CCIF invited the IBU to its meeting that same year, the IBU stressed the utmost importance of the network for thousands of people in Europe who listened to broadcasts already. All developments in long-distance telephony should consider the special needs of radiotelephony, the IBU argued, because in three years, broadcasting had become part of European social life. CCIF received the request favorably. The agreement turned out to be the start of a long-term collaboration between the two bodies to work on the project of a European long distance telephone network that enabled the high quality transmission of music.

The CCIF was an advisory committee that closely managed the building blocks of the European telephone network by establishing a "more precise and universal set of telephone regulations."[76] In 1924 the French Ministry of Post and Telegraphy had given the incentive for the creation of CCIF. Via periodical meetings, telephone experts from various countries would discuss issues and exchange information. Within a year the CCIF became part of the ITU at the request of the Scandinavian countries, but continued in its original composition. Its only link with the ITU was the agreement that all administrations could participate in CCIF studies on request, while CCIF would have to communicate its conclusions and recommendations to the ITU for publication.[77]

According to Léonard Laborie "CCIF is the fine result of a revised articulation between national power, which remains central to all approaches, and European space..."[78] CCIF telephone experts were an interesting mix of the private and public

---

74 "note la plus haute du piccolo ayant une fréquence de 4750, mais des fréquences plus nombreuses encore sont produites par de la musique d'orchestre." "Relais Internationaux," Série 563, 2, box 85, file Relais, Général I, IBU.

75 IBU, *Broadcasting: An Instrument for the Creation of a Better Understanding Between the Peoples* (Geneva: IBU, 1937), 88.

76 Codding, *The International Telecommunication Union*, 36; Laborie, "La France, L'Europe et l'ordre international des communications," 444-449.

77 Codding, *The International Telecommunication Union*, 35-36, 47.

78 "Le CCIF est bien le résultat de l'articulation révisée entre la puissance nationale, qui reste centrale dans les approches, et espace européen, qui se construit avec la stipulation américaine: le cadre international aux contours raffermis que dessine la coopération offre la possibilité de rester souverain et de continuer à poursuivre assidûment ses intérêts." Laborie, "La France, L'Europe et l'ordre international des communications," 465.

sector, operating in a technical and commercial culture. They based their entire *raison d'être* on the efficiency and increased use of the network. In addition to such a pragmatic culture, they believed their approach would contribute to solidarity and peace between Europeans, a generally recognized characteristic of engineers who were active on an international level.[79] Without cooperation, international service would fail to develop.[80] Like the IBU, the CCIF did not regard the national and the international as mutually exclusive processes.

In accordance with its connection to the ITU and the consideration that networks did not stop at Europe's boundaries, the CCIF took a global perspective on matters, with a preference for the American federal model. Laborie stresses the link between CCIF secretary Georges Valensi and the United States that influenced the committee substantially.[81] The CCIF for instance regularly designed inter-continental agreements and allowed American telephone company AT&T to participate in its work on a non-official basis.[82] Nonetheless in practice, the CCIF focused on designing a European telephone network. Much like the IBU, in the 1920s the CCIF brought together cooperators and rivals from the European continent and beyond to work on a European scheme that would benefit a global network. In the European journal *L'Europe Nouvelle*, Valensi once pragmatically compared the work of CCIF to "the blueprint of the nerve system of the United States of Europe."[83] A system of telephone arteries coupled these united states, making Europe the simple sum of the various nation states.

Initially, the IBU and CCIF developed a loosely coupled collaboration. The IBU approached CCIF each time it needed information about relaying music over telephone arteries. The CCIF then performed the necessary study with the IBU Technical Committee. It communicated its final solutions to the ITU who then forwarded the results to the national PTT administrations. They in turn performed the actual construction works.[84] As such, the IBU had a considerable impact on which issues to bring to the attention of the CCIF, after which the IBU Technical

79 Ibid., 464; Johan Schot and Vincent Lagendijk, "Technocratic Internationalism in the Interwar Years: Building Europe on Motorways and Electricity Networks," *Journal of Modern European History* 6, no. 2 (2008): 198.

80 Laborie, "La France, L'Europe et l'ordre international des communications," 464.

81 Georges Valensi was educated as a polytechnician with a long-term interest in transatlantic telephony including its technical applications in radio broadcasting. He also worked in the United States.

82 Laborie, "La France, L'Europe et l'ordre international des communications," 447-449, 466.

83 "l'ébauche d'un système nerveux des Etats-Unis d'Europe." Ibid., 467; Quote by the editors in the introduction to George Valensi, "La téléphonie européenne," *L'Europe Nouvelle* 417 (1926): 204.

84 Among others, please see: "Relais internationaux par fils téléphoniques a grande distance. Rapport élaboré à l'intention de la Commission des Relais Internationaux par le Secrétaire Générale de l'Union Internationale de Radiophonie," Série 1038, 3 août 1928, box 85, Relais, Générale III, 1928, IBU; Laborie, "La France, L'Europe et l'ordre international des communications," 445-446.

Committee could collaborate with CCIF engineers.

In 1930 the two organizations set up a more straightforward cooperation that further increased the IBU's impact on achieving telephone relay network standards. The IBU could now formally join in with CCIF studies. The heads of the IBU Technical Committee and the International Relay Committee assisted at CCIF meetings and other IBU delegates such as Secretary General Burrows, head of the Rapprochement Committee Czeija, and members of the IBU Technical Committee Austrian chief engineer Harbich, British chief engineer Hayes, and French chief engineer Baize also participated. These men could negotiate the best possible terms for radio broadcasting in respect of the general CCIF activities and their European telephony network project.[85]

While the CCIF was working on its European telephone "nerve" system, the IBU had its own ideas about the nature of such long-distance telephone networks. Together they focused on two major aspects. First of all, they had to define standards to improve the technical quality of the musical programs relayed over these networks. In November 1925 the IBU asked the CCIF to study broadening the number of frequencies that the telephone cables and so-called aerial lines – the plain copper wires hanging between poles – were capable of relaying. The CCIF, according to the IBU, had to focus on examining the pupinization of these cables and adapting the amplifier stations. These strengthened the transmitted signals along the way and avoided receiving extremely faint signals at the end of the line. The telephone cables and amplifiers could transmit the limited range of frequencies that characterized speech, but would flatten the musical tones when relaying music by omitting the high and low frequencies that lay beyond the range of speech.[86] The IBU stressed its desire to avoid such inferior quality of music at the end of the line.

Their other focal point was the design of the network. In order to create the best possible transport of programs from A to B, the IBU envisioned a relay network that connected the various European capitals.[87] This design would sustain the underlying vision of enhancement of European social life or the League's ideals of peace and understanding, and also sustain technical and economic lucrativeness. Relays via the telephone network would be exempt from atmospheric disturbances, one of the big problems of wireless broadcasting. From an economic

85 P.V. Conseil Administrative 1930, 270-271, IBU; P.V. Conseil Administrative 1931, 13-14, IBU.

86 "Au Comité Consultatif International des Communications Téléphoniques à Grande Distance, Paris," 4 déc. 1925, 2-3, box 85, file Relais Général I, 1927 et précédents, IBU; "August-September 1928, IBU at Berlin, Transmission radiophoniques sur circuits, Rapport de M. le Conseiller des Postes Sprinck," box 85, file Relais, Général III, 1928, IBU.

87 "Relais Internationaux," Série 563, 3-4, box 85, file Relais Général I, IBU.

perspective, they also increased the efficient usage of the network by employing it for radio broadcasting relays during low traffic hours. The last point especially could resolve the problem facing the CCIF, namely the declining use of the telephone cable network. The CCIF decided to study a possible relay network between several European capitals immediately, along with testing long-distance transmissions of music via cable and plain copper wire.[88] In the meantime, the IBU appointed Braillard as main contact person for the CCIF, and asked its members to open their doors to the CCIF in order to facilitate long distance relay experiments. Clearly the CCIF responded positively to the IBU's ideas. These would serve their economic, technical, and more idealistic interests simultaneously.

Occasionally experts had to justify their choice of network. Constructing an international relay network for music was a long term affair, and did not come about easily. In the eyes of the IBU, high quality broadcasting and Europe-wide transmissions were the Union's principal goals. These could not be achieved without such a relay network. In July 1926 the new Committee for Intellectual, Artistic, and Social Rapprochement (Rapprochement Committee) immediately created a sub-committee dealing only with relay questions. The sub-committee focused on all organizational and legal issues such as the division of costs for renting the telephone lines from PTT Administrations. Oskar Czeija, Director General of Austrian RAVAG, an inventor and entrepreneur, and one of the most colorful people in Austrian radio, coordinated the international relay activities as head of the Rapprochement Committee.[89]

Czeija, who had wide experience with telephone and telegraph technology, realized that progress in international relay over telephone cable or wire could not be speedy. Progress depended on the economic and political circumstances in European countries and on the developments in telephony. Furthermore, the relay ideas went against the grain of telephony engineers' general tendency to replace aerial lines with subterranean cables. Cables were well equipped to transmit speech but by no means suitable for music. Czeija stressed the need to adapt and/or couple existing lines in addition to constructing new connections. Occasionally, Czeija warned, IBU members could be asked to contribute to the costs involved in adapting circuits to music transmission. Swiss broadcasters were asked to contribute to the cost of adapting a special cable for relaying between stations over a

88 Ibid.

89 Reinhard Schlögl, *Oskar Czeija: Radio- und Fernsehpionier, Unternehmer, Abenteurer* (Wien: Böhlau, 2005); Kilian J.L. Steiner, "Buchbesprechung – Oskar Czeija: Radio und Fernsehpionier, Unternehmer, Abenteurer," *Zeitschrift für Unternehmensgeschichte* 52, no. 1 (2007): 120.

period of ten years.[90] An international relay network was thus a technologically complicated, costly, and long-term option.

The CCIF and the IBU thought the economic and technical benefits would outweigh these obstacles in the long run. A relay network would settle all kinds of issues that could not be remedied at a purely local or national level. Relaying music over telephone networks, for instance, would circumvent the wireless long-distance problems such as barriers created by mountain ranges.[91] Furthermore, there were simply no satisfactory recording systems that could broadcast live concerts or important political speeches at another time. It was impossible to exchange programs, a factor that maximized program production costs for each broadcasting organization. An international relay network would reduce these individual production costs substantially.[92] Other economic benefits included the option to ward off parasites listening in without a paid license, whereas all-European and high quality reception in combination with the offer of a high variety of programs would promote radio as an entertainment medium, increasing the number of listeners and the sales of receiver sets.[93] As many members of the early broadcasting society had a high stake in the radio industry, they would benefit enormously from a qualitatively high international relay network. [94]

Within a few years the IBU-CCIF efforts already had effect. Several national PTT administrations had already realized part of the international relay network, a promising prospect to the IBU who wanted to start regular relays in the autumn of 1928.[95] Northern and central Europe realized a considerable amount of the relay network in their countries. The west and south of Europe followed a little more slowly. Each country focused on national improvements to the network, as well as on the "liaison of suitable national circuits with those of neighbouring countries."[96] Scandinavia was one of the first to experiment with this type of relaying. Already in June 1926 the *Ligue des Femmes pour la Paix et la Liberté* organized the first sizeable relay event of music and speech between Oslo, Stockholm, and Copenhagen. Although the music from Copenhagen suffered from distortions due to the submarine cables under the Öresund, the broadcasts from Oslo were excellent as was

90 "Relais Internationaux," Série 563, 4, box 85, file Relais Général I, IBU.

91 IBU, "La radiodiffusion: Instrument de rapprochement international, Numéro spécial," *Radiodiffusion*, no. 4 (January 1937): 94.

92 IBU, "La radiodiffusion," 93.

93 "Rapport général du secrétaire à la commission des relais internationaux," Série 1237, 1 Feb. 1929, 2, box 85, file Relais Général I, 1927 et précédents, IBU.

94 IBU, *Maurice Rambert, 1866-1941* (Geneva: IBU, 1941), 7-11; J. Tomeš a kolektiv, *Český biografický slovník XX. století*, vol. 3 (Praha: Litomyšl; Paseka; P. Meissner, 1999).

95 P.V. Conseil Administrative 1927, 100, IBU.

96 IBU, "La radiodiffusion: Instrument de rapprochement international, Numéro spécial," 88.

expected from the frequently exchanged operas from Stockholm.[97]

On the continent, Germany had been employing all its means to construct a highly efficient telephone cable network and expected to couple with its neighboring European countries on the following dates: Vienna, Austria at the end of 1927 or early 1928; Zurich, Switzerland at the end of 1928; Prague, Czechoslovakia at the end of 1927 or early 1928; Innsbruck, Austria at the end of 1928 and the Netherlands in 1928. Furthermore, Switzerland foresaw additional links with Austria and Italy (Milan) in the course of 1929. Only the BBC, one of the first and most powerful broadcasting organizations in Europe, initially held back from these cross-border relays. To the BBC, the quality of the submarine cables between Great Britain and Belgium could not satisfy their high demands for quality music exchange. In addition to the cables, amplifiers and equalizers would secure modification as well as guarantee a wider band of frequencies. The BBC followed the activities on the continent with interest and considered them full of promise.[98]

These developments reflect the discrepancy between the number of organizations in the IBU and the countries working on an international relay infrastructure at this early stage. In some countries the economic and political situation influenced their participation, whereas in the Baltic States broadcasting was still in its infancy. These countries had not participated in the 1920 IBU-CCIF events and did not become IBU members till the early 1930s.[99] Consequently, the realization of a high quality and Europe-wide relay network for broadcasting was in constant flux. The network did not develop evenly in all corners of Europe.

Central and East European broadcasters immediately took the lead. They joined forces in October 1927 to integrate their national actions into a broader central European regional initiative. At the invitation of the Polish Ministry of Post and Telegraphy and *Polskie Radjo,* the PTT administration and broadcasting organizations from Austria, Germany, Czechoslovakia, and Poland gathered in Warsaw to discuss technical difficulties and create standards in testing, timing, and allocation of costs. These countries had developed their relay networks sufficiently to commence experimental relays of live concerts between their capitals. The meeting had a transnational character. For the first time, international organizations, state and non-state actors as well as telephony and radiotelephony experts, gathered in an official context to negotiate their ideas.

97 "Relais Internationaux," Série 563, 5, box 85, file Relais Général I, IBU.

98 "Remarques de la B.B.C. sur le rapport de M. le dr. Czeija concernant les relais à grande distance, by C.F. Atkinson, 24.6.27," 26 June 1927, 1, box 85, file Relais Général I, 1927 et précédents, IBU.

99 IBU, *Twenty Years*, 66, 69.

In Central and Eastern Europe, international relay construction developed far more quickly than the IBU's ability to control the process. Earlier the IBU Rapprochement Committee had suggested standardizing the new form of collaboration between the radiotelephony and telephony engineers to clarify their division of labor. At that meeting the Polish IBU delegate Chamiec proposed creating a sub-committee to the Rapprochement Committee that would coordinate the developments in international relaying in several central and east European countries, to help them with their first cross-border relay experiments. Chamiec was trying to anticipate the unequal developments in European relaying and their potential problems beforehand. Head of the committee Czeija rejected the idea, and proposed creating a small group within the committee instead. Soon the Rapprochement Committee handed in a request for its reorganization to the IBU Council.[100] The process would take a while and progressed too slowly to deal with the vast growth in international relaying. In the end, the small group of east and central European broadcasters consisting of Czeija, Chamiec, the Czech Sourek, and the German Giesecke, gathered in Warsaw with their respective PTT administrations.

The Warsaw meeting began with a considerable amount of official observations. The opening addresses stressed the unique and international unifying character of straightforward collaboration between radiotelephony and telephony engineers. Superior Councilor Thurn of the Polish Ministry of PTT pragmatically stressed the importance of good relays of music via wire for the artistic domain. Polish broadcasting director Chamiec and Polish Minister of Foreign Affairs, Mr. Libicki highlighted the ideological and cultural benefits. The overall aim extended beyond a unifying network all the way to the cultural impact such relays would have on their audiences. According to Libicki, "the regular exchange of programs by wire will contribute to the rapprochement of nations in the domain of the arts and will deepen in this way the idea of universal peace."[101] The engineers then continued the meeting without the government officials, discussing practical standards for the regional relay network with these unifying visions of society in mind.

100 "Procès-Verbal de la Commission de rapprochement, intellectuel, artistique et social tenue à Ouchy," Série 567, 17 juin 1927, 5, P.V. Commissions des Programmes [anciennement commissions de Rapprochement et Relais], 1926-1939, IBU.

101 "[l]'échange régulier des programmes par fil contribuera au rapprochement des nations dans le domaine de l'Art et approfondira à cette manière l'idée de la paix universelle." "Procès-Verbal de la Conférence des délégués des Administrations Postales et Télégraphiques d'Allemagne, d'Autriche, de la Pologne et de la Tchéchoslovaquie tenue à Varsovie," 14-16 Oct. 1927, 2-3, box 85, file Relais, Général I, 1927 et précédents, IBU.

The PTT delegates discussed the broadcasting organizations' requests. The radiotelephony experts employed by the broadcasting organizations joined them as consultants. They agreed to start a series of regular tests to trace and resolve technology related issues. These tests could start as early as October 27 on the lines between Warsaw, Vienna, and Prague. They agreed that initial testing would be free of charge. Expenses for any further and future relaying should be paid by the broadcasting organizations to their national PTT administrations. There would be a fixed tariff for each telephone line used for international relaying. These transnational agreements paved the way for the first testing of musical relays between capitals in Central and Eastern Europe.

The test results invited further unification activities in European relaying. The initial tests at the end of 1927, performed in collaboration with the IBU's newly established Brussels Checking Center, were positive. The international relay network for music in Central and Eastern Europe was ready to use. The IBU had reorganized its Rapprochement Committee, transforming the relay sub-committee into an independent International Relay Committee. This had two members from the Technical, Legal, and Rapprochement Committees. Furthermore, the Council appointed Zygmunt Chamiec as its head.[102] With the Brussels Checking Center and International Relay Committee, the IBU now possessed a more efficient apparatus to coordinate developments in long-distance relaying.

Relay infrastructure was taking shape all across Europe. In close collaboration with IBU and CCIF technical experts, the national PTTs constructed telephone lines suitable for transmitting music. In spring 1928, regular music relays began. At first the music was relayed via dedicated telephone lines between Germany, Poland, Austria, and Czechoslovakia. On arrival, the music was then broadcast over their publicly accessible home channels.[103] Broadcasting organizations in Hungary and Yugoslavia were able to join the network by 1929 when their PTTs had completed the lines.[104] In 1931, the network diffused further north to Scandinavia and south into France, while Italy was connected to Austria.[105] In the meantime, British, Belgian and German broadcasters and PTTs developed international relaying in Western Europe. The PTTs linked London to Berlin through Belgium. The BBC developed close contact with Radio Belgium. It received its

102 P.V. Conseil Administrative 1928, 139-140, IBU.

103 IBU, *Twenty Years*, 32.

104 "L'avenir des nuits nationales, Rapporteur H. Giesecke, Conseil, no. 4 de l'Ordre du Jour," Série 1592, 4 Nov. 1929, box 85, file Relais, Programmes nationaux (Nuits Nationales), Général, 1926-1931, IBU.

105 "Second talk to be given on April 13 (not checked by Achkridge): Relaying from a distance," 5, box 85, file Relais, Général V, 1931, IBU.

first experimental program of 45 minutes transmitted from Ostend to London on August 27, 1927. In 1928, relays extended from Liege and Cologne to London, enabled by strategically located repeaters in Canterbury, La Panne, Ostend, Bruges, Ghent, Brussels, Tirlemont, Liege, and Aachen. Two years later they had developed the network into the Netherlands and throughout the whole of Germany, foreseeing new high-quality sub-marine cables for the coming year.[106] Cross-border links in Europe now relayed high quality programs on a regional basis, with Germany participating both in the eastern and western initiatives.

The uneven regional development of international relaying moved the representatives from Belgium, Germany and Great Britain to propose the establishment of so-called regional secretariats. The proposal was in sharp contrast to the initial IBU project of an internationally coordinated Europe-wide relay network. At a conference in Brussels in October 1929, the PTT administrations and broadcasting organizations from these three countries considered creating a central secretariat for north-western Europe. Such a secretariat would centralize and provide the interested organizations with all the relevant information for international relaying. The IBU rejected the proposal. It argued that there was no direct need for such a secretariat at that moment.[107] The coordination of international relaying was still expected to become Europe-wide in the hands of the IBU. The vision of a Europe consisting of regional relay networks for music across long distances vanished as quickly as it had arisen.

Shortly after the regional network discussion, the IBU confirmed its role as central coordinator of the international relaying of music over the network owned and established by the PTTs. Network construction efforts were well underway and central European countries had been relaying programs on a regular basis for some years already. The IBU argued that the time had come for a more straightforward collaboration between European broadcasters. At the IBU Council in Budapest in October 1930, the International Relay Committee organized a meeting with representatives from the central-eastern and west European groups to discuss linking up their projects. The east European group had met in Warsaw one month before and agreed to relay five to ten symphony concerts between Germany, Austria, Czechoslovakia, Hungary, and Poland.[108] Other broadcasters had expressed their desire to join the central European group at any rate in part of

106 P.V. Conseil Administrative 1929, 225, IBU; "Second talk to be given on 13th of April," 5, box 85, file Relais, Général V, IBU.

107 "Conférence de Bruxelles, les 8, 9, et 10 octobre 1929: Protocole final, séance plénière de clôture," box 85, file Relais, Général III, 1929, IBU.

108 "Décisions de la Réunion de Varsovie des 25-26.9.30," 1, box 85, Relais, Concerts Internationaux, 1931 à 1939, Général, IBU.

its relays.[109] As a result of these preliminary talks, the Council meeting in Budapest decided in favor of "a complete unification of these two groups..."[110]

The straightforward collaboration immediately led to mutual agreements about future relaying. Their program directors would meet twice in the following year to discuss program contents under the auspices of the IBU Relay Committee. Most importantly perhaps, they anticipated a series of concerts of the highest quality to be relayed each month at fixed dates by all participants to further their ideas about rapprochement between peoples. These concerts would bear the name *Concerts européens* (see Chapter 6) and be part of a large, systematic exchange of artistic programs.[111] All "organizations that have not participated in this cordial exchange" would be encouraged "to adhere to it in accordance with their technical possibilities resulting from the progressive organization of the international telephone circuits..."[112] All parties now voluntarily and formally agreed to create a Europe-wide internationally coordinated relay network. Moreover, they imagined that such a network should be used for an extensive international exchange of national programs as well as unique joint European programs.

These agreements inspired increased uniformity in standards for the international relaying of music via cable and plain copper wire. Although by 1930 the network did not yet encompass the whole of Europe (Figure 3.4), it would expand rapidly in the following years. The broadcasting organizations agreed that there were no costs attached to the musical programs. Each IBU member would at some point organize a concert or produce specific programs. Overall, these activities should ensure an equal division of costs among the various organizations. The cost of renting the relay network from the national PTT administrations, on the contrary, was another matter. The IBU and PTT administrations agreed that special standards should ensure the division of costs and provide a clear set of rules for the act of relaying.

The following outlines the standards specifying the cost for using the international relay network. In all circumstances the IBU member who organized a relay would have to submit a request to its national PTT administration for renting the

109 "Perfectionnement des concerts destinés aux relais: Commission des Relais, Rapporteur M.S. de Chamiec," Série 2130, 7 octobre 1930, 1, box 85, file Relais, Général IV, 1930, IBU.

110 "une union complète des deux groupes..." "Réunion des membres du groupe des échanges de programmes de l'Europe Centrale et de l'Europe Occidentale: Budapest – Octobre 1930," 1, box 85, file Relais, Concerts Internationaux, 1931 à 1939, Général, IBU.

111 "Réunion des membres du groupe des échanges de programmes," 1, box 85, file Relais, Concerts Internationaux, Général, IBU.

112 "...incite tous les organismes qui n'ont pas pu participer à cet échange cordial à y adhérer au fur et mésure que les possibilités techniques résultant de l'aménagement progressif des circuits téléphoniques internationaux..." P.V. Conseil Administrative 1930, 270, IBU.

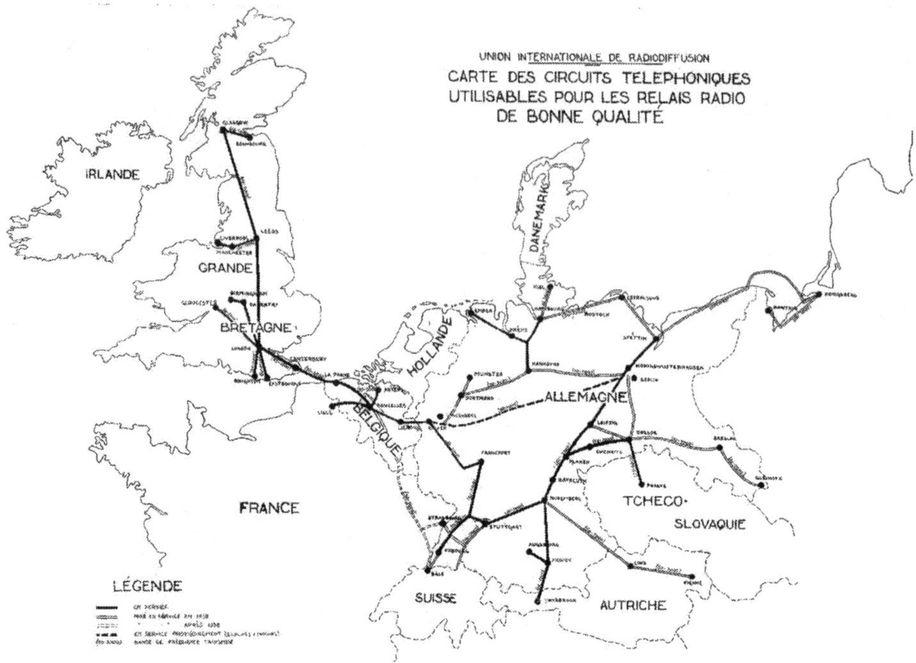

Figure 3.4 – The telephonic circuits usable for high quality radio relays, 1930
Source: Box 85, Relais 1927-1936, IBU. Used by the courtesy of the European Broadcasting Union.

entire circuit required for all participants to receive the relay. Only in cases of local drawbacks of the circuit or an emergency could participants apply to their local PTT. If broadcasters wanted to participate in relays at the last minute, they could only do so with the consent of the organizing broadcasting IBU member. The costs for renting the relay network, however, would have to be divided equally among all broadcasting members using specific arteries of the relay network. For instance, if three out of ten broadcasting members benefitted from one specific artery in the relay network, they would have to split the costs equally.

An exception to this rule was if a country wished to produce and relay a program from a location in another country that did not want to participate in the relay. In such cases, the foreign country would have to apply to the IBU Council representative in the country where the broadcast would take place as well as notify its own IBU Council representative. If the request was not considered harmful to the country where the broadcast would take place, negotiations between the parties could proceed. If favorable, negotiations from then on could be on a bilateral basis if desired, but in general between the IBU Council representatives from the requesting broadcasting organization and their counterpart in the country where the broadcast would take place.[113] The firm integration of the national into the international relay network clearly did not simplify a process like creating standards for cost divisions.

The standards assuring a Europe-wide relay network for broadcasting were a complex and highly nationally oriented set of rules. These rules had to combine locally produced programs with nationally owned telephone and broadcasting networks, as well as international streamlining activities. The standards bound, fine-tuned, and integrated the local, the national, and the international into one set of European standards that consequently had a transnational character. Though contemporaries clearly specified the internationality of the relay network, in the early years these standards only focused on Europe. Together with the wireless network, the relay network enabled the IBU community to create a high-quality and Europe-wide diffusion of music.

Wireless World Visions

The development of the short waves offered visions of a wireless world. Rapid technological innovations by the late 1920s and early 1930s posed new challenges to the community of experts and their European standards. These new challenges made the IBU rethink its 1926 outwardly communicated idea of being an all-*international* society in relation to its recently recognized status of *European* broadcasting expert. According to the director of the Technical Committee "at present the IBU is for all else a European union."[114] The development of the short waves opened up unexpected opportunities to create wireless broadcasting with a global reach via low power and therefore less expensive radio stations. Many organizations therefore quickly employed these waves for so-called empire broadcasting: to overseas possessions, to colonies.[115] The Technical Committee reasoned that the narrowness of the medium and long bands assigned to broadcasting by the Madrid Convention further enticed the explosive employment of the short waves.[116] The short waves had quickly become an attractive complement and alternative to broadcasting.

113 P.V. Conseil Administrative 1930, 246-247, IBU; "Rapport de la Commission des Relais au Conseil, no. 8 de l'Ordre du Jour," Série 1900, 20 mai 1930, box 85, file Relais, Général IV, 1930, IBU.
114 "Radiodiffusion mondiale: Rapporteur: R. Braillard, Président de la Commission Technique," Série 3140, 23 mai 1932, 1, CA Document Series, IBU.
115 Donald R. Browne, *International Radio Broadcasting* (New York, N.Y.: Praeger, 1982); James Wood, *History of International Broadcasting* (London: Peregrinus, 1992).
116 "Recommandation de la commission technique au conseil: Extension de l'U.I.R. Collaboration avec les autres organismes similaires," Série 4840, 22 fév. 1935, CA Document Series, IBU.

In the meantime, further expansion of the European broadcasting network increased the incentive to broadcast on a global scale. Both the wireless networks and the international relay network in Europe continued to diffuse and became denser in the 1930s (Figures 3.5 and 3.6). In particular the swift expansion of the European relay network and subsequent flow of cross-border relays almost immediately attracted the attention of stations in and beyond Europe. In 1929 for instance, the Spanish Union-Radio Madrid asked the French PTT to discuss the availability of telephone lines for relays with France and other European countries.[117] However, it was the inter-continental relay of the American program "Compte Zeppelin" via the short waves to the European relay network that raised a lively interest across the entire globe. The zeppelin, owned by the German Count Ferdinand von Zeppelin made its first ever intercontinental flight. Several European and non-European stations expressed their disappointment to the IBU that they had been unaware of the relay of an event with such worldwide importance.[118] In hindsight Secretary General Burrows recognized the "Compte Zeppelin" experience as the first moment the IBU realistically opted for envisioning an international relay network on a global scale.[119]

Such intercontinental events gave a new dimension to European relaying activities beyond its borders. Broadcasting experts could now project a global, international broadcasting network. Problematically, the global radius of these waves interfered with the frequency standards designed by the IBU, ITU, and PTT administrations, which were only significant to the European region. The problem with the short waves challenged IBU's European focus and the standards for frequency allocation. Consequently, the IBU and other parties also had to renegotiate Europe's role in a globalizing world.

The number of contacts between European broadcasters and worldwide broadcasting stations grew instantly. As early as 1927, National Broadcasting Company (NBC) in the USA, stations in India and Australia, the Peruvian PTT, as well as Nohon Hoso Kyokai (JOCK Broadcasting Station) in Japan joined the IBU as associate members. Initially, inter-continental activities mostly developed between the active European members of the IBU and the major U.S. broadcasters NBC and Columbia Broadcasting System (CBS). Partly thanks to their shared language, NBC, CBS and BBC bilaterally developed a lively program exchange. These exchanges opened up a huge variety of new music to the British and on occasion other

117 "Rapport général du secrétaire à la Commission des relais internationaux," Série 1237, 1 fév. 1929, 3, box 85, file Relais, Général III, 1929, IBU.

118 "Relais internationaux d'évènements d'intérêt mondial, rapporteur: A.R. Burrows," Série 1142, 21 nov. 1928, 1-2, box 85, file Relais, Général II, 1928, IBU.

119 "Rapport général sur les progrès accomplis dans le domaine des relais internationaux, rapporteur A.R. Burrows," Série 1852, 29 avril 1930, 2, box 85, file Relais, Général IV, 1930, IBU.

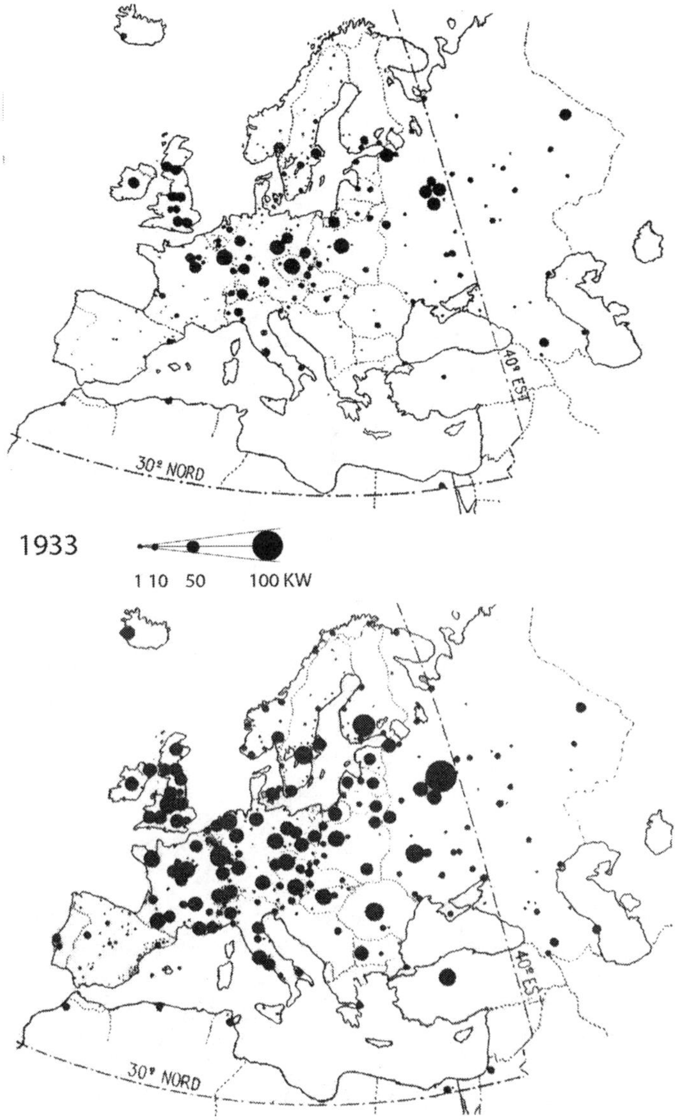

Figure 3.5 – Rise in number and power of European radio stations in 1933 and 1938
Source: "Projet d'attribution des fréquences aux stations de radiodiffusion de la région européenne établi par l'Union Internationale de Radiodiffusion en exécution des stipulations du Protocole Additionnel aux Actes de la Conférence Internationale des Radiocommunications du Caire, 1938 et présenté au Gouvernement de la Conféderation suisse," Doc. 536, Bruxelles, 29 nov. 1938, 37, registry file 9G, box R-4322, file 35525/663, jacket 1, League of Nations Archives, UNOG Library. Used by the courtesy of United Nations Office, United Nations Library, Geneva.

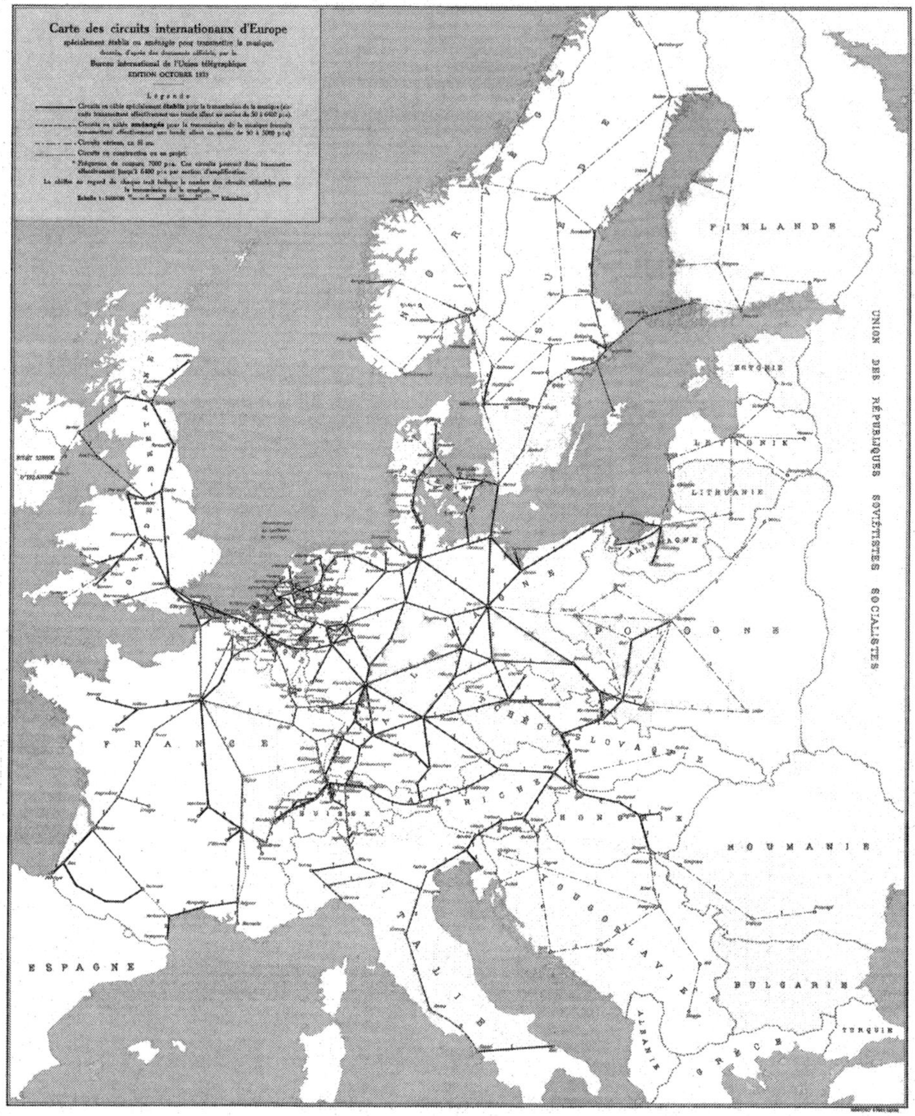

Figure 3.6 – The European relay network in 1933 and 1939 (on page 115)
Source: "Carte des circuits Internationaux d'Europe spécialement établis ou aménagés pour transmettre la musique" (Berne: 1933, 1939), International Telecommunication Union Archive (Hereafter ITU). Used by the courtesy of the International Telecommunication Union, Geneva.

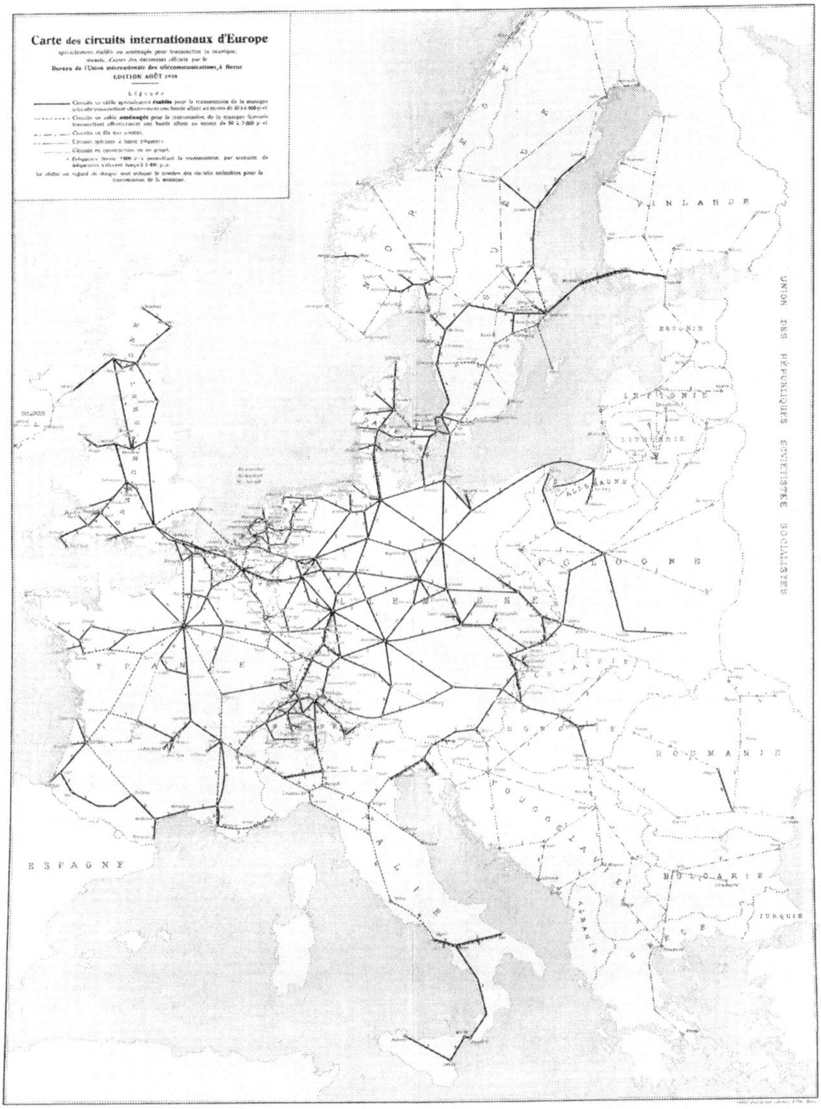

European audiences.[120] Additionally, CBS attempted to establish a multilateral program exchange with several European countries.[121] Both American organizations had become IBU associate members by 1930, appointing directors as their first foreign broadcasting representatives in Europe.[122]

120 "Rapport général sur les progrès accomplis," Série 1852, 2, box 85, file Relais, Général IV, IBU.

121 "Radiodiffusion mondiale," Série 2862, 15 oct. 1931, 1-2, box 85, file Relais, Général V, 1931, IBU.

122 In their roles, César Searchinger for CBS and Dr. Max Jordan for NBC had close relationships with the various European broadcasting stations and frequently attended IBU meetings, not only Council meetings but also those of the various committees. From 1932 onwards they even became actively involved

The IBU also began to collaborate with Japan and established a short wave connection between Europe and Japan to relay the Naval Disarmament Conference in London in 1932. Both the request to Japan to participate in the relay and the broadcast itself made a big impression outside Europe. The broadcast demonstrated the potential of the existing short wave stations in Europe for further developing the international relay network beyond European borders.[123] The growth in such inter-continental contacts soon imposed a sense of urgency on studying worldwide broadcasting via the short waves.

Worldwide broadcasting increasingly came within sight. The first global encounters took place alongside improving international relations. The French Prime Minister Aristide Briand, in his famous address to the League's General Assembly on September 5, 1929, proposed the creation of a United States of Europe.[124] The League had never officially followed up any of the IBU proposals, even when the IBU stressed the importance of European activities to the world as a whole.[125] Informally, however, some members continued to exchange views on the vision of a worldwide organization for broadcasting, of a *Radiodiffusion mondiale*.[126] Discussions took place in particular between the key wireless expert of the League's OCT, Gijsbert Frans van Dissel, and IBU Secretary General Burrows. They agreed that such an organization should coordinate efficiently the evolution of a global and fully interconnected broadcasting space based on global standards for wireless and wired broadcasting. Burrows had expressed his opinion to Van Dissel that "the question of radio broadcasting is a question of worldwide interest." Van Dissel in turn reflected that global broadcasting should be centralized under the auspices of the League.[127] When Briand proposed a United States of Europe, Van Dissel and Burrows considered a global organization for broadcasting beyond European borders.

Van Dissel's thoughts on developing international broadcasting may not have been unexpected considering his personal life, education, and work experience. He grew up in a family that lived for some time in the Dutch East Indies. Although

in the joint meetings of the IBU International Relay Committee and the Rapprochement Committee to exchange ideas about establishing a satisfactory international relay network. "Développement des relais internationaux depuis octobre 1931, rapporteur A.R. Burrows," Série 3167, 20 mai 1932, box 85, file Relais, Concerts Internationaux, 1931 à 1939, Général, IBU; CBS, *Spanning the Seven Seas: A Résumé of Inter-Continental Broadcasting in 1935* (New York; London: Columbia Broadcasting Organization, 1935), 3.

123 "Rapport Général sur les progrès accomplis," Série 1852, 2, box 85, file Relais, Général IV, IBU.

124 LoN, *Verbatim Record of the 10th Ordinary Session of the Assembly of the League of Nations, 6th Plenary Meeting*, LoN doc. ser., A.10.1929 (Geneva: LoN, 1929), 5.

125 "Utilisation commerciale des émissions radiophoniques, rapporteur: O. Czeija," Série 839, 11 juin 1928, 3, box 85, file Relais, Général II, 1928, IBU.

126 "Utilisation commerciale," Série 839, 3, box 85, file Relais, Général II, IBU.

127 13 fév. 1931, 3, P.V. Commission Mixte, Commission Relais, Commission Spéciales, IBU.

his brother and sister were born there, Van Dissel was born in the Netherlands, in the Hague, on October 6, 1891. He studied civil engineering in the Netherlands and specialized in radio telegraphy at the École Supérieure de l'Électricité in Paris, France. In 1921 he published a book on the international organization of radio-electric traffic.[128] Initially, he worked as a civil engineer in the Dutch East Indies on railway, bridges, and irrigation matters. During these years overseas he had become aware of the important role of long distance communication in the "development of relations between the motherland and the colonies." [129] After his specialization in radio telegraphy he returned to the Dutch East Indies to work for the PTT. As chief operations of the large telegraphy and telephony stations in Bandoeng (Malabar) on the island of Java, he gained extensive experience with overseas long distance communications. He was in charge of communications with Europe and other continents. As chief of operations he regularly represented the Dutch East Indies government at international conferences.[130] He married twice. After his divorce from a Dutch woman in 1931, he married a Filipino woman. His life and work suggest a highly internationally oriented mind. When he applied to the League of Nations he stressed his desire to work for a "highly humanitarian institution pursuing such a noble ideal."[131] He died in Geneva on September 29, 1958.

In the 1930s the cordial relationship between Van Dissel and Burrows impacted profoundly on international broadcasting developments. It strengthened the good relationship between the League and IBU's activities. Van Dissel had an important role, and for both the IBU and the League as a whole, he was the key expert on wireless questions in the widest sense.[132] He began as an ordinary member of the OCT regularly representing the League at international conferences, IBU Council meetings, and technical discussions on broadcasting. In 1935, he became Director of the OCT when Robert Haas died.[133] Van Dissel's highly international orienta-

128 Gijsbert Frans van Dissel, *De internationale organisatie van het radio-electrische verkeer: Voordrachten gehouden door comm. Brenot voor de afdeling der radio-telegrafie van de école supérieure d'électricité gedurende het semester 1920-1921* (Leiden: A.W. Sijthoff, n.d.).

129 "Letter from G.F. van Dissel to R. Haas," The Hague, 19 Oct. 1929, Personal Files: G.F. van Dissel, LoN.

130 "Letter from Eysinga, to League Secretary General Drummond," Leyden, 5 Oct. 1929, Personal Files: G.F. van Dissel, LoN.

131 "Letter from Van Dissel to Haas," The Hague, 19 Oct. 1929, 3, Personal Files: G.F. van Dissel, LoN.

132 During the early 1930s Van Dissel focused on electricity issues and was appointed secretary of the OCT's Committee on Electric Questions. Vincent Lagendijk, *Electrifying Europe: The Power of Europe in the Construction of Electricity Networks* (Amsterdam: Aksant, 2008), 90.

133 A few years later the OCT merged with the Economic and Financial section as a result of which the entire directorate disappeared and Van Dissel gave up his position as Chief. BUIT, *Documents de la Conférence Européenne des Radiocommunications Lucerne, MAI/JUIN 1933* (Berne: BUIT, 1933), 18; IBU, *First Intercontinental Meeting of Broadcasting Organisations, Paris, Febrary 27-March 6 1936: Preparatory Documents and Minutes of the Meeting* (Geneva: IBU Office, 1936), 12; BUIT, *Documents de la Conférence Internationale des Radiocommunications du Caire (1938), Tome I: Propositions transmises au Bureau de*

tion must have stimulated the discussions with Burrows further, in particular their exchange of views on a *Radiodiffusion mondiale*. Their contacts seem to have been frequent, informal, perhaps even on the basis of friendship. Van Dissel would give their visions about a *Radiodiffusion mondiale* an official character.

In October 1931 he sent a provocative Memorandum to the IBU Secretariat. In light of international developments across the globe, he proposed that the IBU extend its view and influence beyond the borders of Europe. He mentioned the short wave stations employed in France, the Netherlands, Great Britain, Germany, and North America, and the rapidly increasing interest in radio broadcasts by audiences in far-away countries with different climates and political circumstances who lacked a regular broadcasting service. According to Van Dissel, there was no time to lose in ending this transitory and irregular phase. A collective effort was required on a global scale.

Van Dissel dwelled on his plan for a *Radiodiffusion mondiale*. In addition to an expansion of inter-continental program exchange activities, the IBU, in his opinion, should accelerate the process of worldwide broadcasting in general. Van Dissel believed that the IBU should establish a permanent special committee for short waves in which other interested organizations could participate. Moreover, he considered it of the utmost importance that an independent global institution be created for broadcasting as well, to oversee its efficient development. Members of such an organization should come from the private sector, governments, or international institutions that either already possessed or would soon acquire the technical means to realize broadcasts with a global radius.[134] According to Van Dissel, the IBU should examine as quickly as possible the format of such an organization consisting of members with different nationalities.

In Van Dissel's view, a special study committee should prepare the statutes for this new organization with the help of expertise available in the IBU Legal, Relay and Rapprochement Committees. Since the world already recognized the IBU as European broadcasting expert, it would be fair to place the new organization under direct supervision of the IBU during its first phase, Van Dissel reflected. This construction was designed to be as supportive as possible of the new organization. Supervision here referred to an indirect relationship since many members of the new organization were IBU members already. During a transitory period the new organization should then gain sufficient strength to become an independent

l'Union pour être soumises à la Conférence: Propositions, notes, déclarations, études, remarques soumises pendant la Conférence (Berne: BUIT, 1938), xxi; BUIT, *Documents de la Conférence Européenne de Radiodiffusion, Montreux, Mars/Avril 1939* (Berne: BUIT, 1939), 12.

134 "Radiodiffusion mondiale," Série 2862, 15 Octobre 1931, 1-2, quote: 2, box 85, file Relais, Général V 1931, IBU.

organization. After that it could become an ordinary IBU member like all other broadcasting organizations.[135] Van Dissel had designed a relatively complicated transnational structure for a global broadcasting institution that should realize efficiency in worldwide broadcasting under the supervision and later as member of the IBU. The IBU Council decided to let the Rapprochement Committee study the proposal.[136]

Only one day later Van Dissel withdrew his proposal. He realized that his idea had been too premature given the state-of-the art technology. He proposed that the IBU should focus on the exploration of a possible role for the short waves in international relaying activities instead. There appeared to be a large basis of support for his proposal in the IBU. The IBU Council translated the League memo into a resolution, recognizing the power of broadcasting to improve international relations. The Council encouraged its members to fully cooperation with studies and conferences on topics related to the international nature of the medium. Its members should "develop in their countries radio broadcasting via the short waves." The Council decided to integrate Van Dissel's initial idea of a worldwide organization for broadcasting. Both the Technical and the Rapprochement Committee should study the options to develop "this interesting form of radio broadcasting."[137] Van Dissel's memo was the first time that the League set the IBU's agenda, proposing a European led, global organization of broadcasting. The IBU did not need any further encouragement to pursue such an organization.

The IBU paved the way for a *Union universelle*. Between 1932 and 1936 the IBU Technical and Rapprochement Committees sought to establish international broadcasting on a global level. Initially, they explored options to expand the Union from a European to a universal organization. Such an idea might seem utopian at first, Burrows argued, "but it is highly likely that in some years' time it will be considered the normal development of new possibilities offered by broadcasting."[138] According to Braillard:

> the increasing evolution of broadcasting problems renders increasingly

135 "Radiodiffusion mondiale," Série 2862, 2, box 85, file Relais, Général V, IBU.

136 P.V. Conseil Administrative 1931, 71, IBU.

137 "développer dans leurs pays la Radiodiffusion mondiale à ondes courtes"..."cette forme intéressante de radiodiffusion." P.V. Conseil Administrative 1931, 78, IBU; "Procès-verbaux des réunions de la Commission de Rapprochement tenue à Rome," Série 2924, 25 nov. 1931, 4, P.V. Commission des Programmes [Anciennement Commissions de Rapprochement et Relais] 1926-1939, IBU.

138 "mais il est probable que dans quelques années elle sera considérée comme le développement normal des nouvelles possibilités offertes par la radiodiffusion." "Expansion de l'Union, Rapporteur: A.R. Burrows," Série 3188, 24 mai 1932, 1, CA Document Series, IBU.

> urgent the organization of worldwide broadcasting: this responds to technical necessities on the one hand and, on the other, responds to the general scope that broadcasting has to pursue in its broadest sense, this is to say the rapprochement between peoples.[139]

Whereas Braillard approached the idea from a technical perspective, Burrows in contrast added the political and economic ramifications. Both promoted a global union of broadcasters in light of what they considered the main aim of broadcasting, namely the creation of mutual understanding between peoples.

Braillard discussed the idea of a *Union universelle* in the context of the never-ending increase in inter-continental relaying and the complex character of the allocation of short wave frequencies. Broadcasting over the short wave increased rapidly. Braillard referred to growing colonial broadcasting, to the global broadcasts of the League of Nations and Vatican Radio (Chapter 4), and to increased relay activities between parts of the globe sharing the same languages. As well as the English language, this also applied to regular relays between South America and Spain, he pointed out. Though the need to make arrangements for the regulation of short wave frequency allocations was not urgent at that moment, Braillard stressed that such urgency would soon emerge. Would it not be better to prevent rather than try to cure a disease?[140]

Such standardization required substantial analysis of the characteristics of the short waves. The technicalities of broadcasting over the short waves proved more problematic than broadcasting over all the other waves together. Favorably, broadcasts over the short waves "could be directional if desired, and with quite modest power of perhaps 10 kW, could attain ranges of many thousands of kilometers, relying on reflection from the ionized layers in the upper atmosphere."[141] They were a cheap alternative to the long and medium waves. The short wave transmissions however, rather than following the curvature of the earth like the medium and long waves, beamed against the ionized layers in the air. Varying with the time of day as well as the season, reception depended on the distance between the transmitter and the receiver. As a result, one station needed several frequencies during the day to be able to broadcast its programs.[142] Together with the need to improve

139 "…l'évolution incessante des problèmes de la radiodiffusion rend de plus en plus urgente l'organisation d'une Radiodiffusion mondiale: ceci répond à des nécessités techniques d'une part, et, d'autre part, répond au but général que doit poursuivre la radiodiffusion dans son sens le plus général, c'est-à-dire le rapprochement des peuples." "Radiodiffusion mondiale: Rapporteur: R. Braillard, Président de la Commission Technique," Série 3180, 23 mai 1932, 1, CA Document Series, IBU.

140 "Radiodiffusion mondiale," Série 3180, 1, CA Document Series, IBU.

141 Williams, *A History of Technology: Vol VII The Twentieth Century c. 1900 to c. 1950 Part II*, 1255.

142 Daniel Headrick, *The Invisible Weapon: Telecommunications and International Politics, 1851-1945* (New York: Oxford University Press, 1991), 203.

short wave transmitters, increase power, and fine-tune antennas and improvements in short wave receivers, these frequency characteristics complicated standardization activities substantially.[143] The expanding international relay activities and the technically complicated structure of the short waves required a global approach to the problems of broadcasting. Braillard therefore pleaded for a *Union universelle*.

Braillard and Burrows outlined the idea of a *Union universelle* under the heading of expanding the IBU. In their opinion the IBU had made some remarkable achievements, creating standards to which so many organizations adhered. Burrows pointed out that the European broadcasters had put much research and effort into these standards. Many extra-European broadcasting organizations like those in Japan had always followed these developments with interest, using them to their advantage. This implied a broad interest in and recognition of IBU standards outside the IBU community, even though such standards would not always be integrated one-by-one. It also implied that the greater research and development burden remained with the European broadcasters. According to Burrows, transforming the IBU into a *Union universelle* would further strengthen the global position of IBU standards.[144] Moreover, research and development costs could then be divided more equally among broadcasters globally, a development that would reduce costs for European broadcasters.[145] The new situation would bring economic and technical efficiency under the supervision of a *Union universelle,* the IBU.

The discussion explored several different visions of such a *Union universelle*. The IBU could become a global organization with active membership only, in which all national members had the same rights and obligations including the right to vote. In this case the world would resemble a community of national broadcasters in which Europe would not have a central or exceptional position. The IBU could also become some kind of federation of broadcasters. Supported by his Technical Committee, Braillard favored a simple amendment of the IBU statutes that would delete the Extra-European membership category altogether. Current Extra-European members would then become active members, turning the IBU into a universal organization almost overnight.[146] The alternative was much less favorable, Braillard argued. In a scenario where other parties in the world developed a similar initiative, they might even absorb the IBU in the end. Braillard suggests that with more and more organizations participating across the world, the universality of the IBU would eventually become firmly grounded over time.[147]

143 Fickers, "Broadcasting as Critical Infrastructure," 14.
144 "Expansion de l'Union," Série 3188, 2, CA Documents Series, IBU.
145 Ibid.
146 "Radiodiffusion mondiale," Série 3180, 1, 3, CA Document Series, IBU.
147 Ibid., 3.

Braillard proposed another concept as alternative or step towards a *Union universelle*. To avoid difficulties with extra-European organizations, the IBU could establish sections based on regional definitions like a European section that included Northern Africa or a South American section. These sections could then optimally fine-tune global interests with problems of a more continental nature comparable to issues in the present-day IBU. From time to time qualified delegates would participate in a joint conference studying problems of a global nature.[148] If the proposal took root, the IBU would then become a federation in which the regional sections would have relative freedom to maneuver under the auspices of the general aims and standards of the Union. Although Europe would be only one section among others, the federation would continue in the European tradition of the IBU with Geneva as central hub for the organization.

Burrows posed the question more openly to the other discussants. We have to know "if the Union wants to become a global organization or remain a European organization," he asked.[149] There was possible global support for expanding the Union into a *Union universelle*. Burrows urged the discussants to consider Braillard's proposal, adding that the IBU could also approach the extra-European members to give them the choice of active membership. This might make it easier to achieve a *Union universelle*. Whatever format the *Union universelle* might have, Burrows in any case pleaded in favor of an "inter-continental expansion of the Union."[150] He agreed with Braillard that it was absolutely necessary to envisage worldwide collaboration if only to resolve the increasingly pressing problems with global interference. These problems could not be resolved without a strong spirit of international collaboration.[151]

The discussion ended with the general recognition that no effective study about a specific format for a *Union universelle* could be undertaken until the IBU committees had all the documentation on the topic. The Rapprochement Committee would look at the possibility of more straightforward cooperation with the Extra-European countries on all aspects of broadcasting. The committee sought answers to questions such as whether they should consider expanding the IBU given the actual situation of broadcasting in and outside Europe, and if so, in which

148 "section européenne englobant le Nord de l'Afrique, section Nord Américaine, section Sud Américaine, section du Pacifique englobant le Japon, l'Indochine, l'Australie, etc…Chacune de ses sections traiterait au mieux les intérêts, communs, les problèmes de caractère continental analogues à ceux qui préoccupent notre Union actuelle, mais de temps à autre, des déléguées qualifiés de chacune de ses sections se réuniraient en conférence pour étudier en commun les problèmes ayant un caractère mondial." Ibid.

149 "si l'Union veut devenir un organisme mondial ou rester un organisme européen." "Expansion de l'Union," Série 3188, 1, CA Documents Series, IBU.

150 "expansion intercontinentale de l'Union." Ibid., 3.

151 "Radiodiffusion mondiale," Série 3180, 4, CA Document Series, IBU.

domains and in what form a more straightforward cooperation might emerge. The Technical Committee in turn would carry out a separate study on international relaying over the short waves that should clear the way for future standardization of their usage.[152] It would seek straightforward collaboration with engineers from extra-European countries at the 1932 Madrid Conference. The Assembly agreed but decided instead that the Technical Committee should focus on all the questions that could possibly hinder IBU's freedom of action in the future.[153]

In order to seek a more straightforward collaboration and possible expansion of the Union, the IBU contacted the United States representatives at the 1932 Madrid Conference. Regular relays between Europe and the United States existed already, but their broadcasting standards and networks differed substantially. Braillard noted that "especially in the United States there has developed another technique of which we know too little in general, just as Americans are often poorly informed about what we do in Europe."[154] Their mutual ignorance had led to troubles at times, Braillard found. Straightforward collaboration would facilitate efficient broadcasting. Moreover, it would have a positive effect on their position during international conferences like Madrid regarding other means of telecommunications. Straightforward collaboration would benefit all.

The United States had already been expressing interest in exchange and collaboration. In addition to CBS and NBC, the American National Association of Broadcasters (NAB) had recently applied for IBU extra-European membership. Burrows described the NAB as a kind of American union for broadcasting that represented some 170 American broadcasting organizations.[155] The NAB was the primary lobbying organization for commercial broadcasters in the United States, and consequently the best place to start discussing straightforward collaboration. Although the IBU representatives in Madrid approached engineers from several organizations, they above all seem to have had positive discussions with Louis Caldwell, President of the NAB Committee for Foreign Relations. The IBU asked Caldwell to consider developing an association for all American broadcasting

152 "Procès-verbaux des Réunions de Commission de Rapprochement et de Relais réunis tenue à Montreux," Série 3245, 10 juin 1932, 8-9, P.V. Commission des Programmes [Anciennement Commissions de Rapprochement et Relais] 1926-1939, IBU.

153 "Procès-verbal de la Réunion commune de la Commission de Rapprochement et de la Commission Technique," Série 3263, 27 juin 1932, 1-2, P.V. Commission des Programmes [Anciennement Commissions de Rapprochement et Relais] 1926-1939, IBU.

154 "Parallèlement, et notamment aux Etats-Unis, une autre technique s'est développée dont, en général, nous connaissons trop peu de chose, de même que les Américains sont souvent assez mal informés de ce que nous faisons en Europe (nous avons pu le constater encore récemment à Copenhague.)" "Radiodiffusion mondiale," Série 3180, 2, CA Document Series, IBU.

155 "Expansion de l'Union," Série 3188, 1, CA Documents Series, IBU.

organizations and to establish more straightforward cooperation between any such new organization and the IBU.[156]

Caldwell has been called "one of the most influential figures in the history of broadcasting regulation."[157] The IBU decided it could not wish for a better ally. He "was the most visible [and most important] legal authority on broadcast policy" in the United States. He "worked as a top-level legal adviser on behalf of both NAB and NBC," as chair of the American Bar Association and, moreover, had a successful law practice with the Radio Corporation of America (RCA) as its clientele.[158] In the eyes of the IBU, Caldwell could establish and lead the lobby for setting up an encompassing organization for broadcasting in the USA. Though Caldwell seems to have responded positively to the IBU overture, the initiative stalled before it could get started. Caldwell almost immediately became deeply involved with internal NAB difficulties at home. Even though the IBU continued to correspond with Caldwell, they would not pursue the idea in the following years.[159] The internal U.S. difficulties were an unfortunate set-back for the IBU and its hopes to realize a *Union universelle* in the short term.

In the meantime, the regional organization of telecommunications and broadcasting across the globe took off. The African countries had postponed the regulation of broadcasting to a later date, but several did establish the African Telecommunication Union following the ITU model in October 1935.[160] Efforts in South America as well as in Asia and Oceania also focused on the organization of broadcasting.[161] In Latin and Southern America, broadcasting seems to have

156 "Procès-verbal des Réunions des Commissions de Rapprochement et de Relais tenue à Bruxelles," Série 3550, 10 mars 1933, 1, P.V. Commission des Programmes [Anciennement Commissions de Rapprochement et Relais] 1926-1939, IBU.

157 Michele Hilmes, *Only Connect: A Cultural History of Broadcasting in the United States* (Belmont, CA: Wadsworth/Thomson Learning, 2002), 67.

158 Robert W. McChesney, *Telecommunications, Mass Media, and Democracy: The Battle for the Control of U.S. Broadcasting, 1928-1935* (New York: Oxford University Press, 1993), 131; Hilmes, *Only Connect*, 67.

159 "Procès-verbaux des Réunions des Commissions de Rapprochement et des Relais tenue à Amsterdam," Série 3922, 13 nov. 1933, 2, P.V. Commission des Programmes [Anciennement Commissions de Rapprochement et Relais] 1926-1939, IBU.

160 John D. Tomlinson, "The International Control of Radiocommunications" (Ph.D. diss., Université de Genève, 1938), 251-252.

161 "G. Gallarti to A. Pelt," March 5, 1935, 3, registry file 9G, R-4321, file 15874/663, LoN. Whether their efforts followed up on the IBU initiatives in Madrid to seek straightforward collaboration with these organizations remains unclear from the archives. Neither material on the Madrid Conference collected by the ITU, nor the IBU Relay and Rapprochement Committee documents give any details on the contacts and discussions the IBU had with the engineers and participants of extra-European broadcasting organizations. BUIT, *Documents de la Conférence Radiotélégraphique Internationale de Madrid (1932): Tome II: Travaux de la Conférence (Délibérations, Actes définitifs)* (Berne: BUIT, 1932); Box 14, Conférences et conventions internationales, IBU; Box 85, Relais 1926-1950, IBU; P.V. Commission des Programmes [Anciennement Commissions de Rapprochement et Relais] 1926-1939, IBU; P.V. Conseil Administrative 1932-1935, IBU.

developed in two different ways. Most Latin and Southern American broadcasting stations worked closely with the American radio industry and U.S. advertizing agencies. Moreover, U.S. policy makers worked hard to increase their influence over the development of broadcasting in this region in order to reduce European influence during large International Telecommunication Conferences. These American efforts turned out to be very successful.[162]

When one explores the organization of broadcasting into larger regional agreements and pan-national Southern American organizations, there seems to have been considerable overlap with IBU initiatives. In August 1934, Argentina, Bolivia, Brazil, Chile, Paraguay, and Uruguay together signed the South American Radio Agreement in Buenos Aires. In the eyes of Tomlinson, the new South American Broadcasting Union (USARD) performed many of the functions for South American countries which the U.I.R. [IBU] performed for Europe.[163] Though none of the signatories had links to the IBU, the organization and its focal points resembled IBU activities substantially.[164] Like the IBU, USARD had a permanent center in Montevideo but no technical center. Unlike the IBU, each USARD agreement had to be ratified by the national governments, while countries outside the agreements could choose to adhere to them. The agreement in any case facilitated the technical development of broadcasting networks by for instance creating frequency tables and preventing programs that would damage international understanding.[165] Both aspects of the agreement corresponded with focal points discussed in the IBU. The initiatives in South American broadcasting seemed to combine the tradition of American broadcasting as well as that of the IBU.

Unlike U.S. and South American developments, the initiatives in Asia and Oceania appeared to still be in their infancy. Broadcasting organizations in these areas seemed to want to develop their medium in the direction of a regional organization.[166] Both the recent movement in the east as well as the establishment of USARD west of Europe took place before the IBU took any decisions about its possible global expansion. As a result, the IBU was forced to consider these new developments. The Union created a small sub-committee to examine the possible

162 James Schwoch, *The American Radio Industry and its Latin American Activities, 1900-1939* (Urbana: University of Illinois Press, 1990), 7, 81-83.

163 Their second conference in June 1937 revised the initial Buenos Aires agreement, and went into force January 1, 1938. Tomlinson describes the Union as an organization of a semi-permanent character that consisted of periodic conferences held at least every two years. Tomlinson, "The International Control of Radiocommunications," 212-213.

164 Only Argentina would become an IBU associate member in 1936, followed by Uruguay in 1939. IBU, *Twenty Years*, 73-74.

165 Huth, *La radiodiffusion*, 388.

166 "G. Gallarti to A. Pelt," 3, registry file 9G, R-4321, file 15874/663, LoN.

consequences of these latest developments on their idea of a *Union universelle*.

The IBU concluded that it was less problematic to realize a federal union. The idea of a universal union with only active members who were equally important was abandoned. The IBU decided to pursue the vision of itself as a long-term global federal union for broadcasting, seemingly based on a Eurocentric idea of globalism. The recent developments of a new regional organization might well facilitate IBU's transition into such a federal union, the sub-committee had argued. In fact, the pressing problem of the short waves had already triggered American, South-American and also groups from the Far-East to ask the IBU to take the "initiative to these studies." They even "declared themselves ready to collaborate efficaciously."[167] Apparently, these other groups were seeking more straightforward collaboration worldwide and wanted the IBU to take the lead. The Rapprochement Committee advised its Council it would be better if these extra-European members organized broadcasting by their own choice in groups analogous to that of the IBU and that these groups could meet in a federal cooperation.[168] With this decision the IBU accepted that there should be a worldwide federal organization of broadcasting, preferably based on the IBU's exemplary role.

In response to these requests to take the lead in global studies on short wave broadcasting, the IBU Council decided to organize a world conference on broadcasting in 1936. The conference would have to establish close and straightforward global collaboration between broadcasting organizations on the most pressing issues. In addition, the conference needed to make firm preparations for the important international CCIR conference in Bucharest in 1937 and for the 1938 International Radio Communications Conference in Cairo. The First Intercontinental Meeting of Broadcasting Organizations in Paris in 1936 would have to strengthen the existing bond between PTT Administrations, the IBU associate members, and other broadcasting groups across the globe. The IBU seemed to envisage that Paris would pave the way for a transition of its organization into a global federation for broadcasting in the long term. This vision was facilitated by the request of broadcasters in other regions to take the lead in global studies on the short waves.

---

167 "...l'initiative de telles études, en se déclarant prêts à y collaborer efficacement." "Recommandation de la commission technique au conseil," Série 4840, CA Document Series, IBU.

168 "qu'il serait désirable qu'elles s'organisent selon leur propre choix en des groupements analogues à celui de l'U.I.R. et que ce groupements à leur tour se réunissent en une coopération fédérative." "Procès-verbaux des Réunions de la Commission de Rapprochement tenue à Varsovie," Série 5273, 2 août 1935, 3-4, P.V. Commission des Programmes [Anciennement Commissions de Rapprochement et Relais] 1926-1939, IBU.

During the preparations for Paris, Van Dissel posed a serious warning to the IBU. He had stayed in touch over the years. By 1935 he had become a regular attendee at IBU Technical Committee meetings on short wave broadcasting, something he would continue to do after the conference in Paris in 1936.[169] Van Dissel warned Braillard of the potential problems the IBU could face if it really tried to integrate the standardization of the short waves into its activities. Since short wave transmissions reached out globally, Van Dissel insisted, the IBU should depart from its official framework that was currently nothing more than a European union. Moreover, he foresaw the need to seek a practical definition for the genre of short wave broadcasting "since it does not fall within the idea of national broadcasting anymore. Together with the value given to these special services, this question will raise problems of an international character, some even of a non-technical and political character."[170] Van Dissel's warning merely seems to have stimulated the IBU to find a line of discourse that would justify its taking the lead in the global rather than the European matter of short wave broadcasting.

IBU took the advice and circumvented the problem by discussing a vision of global collaboration based on an equal footing between regions. Simple collaboration between regions would not interfere with the IBU's status as a European union and envisioned a world that would not favor Europe over other regions. The world meeting would have an informal and unofficial character. It would bring together PTT administrations, IBU associate members, representatives of national and international unions, associations and broadcasting groups created outside Europe as well as extra-European organizations operating where there were still no such international groups. Together they would have to find ways of convincing governments of the necessity to reach international agreements on short waves. This was because, even though "broadcasting over the short waves, on short notice, has become a true public service," its technological expansion "continuously and more vigilantly holds the attention of governments," Braillard argued.[171] The

169 Though most of the exchanges of ideas must have happened in the hallways or over dinner, the archives indicate that these discussions continued even after the League Memorandum before the IBU Council in 1931. "Letter from Burrows to Van Dissel," 2 May 1932, registry file 9G, box R-2601, file 13698/2081, LoN.

170 "ne rentre plus dans le cadre d'une radiodiffusion nationale, et cette question, ainsi que la valeur à donner à ces services spéciaux, soulèveront, des problèmes d'un caractère international, certains même d'un caractère non technique et parfois politique." "Rapport sur la session du Conseil de l'Union Internationale de Radiodiffusion tenue à Genève du 20 au 26 février 1935," 11 mars 1935, 4, registry file 9G, box R-4321, file 15874/663, LoN.

171 "...la radiodiffusion sur ondes courtes est devenue en très peu de temps un véritable service publique qui retient l'attention sans cesse plus vigilante des gouvernements." Raymond Braillard, "L'avenir de la radiodiffusion sur ondes courtes et la conférence du Caire," *Radiodiffusion*, no. 5 (November 1937): 13.

Table 3.1 – List of inter-continental circuits spreading out from Europe, 1935-1936

1. France	Paris	Algiers – Rabat – Cairo – Saigon Maracay – Rio de Janeiro – Buenos Aires
2. Germany	Berlin	Bandung – Bangkok – Mantilla – Tokyo Cairo – New York – Maracay – Buenos Aires – Rio de Janeiro – certain ships at sea
3. Great Britain	London*	Canada – New York – Rio de Janeiro Buenos Aires – the Cape – Australia Tokyo – British India – Egypt Palestine – certain ships at sea
4. Italy	Rome	New York – Tokyo – Buenos Aires Rio de Janeiro – Cairo
5. League of Nations	Prangins	New York – Buenos Aires – Rio de Janeiro – Shanghai – Tokyo – Sydney Melbourne
6. Netherlands	Hilversum	Dutch East and West Indies – Japan British India – South Africa – Egypt – Palestine – United States Central and South America – Australia New Zealand – other regions by special arrangement
7. Spain	Madrid	Buenos Aires

* "These circuits can be used for broadcasting during the hours announced for public service; the indication of these hours can be obtained from the General Post Office. Broadcast transmission may, however, sometimes be arranged outside of these hours; applications for the use of these circuits for broadcasting must be made to the local telephone authorities who will put themselves into communication on the subject with the British General Post Office."
Source: Brochure des Relais: corrigendum à la série 5652, Chapter VII, Série 5779, 22 avril 1936, box 85, file Relais, Brochures des Relais 1927-1936, IBU.

relatively small short waveband between 6,000 and 28,0000 kc/sec, assigned to broadcasting by Madrid, only guaranteed space for about 91 available channels, or even as few as 85.[172] The real number of channels in operation exceeded this number by far. In its plea, the IBU merely stressed its advisory role.

The IBU could plan the conference program as it wished. Consequently, the IBU Bureau, together with the heads of its committees, put all topics of a technical, juridical, and programming nature on the agenda. Each issue served to clarify the mechanism for establishing effective worldwide cooperation with a permanent character.[173] In other words, the IBU set the agenda to visualize global broadcasting in all relevant domains. The issue of short waves was one of the most pressing

172 Tomlinson, "The International Control of Radiocommunications," 219.

173 "Procès-verbaux des Réunions de la Commission de Rapprochement," Série 5273, 4, P.V. Commission des Programmes, IBU.

technical questions. As the IBU could also set the agenda for the technical discussions, it could actively pull the use of short waves into the domain of public broadcasting and towards the ideological goal of driving the improvement of international relations, of mutual understanding, and neighborliness between people. The Paris meeting in 1936 was a unique event in broadcasting history. It was the first time broadcasting organizations joined in a world-wide meeting dealing with all aspects of broadcasting. Because of the IBU's dominant role, the meeting seemed to have been inspired by European expertise and visions of society.

In preparation for Paris, the IBU and ITU experts on the IBU technical committee studied the issue of short waves from the perspective of the European region. They collaborated with organizations like the Bureau of Standards in Washington D.C. In order to perform a field survey, the IBU Brussels Checking Center opened up a station to measure short wave transmissions in 1935. The field survey provided a wealth of new information on the status quo of short waves in Europe. The short waves expanded extremely rapidly (Table 3.1). They mostly reached out to colonies like the Paris connection with Algiers, the London connection with British India, and the Netherlands Hilversum connection with the Dutch East and West Indies. Other connections had been constructed only recently either to facilitate bi-lateral or multilateral inter-continental program exchange or in response to the call from the IBU and the League of Nations in 1930 to develop short waves. The connections with the United States, Japan, and China had been constructed for this purpose. The League of Nations' short wave station at Prangins could also be employed as an intercontinental connection in the inter-continental relay network.[174]

Many broadcasters in Europe employed short waves to communicate with specific audiences. At times when they exchanged programs with other organizations or continents, their short wave stations functioned instead as hubs in regional relay networks. These short waves then connected regional relay networks across the globe, creating a global relay network. At the receiving end, broadcasting organizations

174 The idea to make the LoN short wave station at Prangins available to the IBU coordinated international radio broadcasts seems to have been discussed only recently in light of the broadcast of "Youth Sings over the Frontiers" (Chapter 6). The League wondered if such a broadcast "could be considered important enough and of such character that it could be relayed by Radio Nations." "ne pourrait pas être considérés assez importants et d'un caractère tel qu'ils puissent être relayés par 'Radio Nations'." "Rapport sur la session du Conseil de L'Union Internationale de Radiodiffusion," 9, registry file 9G, box R-4321, file 15874/663, LoN. One month later the League again considered the employment of Radio Nations for relaying the IBU international programs. The Committee considered the idea in light of Radio Nation's role in times of crisis or peace, and ultimately decided that a *Comité des Telecommunications* should examine the issue. Van Dissel in any case seems to have been a proponent of the idea. "Metternich to Pilloti," 12 April 1935, 3-5, registry file 9G, box R-4321, file 15874/663, LoN.

would broadcast the shared programs over their home channels. Such a construction allowed the IBU to negotiate its European standards on a global scale. Since the relayed programs would ultimately be broadcast over national channels, the short wave relays served as a means to an end in a similar fashion as the European standards for wired and wireless broadcasting created by the IBU and ITU.

Creating a global broadcasting network turned out to be far more complicated than an interconnected regional network consisting of wireless broadcasting over the medium and long waves. The technicalities of short waves, in particular the lack of knowledge about them, prevented a simple employment of these stations for the IBU relaying activities. Short waves gave no guarantee that inter-continental hubs could be used for broadcasting at every hour of the day. Some circuits could adjust wavelengths according to the time day, but for others there were certain moments in the daytime when the regularity of services over long distance radiotelephonic circuits could not be assured. In these cases, parties would have to contact the responsible radio-telephonic authorities before they decided in favor of relays that included hiring an inter-continental radiotelephonic circuit.[175] The Paris meeting would have to provide additional knowledge about the technicalities of short waves in other than European areas of the globe in order to deal with these problems and uncertainties.

The Technical Section of the 1936 Paris meeting established a special sub-committee to deal with these extra-European issues of short wave broadcasting.[176] The committee considered the current situation of these extra-European regional networks, within what limits and bands they wanted an extension of their wavebands, and what local or technical reasons such as geography, demography, climate, or parasites they thought would justify such extension.[177] The sub-committee consisted of representatives from the French PTT, the Ente Italiano per le Audizioni Radiofoniche, the Portuguese PTT, the Russian PTT, the German Reichspost, RCA, the Argentine Association of Broadcasters (the initiators of USARD), the Papal State, the Dutch Philips Gloeilampenfabriek [light bulb company], as well as the League's wireless expert Van Dissel. The committee had a fair representation of engineers from public broadcasting organizations, PTT administrations, and private radio industry enterprises, but a less balanced geographical representation. Europe dominated the committee, supported by representatives from Southern

175 "Brochure des Relais: Corrigendum à la Série 5652," Série 5779, 22 avril 1936, Chapter VII, box 85, file Relais, Bruchures des Relais 1927-1936, IBU. The IBU officially developed the Brochure in 1935 in preparation for the Paris World Conference to be held early 1936.

176 It also focused on the status quo of broadcasting in general, on interference, on the drafting of technical grounds for creating an intercontinental agreement on the distribution of available short wave frequencies. IBU, *Documents of the Paris Meeting*, 183.

177 Ibid., 229.

and Northern America, as well as global wireless broadcasters like Vatican Radio and the League's Radio Nations. The east did not participate at all. The committee that was supposed to focus on broadcasting questions outside Europe had a transnational character, but an unbalanced Europe-centered global representation.

No international meeting had been held about the technicalities of short waves on such a global scale before. The committee obtained a clear status quo of short wave activities by compiling a list of all the short wave stations in operation across the world, their frequencies, and the regularity with which they broadcast (all day or less regularly). The few short wavebands assigned to broadcasting by the 1932 Madrid Convention failed by far to meet the demand in 1936. The frequent use of short waves for global broadcasting from Europe already caused problems, but the true problem was in the tropical countries close to the equator. The geographical location and mountainous terrain meant that medium and long waves could not secure a nation-wide broadcasting service like they could in Europe or Russia. The Columbian PTT stressed the dire need for using short waves for national broadcasting, let alone for realizing regional broadcasting.[178] Global interaction on short waves should secure a global connection of regional broadcasting networks as well as the idea that national broadcasting services existed at the receiving end.

The Paris meeting concluded in favor of creating an "intercontinental wave plan" based on the cordial relations established there. It was hoped that such a plan would serve "effectively the cause of better international understanding and the cause of peace" as the French host Pellenc expressed it.[179] Non-European participants unanimously agreed that the IBU should take the lead in the realization of such a plan, to which end the organization would have to arrange future follow-up meetings.[180] The Paris meeting seems to have paved the way for the IBU's transition into a global federal organization for broadcasting, and the IBU was appointed to centrally coordinate the process. European values and traditions would form the basis of the federation, creating a Eurocentric global organization of broadcasting. In practice this meant that the IBU, ITU, and European PTT administrations, via their statutory collaboration, were closely involved in the process.[181]

178 Ibid., 209-210.

179 Ibid., 183, 220, 235.

180 Ibid., 240-241.

181 Where the overseas countries asked the IBU to take the lead in coordinating technical aspects, at the same time the section dealing with programs accepted a resolution with similar content. The resolution appointed the IBU Office "as the centralising office for documentation relating to intercontinental programme questions and to act as a liaison office between interested organisations for the time being." Ibid., 223, 237.

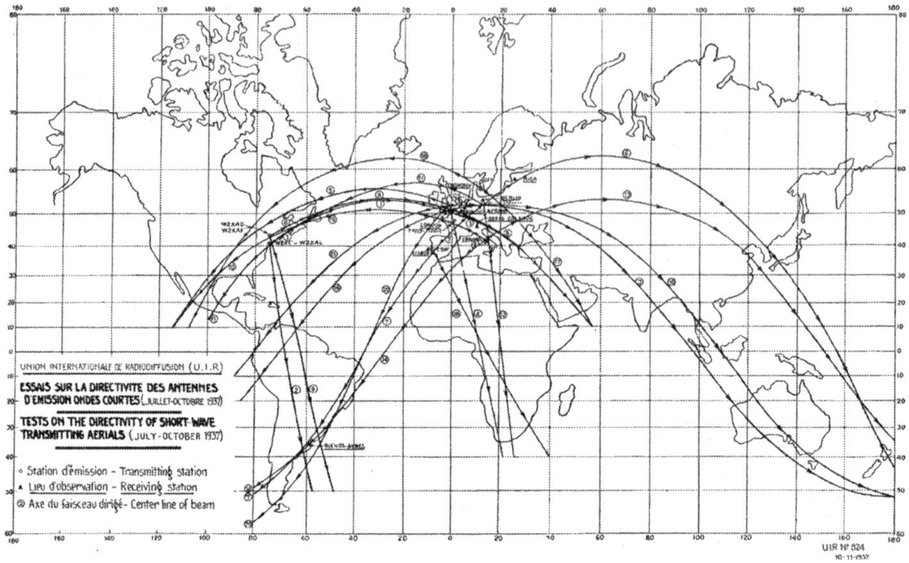

Figure 3.7 – Testing the directivity of short-wave transmitting aerials, July-October 1937
Source: BUIT, Documents de Caire Tome I, 408. The test has been published by the International Telecommunication Union, but has been originally made by the International Broadcasting Union. Used by the courtesy of the European Broadcasting Union.

The organization of global broadcasting became a loose collaboration officially. In the years leading up to the 1938 International Radio Communications Conference in Cairo, broadcasters conducted further regional tests with short waves as arranged in Paris (Figure 3.7). As agreed, the IBU collected all test results. Though the tests immediately increased the efficiency of the global wireless short wave network, the improvements could not compete with the incessant growth of short waves during these years. In January 1937 only half of the more than 200 short wave stations from 48 countries could operate within the short wavebands assigned for broadcasting by Madrid. No fewer than 94 stations operated outside these bands, 17 in Europe, 8 in Asia, 68 in Central and South America, against 1 in Africa between the 50-10 meter band.[182] Less than five months later the number of short wave stations had risen to 217, of which some 99 functioned outside the Madrid wavebands.[183] As these were minimum figures, the actual situation was far more pressing.

182 "Proposition pour la Conférence Internationale des Télécommunications du Caire (1938)," Doc. 366, Bruxelles, 25 janv. 1937, 6-8, registry file 9G, box R-4321, file 27520/663, LoN.
183 "Sous comité des ondes courtes," Doc. 418 corrigée, Bruxelles, 22 juin 1937, 3, registry file 9G, R-4321, file 29167/663, jacket 1, LoN.

When the IBU discussed the new status quo in 1937, it established a special subcommittee to deal with short waves. As a result, the IBU decided that the time was right to create standards for short wave broadcasting like the separation of waves of 10 kc/sec, exploration of simultaneous broadcasting over short waves, and the possible employment of directional antennas. The IBU decided that standardizing the power of stations would be too premature given the state-of-the art technique.[184] On the basis of these globally conducted tests, the IBU designed a draft plan as starting point for discussion at the Cairo Conference.

The plan urged Cairo to create larger wavebands for broadcasting over long, medium, and short waves, as well as in the fixed services (since the broadcasting relay activities went over the telephone lines). The different regions would continue to allocate medium and long waves according to their own regional standards for wireless broadcasting. Regarding the short wave bands, Cairo should distinguish between "intercontinental services" and "regional services in the tropics." Both categories agreed with the European IBU standards of cross-border connections via relay links, complemented with nation-wide wireless broadcasting services. Due to a lack of short wave regulations, the IBU, on behalf of all broadcasting stations in the world, recommended organizing a world conference on short wave broadcasting. Such a conference would need to create a rational global plan.[185] The broadcasters imagined a loosely coupled global broadcasting network of regions linked via the short waves whereby program exchange would enable international programs to be broadcast over the national home channels of countries worldwide. The plan suggests that the allocation of short waves happened along the same lines with similar visions of society as with the European standards. Nonetheless, the global plan integrated the concept of neighborliness between regions as well as between nation-states.

Even though Cairo was a significant improvement compared to Madrid, the 1938 Convention failed to meet all the global broadcasting society's recommendations. The short waves developed far more quickly than official regulations could manage.[186] Cairo nevertheless recognized the problems of short wave broadcast-

184 Like broadcasting over medium and long waves, the short waves could be used for "simultaneous broadcasting" as well, the activity in which several broadcasting stations within one geographical area transmitted the same program via the same frequency. Simultaneous broadcasting improved usage of medium and long waves, and did the same for short waves. "Rapport sur les travaux des commissions de l'Union Internationale de Radiodiffusion tenues à Ouchy du 17 au 25 juin 1937," 5-8, registry file 9G, R-4321, file 29167/663, jacket 2, LoN.

185 "un plan mondial rationnel." "Proposition," Doc. 366, Bruxelles, 6-8, 10, 12, registry file 9G, R-4321, file 27520/663, LoN.

186 "Rapport au Conseil sur réunions de la Commission Technique à Ouchy du 20 au 27 juin 1938, Rapporteur: Raymond Braillard," Doc. 501 (remplaçant le no 494 corrigée), 7 juillet 1938, 2, box 80ter, file

ing, and agreed to organize a world conference to create a global standard for allocating short waves.[187] Cairo entrusted this to the IBU who had to arrange this alongside its regular European conference for broadcasting in Montreux in 1939. According to Van Dissel, preparations for this short wave conference would have to start in Montreux.[188] This assignment further indicates that the IBU was becoming the "leader" of the regional broadcasting organizations. Cairo herewith unofficially acknowledged the IBU as a kind of global federal expert on broadcasting without interfering in regional independence. Officially broadcasters would not manage to create a global federal *Union universelle*. In reality however, such a structure seems to have been slowly emerging.

The 1930s would not witness the creation of a global wireless frequency plan for short waves. The preparatory work done by the technical experts of the IBU, ITU, and PTT administrations did result in a new frequency plan for the European region in Montreux in 1939. Just before the Montreux plan was to go into effect, in March 1940 the IBU Technical Committee recommended that European governments should not implement the Montreux Plan "until the moment the existing conditions in Europe will allow it."[189] Furthermore, with the outbreak of World War II, the preparations for organizing a world conference on allocating short waves came to a standstill.

The results of the global studies on short waves had increasingly been finding their way into short wave broadcasting routines across the globe in the 1930s. Though these results would never be recognized as legally binding standards in the interwar years, many organizations integrated them into their activities, following the recommendations for shared waves, directional antennas, and separation in kc/sec. With respect to frequency allocations, the various stations did not necessarily follow the European model, and mostly continued along the lines they had developed over the years.[190] By the time war broke out, the IBU's vision of becoming a loosely coupled Europe-centered global federation based on a European broadcasting tradition was far from reality. However, the European

---

Rapports Généraux de la Commission Technique au Conseil, de 1932 à 1946, IBU.

187 "The wish expressed by the Cairo Telecommunication Conference of 1938 concerning the Meeting of a World Conference for Shortwave Broadcasting proposed by the Committee for the European Broadcasting Conference," Série 7610a, 9 July 1938, CA Document Series, IBU.

188 "Letter by Van Dissel to M. Metternich," 17 Feb. 1939, 1, registry file 9G, box R-4322, file 36772/663, jacket 1, LoN.

189 "Rapport au conseil sur l'activité de la commission technique et du centre de contrôle pendant l'exercice 1939," Doc. 633, 18 mars 1940, 2, box 80ter, file Rapports Généraux de la Commission Technique au Conseil, de 1932 à 1946, IBU.

190 Tomlinson, "The International Control of Radiocommunications," 215-218; IBU, *Twenty Years*, 73-74.

expert community did achieve an unofficial leading position that over time might have developed into such a federation if the war had not intervened.

Conclusion

Over the years, the transnational expert community standardized wireless and wired broadcasting. The aim was to resolve the problem of electromagnetic interference across borders as well as create an international long distance music relay network. This was not a straightforward process. It was one of negotiation, of redefinition, of catching up, of falling behind, and of visions that failed and visions that were challenged. Experts continually had to keep pace with the technology.

In response, they created a community consisting of IBU broadcasting experts, wireless experts of the League of Nations, and international radiotelephony engineers in the ITU and CCIF, as well as radiotelephony engineers from the various national PTT administrations. Together they tried to match their standardization efforts to the fast developing broadcasting technology. First of all, they developed specific transnational skills for this purpose. These skills were geared towards reconciling national interests with the political and technical aspects of their activities. Their expertise can be seen as a kind of transnational techno-political diplomacy, useful for anticipating high-speed and complicated transnational developments.[191]

Secondly, the expert community kept the structure of their international organizations relatively open, informal, and flexible. At times their easy cross-institutional collaboration could not effectively resolve problems due to inconsistent developments in the organization versus broadcasting technology. Persistent problems with the allocation of frequencies made the IBU, the ITU, and national PTT administrations decide to form an official statutory partnership. Although this solved some of the problems, it also limited future freedom for organizational action.

Initially, the expert community decided to focus on Europe to resolve their technical issues. This was an obvious choice since the problem of electromagnetic disturbances caused major problems, particularly in Europe with its large number of nation states. Although the IBU experts defined Europe geographically for the first time in 1926, this was not a topic which aroused emotion and interest. Europe's spatial definition only became of interest to the transnational

191 For a discussion on the idea of techno-politics see: Andreas Fickers, "The Techno-politics of Colour: Britain and the European Struggle for a Colour Television Standard," *Journal of British Cinema and Television* 7, no. 1 (2010): 95-114.

broadcasting community when what were defined as short waves opened up the option of worldwide broadcasting. Agreed European standards gained more explicit recognition once they were fine-tuned with other standardization efforts across the globe. The broadcasting community thus determined the spatial definition of Europe in the encounter with the global and non-European. By the time World War II broke out, the community's push for a global broadcasting system resulted in an unplanned regionalization of broadcasting.

In the course of these processes, the broadcasting experts pursued a wide variety of visions about who was included in this system and who was excluded. Initially the community projected a transnational system with a place for local participants such as radio amateurs. The wave of nationalizing broadcasting organizations, however, eventually resulted in an infrastructure that interlinked national stations, mostly at the expense of amateurs. The relationships between these national broadcasting stations were not only envisioned via proposals like the international convention for broadcasting over long distances, which was proposed by the League OCT. They were also envisioned via transnational organizational formats like *Radiodiffusion mondiale* or *Union universelle.* These had a global focus, but were mostly designed by the IBU, which had earned its reputation in the 1920s as a European expert. Such global visions could only be realized in the long term. In the short term, the search for global solutions outlined and stimulated the search for practical solutions to inter-organizational collaboration.

The expert community was not striving for a specific European solution. Of course, the experts brought their Eurocentric visions and traditions to the negotiating table. They also believed that building networks would contribute to international peace and rapprochement between nations and states. These developments revealed the impact of visions defining the linking, or non-linking of networks and broadcasting organizations. In the interwar years, the IBU visions were translated into standards which then became dominant but did not remain unchallenged. Other actors with transnational interests were also building broadcasting networks: Radio Moscow, Radio Nations, Vatican Radio and Radio Luxembourg. These were grounded in other visions.

Chapter 4
Battles over Europe's Borders

"Dui arcano Dei consilio succedimus loco Principis Apostolorum..."

It was the first time in history that a pope's voice was heard by the world at large. Beyond the borders of the Vatican, in every country in Europe, in all of the five continents, a multitudinous audience, the greatest that had ever listened to a single man, listened in devout silence to words which only very few could understand. In many places through the far-flung Christian world men and women knelt in streets and public places, listening with feelings of bliss and awe. A maze of radio circuits carried the words around the earth.

César Searchinger, European Director of CBS[1]

Thus César Searchinger describes the mind-blowing global relay event on February 12, 1932 that fired everyone's imagination: a radio speech given by his Holiness the Pope. It was a speech by "the Old Gentleman," the code word for this unimaginable event. Up till then Searchinger had been working as the first foreign and overseas radio representative of the U.S. Columbia Broadcasting System (CBS) in London. He had mostly worked on two-way relays between the United Kingdom and the United States, but occasionally arranged the relaying of an event from the European continent over the nation-wide CBS broadcasting network. A few weeks earlier, CBS had been taken by surprise when its largest competitor, National Broadcasting Company (NBC), had managed to relay the one radio speech Italy's Benito Mussolini would ever give in English during his entire life. To Searchinger this "was an awful blow."

But only a few days later he was given the lead of his life. Marchese Guglielmo Marconi had been working on a wireless short wave station in Vatican City. The whole project had been shrouded in secrecy, not least because of its technical capabilities which people predicted to be no less than scientifically almost perfect. "No one had ever dared to suggest that the Holy Father himself would engage

1 César Searchinger, *Hello America! Radio Adventures in Europe* (Cambridge, Mass.: Houghton Mifflin, 1938), 78.

in anything so mundane as broadcasting...But fools rush in where clerics fear to tread. And we Americans have the imagination of fools," as Searchinger vividly described the event.[2] Since he was the only CBS representative in Europe at the time, his arena suddenly expanded to the whole of Europe. Running ahead of developments he placed "European Director" on his letterhead, hoping for the best. He rushed to Vatican City, and few weeks later he had the scoop of a lifetime, the first worldwide speech by the Pope. It was made possible via Vatican Radio's short wave station and a maze of relay circuits, among them that of CBS, all of which had been connected for the occasion. It was this global broadcast that quite by chance transformed Searchinger's job into that of European director.

Searchinger's anecdote reveals that there were different plans for one particular broadcasting space. Although the Vatican broadcast made Searchinger the first European director for CBS, Vatican Radio did not consider itself European at all. It targeted a global rather than a European Catholic community. Apparently not every station on the European continent thought of itself as European, whereas the overseas CBS considered every station on the European continent to be European.

Vatican Radio and stations like Radio Moscow, Radio Nations, and Radio Luxembourg created their international broadcasting networks according to the kind of society they desired to create, or in other words, the audience they wanted to reach. With different broadcasting systems being constructed in the European Region, the broadcasters would have to deal with one another to secure the efficiency and the operability of their stations. They had to fine-tune their efforts. At times these negotiations could be harsh and led to true battles over Europe's borders. Standards that would make viable Vatican Radio's global Catholic community, for instance, would not necessarily have to coincide with International Broadcasting Union (IBU) and International Telegraph Union (ITU) European standards for frequency allocation that little by little found their way into "the official conferences of Prague, the Hague, Copenhagen, of the CCI, etc..."[3] Several proposals ignored Europe as a category altogether, whereas others wanted to construct an entirely different kind of European system. These different broadcasting projects each in their own way challenged the evolving European standards as discussed in Chapter 3.

This chapter shows how the USSR designed a global broadcasting network for spreading the socialist message. Though the socialist view rejected the idea

2 Searchinger, *Hello America!*, 65-68.

3 "Peu à peu une technique général s'est créée, et elle a pu se faire reconnaître dans les conférences officielles de Prague, la Haye, Copenhague, du C.C.I., etc..." "Radiodiffusion mondiale: Rapporteur: R. Braillard, Président de la Commission Technique," Série 3180, 23 mai 1932, 2, CA Document Series, IBU.

of nation-states in the long run and the category of Europe altogether, by the late 1920s the USSR did seek practical collaboration with the IBU. The chapter continues with an examination of Vatican Radio and Radio Nations. Whereas the first was aimed at all Catholics worldwide, the League of Nations had established Radio Nations primarily to facilitate worldwide communications in times of emergency. The two stations aimed for a global audience, but for various reasons focused on Europe as a distinctive category. The chapter ends with an examination of Radio Luxembourg, generally described as the main nationally-owned broadcasting station that deliberately aimed at a Europe-wide audience for commercial motives. Radio Luxembourg initially saw itself as part of Europe, as a niche in the IBU and League of Nations joint initiatives. Nevertheless, it was brutally rejected by this community partly on the grounds that it had not been a problem for this community to collaborate with Radio Moscow, Vatican Radio, and Radio Nations. These cases give us an insight into the negotiating of a Europe on the threshold of differing broadcasting infrastructures.

Radio Moscow and Confining Europe in the East

Radio Moscow's activities were confined to the east of Europe. The USSR already had its own established broadcasting network for some years when broadcasting organizations in Europe began collaborations based on neighborliness between nations. The USSR did not show much interest in IBU developments, preferring a broadcasting system that suited its needs. In his seminal work *The Age of Extremes*, Eric Hobsbawm writes "…the history of the Short Twentieth Century cannot be understood without the Russian revolution and its direct and indirect effects."[4] The same holds true for understanding the way early building of broadcasting systems played out between the USSR and the IBU.

As early as 1918, Russian engineer-inventor M.A. Bonch-Bruyevich began developing long-range radio transmission in a small laboratory at the Tverskaya Central Receiving Radio Station. Though he and his team initially had a certain degree of freedom, Soviet leader Vladimir Il'yich Lenin soon interfered. Directly after the end of the war, the USSR experienced a revolution in which a socialist regime came to power under the leadership of Lenin. The USSR aimed at a global revolution of workers. Lenin, who had an interest in radio engineering, immediately wanted to create a global broadcasting network to ensure worldwide

4 Eric Hobsbawm, *The Age of Extremes: A History of the World, 1914-1991* (London: Abacus, 1994), 84.

diffusion of the Socialist Revolution's ideals. To Lenin, radio was a "newspaper without paper and without distances."

Such a socialist world had a highly specific outlook. Already in 1915 Lenin had stated that a "united states of Europe" was either impossible or reactionary if it should come about under capitalism.[5] Lenin argued that only a successful and worldwide Socialist Revolution would lead to some kind of "united states of the world." Then the whole idea of a state would wither away.[6] Lenin rejected Europe both as a geographical entity and as an ideal. He focused on a large movement of socialist workers with no relation to any kind of "Europe." Nation-states would only play a role initially. As soon as Lenin decided to interfere with Bonch-Bruyevich's efforts, broadcasting in the USSR developed along socialist lines. Russian broadcasting focused on the world rather than on Europe or nation-states.

Lenin tried to speed up the development of broadcasting. He set up a large state scientific research center for radio engineering in Nizhniy-Novgorod with Bonch-Bruyevich in charge. In March 1920 the Council of People's Commissars, the official name for the Russian Soviet Council, charged the Commissariat of Posts and Telegraphs (NKPiT) to construct a Central Moscow Radiotelephone station as well as stations in other important points in the Republic.[7] In 1922 Russia inaugurated its first radio station with long wave transmitters. The station was formally owned by COMINTERN, a private global organization of national communist parties, and bore the name *Imeni Kominterna*. The station would diffuse socialist ideals and facilitate the social revolution.

The western world feared the idea of a large socialist broadcasting network operated by COMINTERN, associating it directly with the Russian Communist Council. Such a scenario could allow the unbound diffusion of the Russian socialist message firmly into the western world. The Russian government rejected the idea of state-interference by arguing that COMINTERN was a private organization in whose affairs it could not intervene.[8] Less than two years later the Council officially strengthened its influence over broadcasting. It voted in favor

5 Peter Bugge, "The Nation Supreme: The Idea of Europe 1914-1945," in *The History of the Idea of Europe*, ed. Kevin Wilson and Jan van der Dussen (London: Routledge, 1993), 112.

6 Bugge, "The Nation Supreme," 112. The information comes from: Vladimir I. Lenin, "On the Slogan for a United States of Europe," in *Lenin Collected Works. Originally in Sotsial Demokrat, No. 4, 1915*, ed. Marxist Internet Archive, vol. 21 (Moscow: Progress Publishers, 1974), 339-343.

7 The NKPiT would later become the Ministry of Communications. S.V. Kaftanov and Soviet Union. Gosudarstvennyi Komitet po radioveshchan i iu i televiden i iu, *Radio and Television in the USSR* (Washington: U.S. Joint Publications Research Service, 1961), 6-11.

8 Jennifer Spohrer, "Ruling the Airwaves: Radio Luxembourg and the Origins of European National Broadcasting, 1929-1950" (Ph.D. diss., Columbia University, 2008), 315-316; Vernon van Dycke, "The Responsibility of States for International Propaganda," *The American Journal of International Law* 34, no. 1 (1940): 59-60.

of opening "the way for a development of amateur radio and offered extensive possibilities for the utilization of radio by the population for economic, scientific and cultural requirements."[9] Furthermore, the Council aimed to facilitate "the development of the radio engineering industry and the dissemination of radio engineering knowledge."[10] The Council created Radio Transmissions, a joint stock company that served as an umbrella for all kinds of broadcasting activities.[11] Radio Transmissions had its headquarters in Moscow. The fear of the western world seemed well-founded.

In 1933 the Council formally standardized the institutional structure of Russian broadcasting when it changed a 1926 resolution that "assured a systematic party guidance of radio broadcasting" into a firm and state-owned organization of radio. With this decision the Council ended all discussions on the issue of private or government ownership and control. Little by little the Council was increasing its control over broadcasting and created a centrally organized and state governed socialist broadcasting network.[12]

Right from the start, Bonch-Bruyevich envisioned a Russian broadcasting network with long-range radio possibilities. Wireless broadcasting should assure both global and domestic coverage, whereas a relay network would have to connect the various Russian regions with the major centers for Soviet culture. In 1925 already, Bonch-Bruyevich was able to equip *Imeni Kominterna* with the world's first shortwave transmitter. This was a unique achievement since short wave broadcasting was still in an experimental phase in the West. The new transmitter could spread the socialist message for international unification of workers, and came to be known as Radio Moscow. In addition to its global outreach, the new transmitter enabled full domestic coverage, something that had been a big problem given the size of the country. By 1925 already, Radio Moscow's shortwave transmitter had become an important node in the development of a domestic relay network as well as a means to reach out to all supporters of socialism across the globe.

NKPiT and Radio Transmissions together developed the domestic relay network. Radio Transmissions helped cities like Khar'kov, Minsk, Voronezh, Rostov-on-the-Don, Krasnodar and others to construct long wave radio stations. Together

9 Kaftanov and Soviet Union. Gosudarstvennyi Komitet po radioveshchan i iu i televiden i iu, *Radio and Television in the USSR*, 12.

10 Ibid.

11 The initial name of the organization was Radio-for-All. For a discussion of its organizational structure see Arno Huth, *La radiodiffusion: Puissance mondiale* (Paris: Librairie Gallimard, 1937), 144-146.

12 "In view of the fact that radio is acquiring exceptional significance in the entire economic and political life of the country..." the Party centralized all radio activity within an All-Union Radio Committee directly under the Council of People's Commissars USSR. Kaftanov and Soviet Union. Gosudarstvennyi Komitet po radioveshchan i iu i televiden i iu, *Radio and Television in the USSR*, 14.

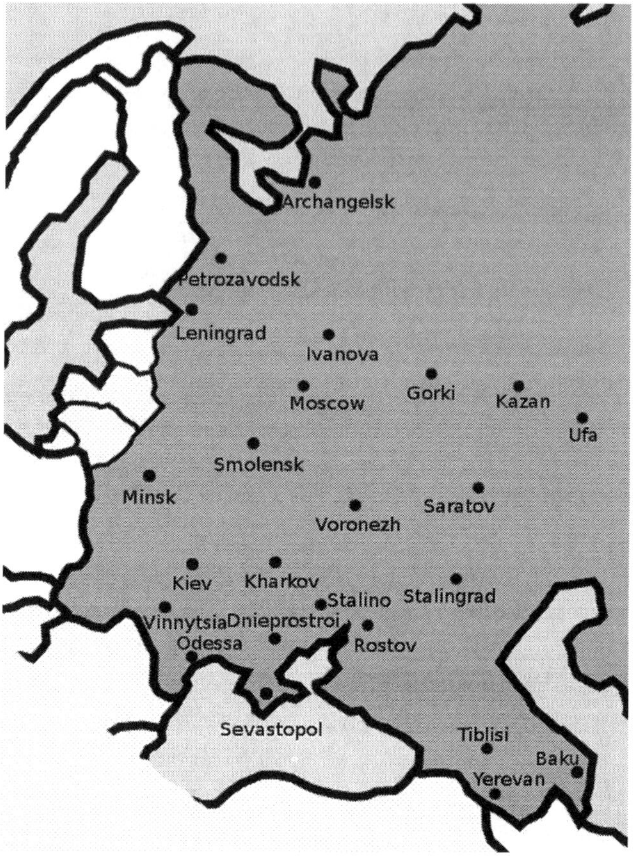

Figure 4.1 – Main Soviet radio broadcasting stations situated within Europe, 1936-1937
The names of a number of these Russian cities have changed over the years. Leningrad is present-day Saint Petersburg, Gorky was the name of Nizhny-Novgorod in between 1932-1990, Stalingrad nowadays is Volgograd, and Stalino is renamed Donetsk. Source: Arno Huth, La Radiodiffusion: La puissance mondiale (Paris: Librairie Gallimard, 1937), 143.

with NKPiT it then arranged these stations into a single relay network.[13] The relay connections served to link "a number of theaters, a conservatory and Smol'nvy for the rebroadcasting of plays, concerts and conferences."[14] Transmissions often

13 Both parties extended their collaboration with a contract enabling them to give shape to this national relay network. Radio Transmissions obliged itself to construct a number of new radio transmitters in the largest centers of the country. NKPiT would allow Radio Transmissions the usage of its transmitters for several hours a day whilst constructing a number of new stations as well.

14 Kaftanov and Soviet Union. Gosudarstvennyi Komitet po radioveshchan i iu i televiden i iu, *Radio and Television in the USSR*, 12.

focused on "political information, reports, transmissions for workers, for the village, educational programs, broadcasts for national republics and oblasts, regular concerts, radio plays and radio dramatizations, [as well as] rebroadcasts from theaters."[15] The relay network diffused these programs, statements of new socialist culture and way-of-life, to the various long wave stations which broadcasted these programs domestically, and to the short wave station which in turn relayed programs across the country and further abroad.[16]

Throughout the 1930s, the domestic relay network expanded with one new short wave station and a large number of regional long wave stations.[17] By 1936 the USSR had arranged its "local" relay activities effectively into three territories covering the central Russian provinces, the areas with national populations like Ukraine, Uzbekistan, Turkmenistan, and Transcaucasia, as well as the Ural and Siberia.[18] With this domestic relay network and wireless global outreach, the USSR had created a centrally organized global broadcasting network without regional, national, or continental boundaries. They foresaw no place for "Europe" in whatever form.

Geographically speaking, however, the USSR bordered many European countries. People living in the border regions experienced annoyance due to interference between their own national and Russian broadcasts. Only by fine-tuning their activities would the USSR and other European countries be able to deal with the problem.

While the USSR ignored IBU efforts, in 1926 the IBU community tried to reduce interference by creating a Geneva Plan integrating those parts of Western Russia that bordered Europe. Only then could interference truly disappear. The IBU had not invited Russia to its conferences. Russia was not and would never become an IBU member. The IBU simply decided to assign the Russian stations in the border regions five medium wave frequencies and informed the Russian PTT of their decision after the conclusion of the Plan.[19]

15 Ibid., 13.

16 On Soviet cultural expressions see: Orlando Figes, *Natasha's Dance: A Cultural History of Russia* (New York: The Penguin Press, 2002).

17 Huth, *La radiodiffusion*, 146-147.

18 Ibid., 145.

19 The Conference participants assigned to the USSR five frequencies with such casualness that the designers of the Geneva Plan might well have been in touch with the USSR before the Conference to discuss the border frequencies. Such informal contacts are characteristic of the international climate since the late nineteenth century, and they remain largely invisible in archives of international organizations like the International Broadcasting Union. Notwithstanding, it could also well be that Russia did not use these medium waves.

The USSR in the meantime prepared itself well for the first International Radiocommunications Conference in years. It handed the organizers of the Washington Conference a complete draft of service regulations that could serve as a points of discussion during the negotiations.[20] However the United States, which organized the conference on behalf of the ITU, declined to invite the USSR since it refused to acknowledge the Russian Soviet Council. The decision placed the ITU in an awkward position, questioning its alleged political neutrality.[21] The USSR had been an ITU member ever since 1866 and had taken part in the previous 1912 London Convention. Article 12 of that convention even stated that these "conferences are composed of delegates of the governments of contracting countries," an agreement the USSR lived up to.[22] Needless to say, the USSR felt disturbed by the Washington developments which, it considered, indicated the subordination of "the interests of international telecommunications to purely political considerations."[23] Apparently, ideological beliefs greatly influenced the negotiation of frequency standards.

After the Washington Conference, the USSR gradually took a more active interest in the practical fine-tuning of frequencies to secure the efficiency of its own network. The USSR and the IBU intensified their collaboration. In its 1928 Brussels Plan, the IBU still excluded Russia, defining a European zone without the USSR "with the exception of a band very closely situated to the west of this country."[24] One year later the Czech PTT, at the request of the IBU, invited the Russian PTT to the Prague Conference to fine-tune the medium and long waves. The USSR accepted the invitation. Both the USSR and the IBU appeared to share a strong discontent over the course of events in Washington that according to the Russian PTT "has not sufficiently taken into account the needs of radiotelephony."[25] This shared discontent facilitated their collaboration substantially.

These exploratory overtures in frequency allocations were a turning point that would affect the boundaries of IBU's Europe and Russia's relation to and position

20 There are documents that suggest that in preparation for Washington, Russia and the IBU for the first time joined forces. These nonetheless do not give any details. John D. Tomlinson, "The International Control of Radiocommunications" (Ph.D. diss., Université de Genève, 1938), 60; IBU, *Twenty Years of Activity of the International Broadcasting Union* (Geneva: IBU, 1945), 19.

21 Nina Wormbs, "Technology-Dependent Commons: The Example of Frequency Spectrum for Broadcasting in Europe in the 1920s," *International Journal of the Commons* (submitted): 7.

22 Tomlinson, "The International Control of Radiocommunications," 59-60; George A. Codding, *The International Telecommunication Union: An Experiment in International Cooperation* (Leiden: Brill, 1952), 31; Quote from BUIT, *Documents de la Conférence Radiotélégraphique Internationale de Washington (1927): Tome II: Travaux de la Conférence (Délibérations, Actes définitifs)* (Berne: BUIT, 1928), 731-732.

23 BUIT, *Documents de la Conférence Radioélectrique européenne de Prague 1929* (Berne: BUIT, 1929), 84.

24 "Rapport de la Commission Technique au Conseil," Série 1268, 20 fév. 1929, 3, box 80ter, file Rapports généraux de la commission technique au Conseil à fin 1931, IBU.

25 "la Conférence n'a pas assez tenu compte des besoins de la radiophonie." BUIT, *Documents de Prague*, 84.

within the international but Europe-oriented society of broadcasters. Since the IBU was increasingly recognized worldwide as the European broadcasting expert, the USSR had to deal with this "Europe" if it wanted to improve the efficiency of the broadcasts over its long waves in the western part of its broadcasting space. From the USSR's point of view, Europe was useful on the sidelines for reasons of practical fine-tuning.

Europe became a politico-economic strategy to the USSR in 1927. In addition to the technological motivations, changes in Russia's political and economic situation affected its decision to collaborate with "Europe." Stalin replaced Lenin as the new Soviet leader. The USSR now "pursued a policy of reconciliation with the west, insofar as possible, while preparing Russia for isolation and defense." Stalin's commissar of Foreign Affairs Maxim Litvinov tried to ease this policy. He argued that safety for his country could only be achieved "through some form of integration into the European System. First and foremost he had to create a breathing space for the USSR…"[26] For economic reasons, Moscow could not permit itself to become isolated internationally. The USSR needed the Europeans to boost its economy in case it wished to effectuate Stalin's second Five Year Plan.[27] Consequently Russia pursued a general "effort to improve its relations with the outside world at a time of domestic upheaval."[28] The new politico-economic strategy opened the doors to a more intensive collaboration.

One point of discussion in Prague was the Brussels Plan, which according to the IBU, had to be adapted to integrate Russian radio stations. Czech chief engineer Steinbach, who had been working on IBU technical activities since the founding days of the Union, mentioned that the "delimitation of the territory on the basis of which the plan must be applied and the solution of radio broadcasting wave distribution on nearby territories" deserved special attention, in particular long waves in the USSR.[29] All parties could have agreed to design a plan in which the USSR formed the border of the IBU European zone, similar to the Brussels Plan. In Prague the participants however decided to relocate the Eastern border of the European zone to include part of the USSR. With this definition they returned to the ideas of the 1926 Geneva Plan. They adapted their definition of Europe

26 Zara S. Steiner, *The Lights that Failed: European International History, 1919-1933* (Oxford; New York: Oxford University Press, 2005), 546.

27 Russian economic strategies and economic targets were communicated via the so-called Five Year Plans. A state planning commission saw to the conclusion and development of these plans.

28 Steiner, *The Lights that Failed*, 526.

29 "délimitation du territoire sur lequel le plan doit être appliqué et la solution de la distribution des ondes de radiodiffusion sur les territoires voisins." BUIT, *Documents de Prague*, 24.

eastwards to the 40° meridian East of Greenwich, a little further to the east than the initial Geneva Plan definition. The new meridian ran right across the west of the USSR, directly through Moscow. Suddenly the western part of the USSR voluntarily and in agreement became part of a newly IBU defined European infrastructure. The participants suitably termed the meridian, the Moscow meridian.[30]

Though the Prague meeting raised the question of defining the territory for the Plan, the documents make no reference to discussion about any such new definition. Most likely the engineers did not consider the discussion important enough, or they addressed the issue over coffee, lunch, or dinner. Only during the last meeting did the NKPiT stress that it "reserved for itself total freedom in the choice for waves east of the Moscow Meridian."[31] A new "Europe" that integrated part of the USSR had clearly been defined.[32] The lack of discussion on the definition in the Prague documents either suggests the minor importance attached to the definition or the obvious need for it. In any case the new definition seamlessly matched the USSR's new politico-economic strategy and it managed to achieve some kind of integration into the European system.

As a result of the growth in European broadcasting stations and the participation of the USSR in the Prague meeting, the Prague Plan integrated an agreement made in Washington on the practice of derogation. The idea was that stations could have a waveband between the medium and long wave bands assigned to broadcasting which were used for maritime and air traffic. Derogation was allowed as long as stations did not disturb the existing traffic. The Prague Plan placed about ten European and six Russian stations between the wavelengths assigned to two stations, separated by 4.5 kc/sec. This practice worked well since most of the stations were located at a long distance from other stations.[33] Even though the USSR had not signed the Washington Convention, it did adhere to the convention when collaborating with the European broadcasting community on material infrastructure related issues.

30 Ibid., 142.

31 "se réserve toute liberté dans le choix des onders à l'est du Méridien de Moscou." Ibid.

32 The only reference that the archives provide is that the decision to move the borders of Europe eastwards to the Moscow Meridian happened on juridical grounds. The document literally states: "Cette extension a été basée, du point de vue juridique, soit sur l'article 5, paragraphe 1er, du Règlement, soit sur l'arrangement intervenu à Prague en 1929 entre les Administrations européennes. Il est à noter que l'U.R.S.S., qui n'a pas adhéré à la Convention de Washington, a participé à la Conférence de Prague." "Commission Mixte (Conférence de Madrid) Rapport provisoire de la Commission Mixte concernant les amendement a apporter a la Convention de Washington (1927) et au reglement general annexe," Série 2291, 5 janv. 1931, 9, registry file 9G, box R-2601, file 13698/2081, LoN.

33 Wormbs, "Technology-Dependent Commons," 8; Nina Wormbs, "Invisible with Global Reach: Radio Spectrum for Broadcasting in Europe" (presented at the the Fourth Tensions of Europe Conference, Sofia, Bulgaria, 2010), 6-7.

With the Prague Plan in place, the USSR continued to join in discussions about the projection of European broadcasting with the IBU Legal, Technical and Program Committees as well as European PTT administrations. Preparing for the 1932 Madrid conference, they speculated about the "Europeanness" or form of a European region for telecommunications in general and for broadcasting in particular. The Mixed Committee established by the IBU for the occasion would have to devise a plan that was advantageous to broadcasting. In June 1930, to improve collaboration, the IBU had begun sharing technical information with the Russian Council. The IBU exchanged lists of long wave partition, its monthly bulletin with all sorts of broadcasting related information, radiotelephony laws, statistics, and technological information.[34] The USSR accepted the request. The Washington agreement had to be revised for it contained errors, head of NKPiT Hirschfeld pointed out.[35] The NKPiT helped to draft the Mixed Committee Report that served as basis for the discussions on broadcasting frequencies in Madrid.

The final Mixed Committee report made two requests concerning the defining of a European region. The Committee argued it was necessary to examine whether the European border in the east defined by the Moscow Meridian in the Prague Plan should not move a little further to the east, all the way to the Ural Mountains. Furthermore, the Report proposed a differentiation between affairs that could be resolved by regional arrangements for Europe only, and those that specifically required arrangements of a more universal character.[36] The USSR collaborated on a proposal that made a clear distinction between the world and a Europe that integrated an even larger part of its territory into what Madrid would call "the European region." Over the years, the USSR actively pursued its politico-economic strategy. By participating in Prague and in the Mixed Committee, it realized a form of integration into "the" European system within the material and organizational domain.

During the mid-1930s, the USSR increasingly bailed out of active European collaboration. Its shift in politico-economic strategy had clearly opened up the possibility to fine-tune technical standards. However ideological differences were a considerable problem especially regarding organizational issues in light of the vastly deteriorating international relations. The ideological differences between the IBU that was formed in a capitalist system and socialist Russia, became near

34 "Rapport du Directeur sur les mesures prises pour exécuter les décisions adoptés par le Conseil en mai 1930," Série 2061, 24 Sept. 1930, 1, CA Document Series, IBU.
35 "Procès-verbal de réunion de Commission Mixte," Série 2370, 9 fév. 1931, 4, P.V. Commission Mixte, Commission Relais, Commission Spéciales, IBU.
36 "Commission Mixte," Série 2291, 9, registry file 9G, box R-2601, file 13698/2081, LoN.

to insurmountable (see Chapter 5). The success of the technological collaboration ultimately depended on how well all parties were able to leave ideology out of their discussions.

During the Prague Conference, when socialist Russia and the IBU collaborated for the first time, the various parties tried to make the collaboration work. The dominant discussions in Prague were the role and responsibilities of the ITU regarding the IBU. The conference agreed that final responsibility remained with the ITU whereas the IBU would become the official European expert for broadcasting issues. Initially the IBU had tried to obtain final responsibility. The Russian delegation objected, tenaciously debating that the ITU must have final responsibility. The ITU did not object and the Russian delegation got its way in the end.[37] Thus the USSR achieved an agreement that the final decision power remained with an international organization that had no interest in the cultural and political side of broadcasting. Ideology would stay out of material network discussions.

Nonetheless, the question of trust between the USSR and other European countries lay dormant during the ensuing years. The 1932 Madrid Conference opened up the opportunity to discuss interference between Russian and other European stations in which the trust issue came to the fore.[38] Germany expressed fear of possible interference with a frequency which had been assigned jointly to Germany and the USSR. The Russian delegate replied in pro-European terms that the USSR had always collaborated in the best possible spirit with the other European countries to avoid possible interference. The USSR would continue to do so in the future and "foresaw a way to treat the issue in question separately with the German Administration."[39] The USSR was positive about collaboration

---

37 BUIT, *Documents de Prague*, 88-93.

38 In preparing for the Fourth International Radio Telegraphic Conference in Madrid the USSR had organized a preparatory Conference in Moscow to discuss all its telecommunication technologies together currently in use in radio electricity. From that Conference followed a path-breaking proposition for frequency allocations in telecommunications. The Madrid Conference quickly rejected the proposition because it proved too costly, yet worthwhile considering in a small sub-committee. A German delegate mentioned the proposal was too new and the time was too short too fully grasp the idea and implement it now. IBU, *Extrait des documents de la Conférence Radiotélégraphique Internationale de Madrid (1932): COMMISSION TECHNIQUE* (Geneva: Selection of all formal Madrid proceedings on broadcasting, 1932); BUIT, *Documents de la Conférence Radiotélégraphique Internationale de Madrid (1932): Tome I: Propositions transmises au Bureau de l'Union pour être soumises à la Conférence: Propositions, notes, déclarations, études, remarques soumises pendant la Conférence* (Berne: BUIT, 1932), 881.The examination of this preparatory meeting in Moscow might open up Russia's point of view not only on global and European telecommunications in general but also on its own position in this global and European world of telecommunications.

39 "envisagera la manière de traiter cette question particulière séparément avec l'Administration allemande." IBU, *Extrait des documents de Madrid*, 1065.

with other broadcasters which up until now had only happened outside the sphere of official frequency standard debates. The delegate affirmed his country's commitment more strongly by adding that the USSR was ready to collaborate with the other European countries individually as well as in European conferences to create a better organization of wireless services in the European region.[40]

Ultimately the parties agreed to integrate the USSR into their European broadcasting space, but under a special heading. The fact that the USSR was not obliged to adhere to Washington and thus had designed a different network than the Washington standards would need to be specified, according to the UK. Though the overall tone could be a little harsh at times, the participants finally agreed that the convention should maintain an extra clause that stressed the exceptional status of the USSR "for reasons of individual rights."[41] The USSR could use specifically assigned frequencies. This exceptional position, as agreed in the Madrid Convention, allowed both the USSR and the IBU to realize the fine-tuning of their wireless broadcasting networks over the long and medium waves. Moreover, the formulation indemnified both parties against any liability of the other party whether of an ideological or a practical nature.

The discussions in Madrid show that even though the USSR remained outside the official European institution for broadcasting, it discursively portrayed itself as a European friend. In practice this meant that Russian experts actively participated in the European broadcasting community. Madrid adopted the Prague definition of the border of Europe in the east determined by the IBU, PTT administrations, ITU and Russian delegation.[42] The friendly collaboration of Russian experts as part of both this physical Europe and this European broadcasting community rather surprisingly extended to the short waves as well. In spite of ideological differences that formed a political complication when it came to global broadcasting, the USSR obtained one of the Vice-Presidency seats in the Technical Section of the 1936 Paris Meeting that studied the short waves.[43] It is highly likely that the technical nature of the discussions as well as the fact that the IBU did not have the power to allocate the short waves facilitated this surprising development. By then the USSR had already been collaborating with the European community of broadcasters for over seven years. The relatively stable and peaceful status quo remained almost until the outbreak of World War II. In 1938 the USSR still signed

40 Ibid., 1068.

41 "a de ce fait des droits individuels et est justifié à présenter à ce sujet des réserves en ce qui regarde l'utilisation spéciale, par ses services, des fréquences suivantes :..." Ibid.

42 Codding, *The International Telecommunication Union*, 157, note 113.

43 IBU, *First Intercontinental Meeting of Broadcasting Organisations, Paris, Febrary 27-March 6 1936: Preparatory Documents and Minutes of the Meeting* (Geneva: IBU Office, 1936), 181.

the International Telecommunications Convention in Cairo, that like Madrid, made up for Russia's exclusive position.[44] But shortly after, international tensions would brutally disturb their mutually developed contact.

Part of these tensions arose because of the vast increase in broadcasting stations in combination with the scarcity of available frequencies. The problems started with the drafting of the 1939 Montreux Plan. The new frequency plan assigned the USSR far fewer frequencies than the 1933 Luzern Plan. The USSR felt betrayed for not being taken seriously by a plan that was "absolutely unacceptable to the USSR."[45] The final plan would reduce the density of the Russian wireless network considerably. It assigned the USSR far fewer exclusive waves, and obliged the country to reduce considerably the power of many of its stations to avoid interference with other European ones. Consequently, many stations would have to be taken off the air for quite some time to apply technical adjustments to their transmitters.[46] It would be a costly affair. The documents suggest that the USSR felt insulted by the shrinkage of the wireless network in western Russia as proposed at Montreux. The USSR had the feeling that it was sacrificing far more than other countries.

Though the large growth in stations made it difficult to create a plan favorable to all parties, it is highly likely that political considerations decisively influenced Europe's attitude towards socialist Russia. The USSR could not understand why it had to make concessions to such a degree when other countries who broadcasted solely in one language obtained extensive favors (see Table 4.1). The USSR insisted on obtaining 18 exclusive waves as proposed in the draft plan in contrast to the now intended 8 exclusive waves.[47] It proposed several changes to the plan as a whole, but concluded that the conference had not been very forthcoming with its remarks and requests. The USSR refused to sign the Montreux Plan.

The Russian delegation disappointedly and harshly declared that it reserved the right to assure itself of a satisfactory broadcasting service on Soviet territory, and resolve broadcasting problems conformable to its national needs. Furthermore, in case of interference with foreign stations on Soviet territory, the USSR "reserved for itself the right to protect by all means necessary the interests of its services

44 BUIT, *Documents de la Conférence Internationale des Radiocommunications du Caire (1938), Tome II: Travaux de la Conférence (Délibérations, Actes définitifs)* (Berne: BUIT, 1938).

45 "absolument inacceptable pour l'U.R.S.S." BUIT, *Documents de la Conférence Européenne de Radiodiffusion, Montreux, Mars/Avril 1939* (Berne: BUIT, 1939), 101.

46 Russia would lose one exclusive wave within the band of 150 to 285 kc/sec. lowering its number from four to three, and additionally would receive far too few exclusive waves in the band between 600 to 192,3 m.

47 BUIT, *Documents de Montreux*, 101-103.

Table 4.1 – Excessive favors to countries according to the USSR in the draft Montreux Plan, 1939

The increase in number of shared waves (very low power stations) referring to Luzern		
Italy	+7	
Great Britain	+9	
Germany	+8	
Sweden	+3	
Poland	+5	
France	+12	
The increase in number of exclusive waves referring to the 1933 Luzern Plan		
	Luzern	Draft Montreux
France	5	+1
Norway	1	+3
Italy	4	+3
Belgium	1	+1
Poland	3	+2
Yugoslavia	2	+1
Netherlands	1	+1
Finland	1	+1

Source: BUIT, Documents de la Conférence Européenne de Radiodiffusion, Montreux, Mars/Avril 1939 (Berne: BUIT, 1939), 102-103.

against disturbances caused by the stations placed in infringement."[48] The USSR refused any further collaboration with the IBU members. Technologically the USSR again isolated itself, returning to a policy it had pursued at the end of World War I. International tensions and rapidly increasing distrust towards communism by the end of the 1930s had created too harsh a climate for the frequency standard debates between the USSR and other European countries. The USSR bailed out of friendly collaboration with the European broadcasting community as coordinated by the IBU, but would continue to officially form part of the Madrid definition of the European Region.

For the major part of the interwar years, the Russians and the IBU community managed to fine-tune their material broadcasting networks relatively well. They could collaborate technically as long as they held no responsibility for one another and avoided ideology. The USSR managed to place final responsibility for the frequency plans with the ITU during the Prague Conference. Moreover, the Madrid Convention confirmed that neither of the two parties could be held responsible for each other's actions. Initially the USSR had envisioned a global broadcasting network based on socialist ideologies without intending to place

48 "se réserve le droit de protéger par tous les moyens utiles les intérêts de ses services contre les perturbations causées par les stations placées en dérogation." Ibid., 261.

emphasis on "Europe." Practical problems of interference and a shift in Russia's politico-economic strategy laid the foundation for some form of Russian integration into the IBU's European broadcasting system. Though western Russian territory formed part of the European Region as defined in Madrid, organizationally the USSR portrayed itself as a friendly collaborator in the European broadcasting community. In the end, nonetheless, ideological upheavals hindered the relatively stable, fruitful, and peaceful collaboration.

Vatican Radio, Radio Nations, and Established Standards

Vatican Radio and Radio Nations did not suit established European broadcasting at all. By the early 1930s, both Vatican City, in other words the Roman Catholic Church, and the League of Nations had created a transnational broadcasting network. Their stations transmitted over the short waves from within the newly defined European Region to audiences in and beyond the borders of Europe. Their institutional standards as well as the geographical scope of their networks deviated substantially from the stations represented by the IBU. In the course of their attempts to realize global broadcasting networks, both organizations found themselves hindered by technological deficiencies and ongoing European developments in frequency standardization. As a result, they would have to collaborate with the IBU and ITU and be forced to distinguish Europe in their developing broadcasting networks.

The League projected its own global station. This idea goes back to the discussions about the establishment of the League itself. At the Paris Peace Conference in 1919 U.S. officials proposed "a project that equips the future organization for world peace with an instrument for diffusion and communication of its ideas and its activities." The project integrated "the possibility of erecting in Switzerland a radio electricity station in the proximity of the terrain where the League of Nations will be set up."[49] Such a radiotelegraphic station would not only stimulate the development of the League but also trigger "the peaceful future of mankind."[50] When shortly afterwards the USA decided not to enter the League of Nations, the idea died prematurely.

49 "un projet de doter la future organisation de la paix mondiale d'un instrument de diffusion et de communication de ses idées et de ses activités…la possibilité d'ériger en Suisse une station radioélectrique à la proximité des terrains où s'installera la SdN." Antoine Fleury, "La Suisse et Radio Nations," in *The League of Nations in Retrospect: Proceedings of the Symposium* (Berlin; New York: Walter de Gruyter, 1983), 196.
50 "l'avenir pacifique de l'humanité." "Relais internationaux, Résumé de la situation présenté par le Secrétaire Général," Série 562, 15 juin 1927, 4, box 85, file Relais, Général I, 1927 et précédents, IBU.

The plan for a wireless radiotelegraph station to diffuse its message globally continued to lie dormant within the organization. The League's Communications and Transit Committee witnessed the creation by Radio-Swiss S.A of a Swiss broadcasting network, as well as the establishment of the IBU some three years later. When the IBU asked the International Telephone Consultative Committee (CCIF) in 1925 to examine the possibility of relaying music over the telephone cable network, the League approached the IBU to let Geneva be treated as a European capital in this proposal. A direct connection would make Switzerland the center of international broadcasting. A central office in Basel or Zürich would be working under the auspices of the Geneva headquarters and would serve its ideals of peace and of the prevention of war.[51] This alternative plan for a Europe of national capitals would allow the League to contact other European capitals in case it did not realize its proper radiotelegraphic station.

When in 1926 the League discussed for the second time building its own radio station, existing regulations narrowed down their options of realizing the global network the way they projected. In December 1926 the League Council asked the OCT to examine having "at its disposal a radio-telegraphic station of its own, sufficiently powerful to enable it to communicate independently with the greatest possible number of State Members of the League."[52] Ownership of an independent radio station seemed to be the most advantageous option. Participation in the relay network as envisioned by the IBU would subject the organization to various systems of national communication. The importance of having access to independent means of communication could prove essential during "times of emergency." In time of war, the League would have to be able to contact all European governments while the League Secretary-General could contact member countries (especially its Council members). In the eventuality that it had to use existing stations for its communications, the League would be hampered by the International Telegraphic Convention which reserved each Government "the right to suspend the international telegraphic service for an indefinite time if it thinks necessary, either as a whole or only on certain lines and for certain classes of correspondence, provided that it immediately informs each of the other contracting Governments."[53] To live up to its Covenant for the maintenance of peace the League wanted to establish its own radio station. The station must have global radius and would have to fall outside the sphere of influence of national governments, in particular that of the Swiss.[54]

51 "Relais internationaux," Série 562, 4, box 85, file Relais, Général I, IBU.

52 LoN, *Establishment of a League of Nations Radio-Telegraphic Station*, LoN doc. ser., A.22.1928.VIII (Geneva: LoN March 1928), 1.

53 LoN doc. ser., A.22.1928.VIII, 2.

54 Ibid., 1.

This plan for a League radio communications network connecting with nation-states on a global scale did not refer to Europe in any way. In March 1928 the OCT advised the Council to build a short wave station that could transmit over two wavelengths simultaneously, supplemented, if necessary, by a medium wave transmitter. For financial reasons it dismissed the idea of a long wave transmitter altogether. An additional medium wave transmitter would only be required in case the short wave transmissions could not be received within Europe. The League would then have to distinguish between Europe and the rest of the world for technical reasons. The station would allow the League to communicate via wireless telegraphy and live broadcasts with all countries in and outside Europe. The station would strengthen relations between the League, all governments, and all people worldwide.[55]

The Swiss intervened immediately. Radio Swiss S.A. and the Swiss government were not at all pleased to have such a global and pan-national broadcasting station within Swiss borders. Though they shared the basic ideological aims of the League, they had economic and political objections. Radio-Suisse needed to follow the League project closely. Director Fritz Rothen remarked that "Switzerland should stay vigilant towards this project; it cannot oppose it, but it must make the League understand the difficulties and inconveniences of an independent station."[56] Rothen was afraid of the competition from another radio station in his country. The Swiss government on the other hand foresaw potential problems with the country's strict position of neutrality including possible effects for its national defense.[57] "Switzerland must avoid being faced with a fait-accompli."[58] As a result, a few months after the OCT report, the Swiss federal political department made a counter-proposal that set the tone for a period of intense deliberation between the Swiss authorities, Radio-Suisse, and the League.

The Swiss counter-proposal gave the League three options. Firstly, the League could establish its own radio station according to its initial plan, to the dismay of the Swiss. Secondly, the Swiss proposed a mixed solution whereby the Swiss and the League equipped a radio station together. The Swiss would administer and

55 Ibid., 2-3, 5.

56 "la Suisse doit rester attentive à l'égard de ce projet; elle ne peut s'y opposé, mais elle doit faire comprendre à la SdN les difficultés et les inconvénients d'une station indépendante..." Fleury, "La Suisse et Radio Nations," 200.

57 LoN, *Radio-Telegraphic Station for the Purpose of Ensuring Independent Communications for the League of Nations at Times of Emergency, Note and Memorandum from the Swiss Federal Government*, LoN doc. ser., A.31.1928.VIII (Geneva: LoN 1928), 1.

58 "la Suisse doit éviter d'être placée devant un fait accompli." Fleury, "La Suisse et Radio Nations," 201.

exploit the station in normal times while the League would take over in times of crisis. Though this solution was built on mutual trust and collaboration, the Swiss government wanted to secure its neutrality by agreeing to a written consent that Switzerland could not be held responsible for any of the League's actions or transmissions. The third option assumed that the Swiss would administer and exploit the station at all times even in times of crisis. The Swiss then would not only guarantee the League independence and security of communications, but would also allow a League delegate to work as observer in the station.[59] Though the Swiss preferred the last option, it did not play an important role in the discussions.

Each option had its advantages. Whereas the second option would enable the League to make great cost-efficiencies, the first option would exclude complicated collaborative efforts. Both options enabled the League to communicate with the largest possible number of members in and outside Europe, downplaying any future dependency on re-transmissions that were currently necessary for communication with extra-European countries.[60] Regarding the mixed option, the League would have to agree to the technical specifications jointly with the Swiss wireless telegraphy and telephony joint stock company Radio-Suisse.[61] The pros and cons expressed by the League's wireless engineers led to extensive behind-the-scene lobbying between the Swiss and League Secretary-General, Drummond. In September 1929 the League Assembly finally voted in favor of the mixed solution.

The Assembly considered the new station to be one of worldwide importance and Rothen confirmed that Radio-Suisse fully backed the League's ideals. He mentioned his delight at participating in a collaboration that would satisfy needs beyond the reach of a national station alone:

> 'It is because we are friends of the League and we consider it a privilege to facilitate its mission that Radio-Suisse…has the honor to establish the widest possible number of international links.' On balance, it declares itself happy to collaborate with the LoN to construct a station that has to satisfy the needs more than a national station could have done alone.[62]

---

59 Ibid., 207-208.

60 LoN, *Radio-Telegraphic Station for the Purpose of ensuring Independent Communications for the League of Nations at Times of Emergency, Supplementary Report of the Advisory and Technical Committee for Communications and Transit*, LoN doc. ser., A.23.1928.VIII (Geneva: LoN 1928), 2-3.

61 LoN doc. ser., A.31.1928.VIII, 2. In the mixed solution operations could consist of 1) radio-telegraphic connections for correspondence between the League and other parties like government and press; 2) weekly exchanges of traffic on appointment; 3) "Daily broadcasting of one or more League *communiqués* by means of suitable transmitting apparatus in order to establish contact daily with the stations intended to receive messages from Geneva"; 4) a listening-in program consisting of other stations sending information to Geneva. LoN doc. ser., A.23.1928.VIII, 6.

62 "'C'est parce que nous sommes des amis de la Ligue et nous considérons comme un privilège de faci-

Figure 4.2 – Radio Nations, Prangins
Source: League of Nations Archives, UNOG Library. Used by the courtesy of United Nations Office, United Nations Library, Geneva.

Though all parties seemed to be happy, the League would soon experience limitations when seeking to influence the development of its station's network.

The League had agreed to construct a network that radiated globally, but for its technical qualifications and collaboration with the Swiss broadcasters, distinguished between Europe and the world. The construction of the global network over the short waves would have to become a truly international project to avoid

liter sa mission que Radio-Suisse…s'est fait l'honneur d'établir le plus grand nombre possible de liaisons internationales.' En fin de compte, il se déclare heureux de collaborer avec la SdN à la construction d'une station qui doit satisfaire des besoins au-delà ce qu'une station nationale eût pu assumer seule." Fleury, "La Suisse et Radio Nations," 208.

political quarrels. The League ordered that all major radio industry companies including Telefunken, Marconi, and RCA should take part in the construction of the station.[63] The short wave connection would assure global communication, with a special desire to communicate with South America, Argentina and Brazil in particular.[64] Within Europe, the new station could rely on the already existing medium wave transmitter of Radio-Suisse that allowed broadcasting to all stations within IBU's Europe including those in large parts of the USSR, in North Africa and the Near East. The League of Nations in its political affairs was thus dealing with a smaller Europe than the one to which it broadcasted over the European wireless network of medium waves.

The League's decision to collaborate with the Swiss limited its influence substantially. The League could operate its short wave network freely as to date there was no regulation of the short waves. The medium waves were a more difficult matter. The Swiss PTT owned the Swiss network. Moreover, it was Radio Suisse that had a seat in the IBU and consequently could negotiate and influence the allocation of the medium waves and the designing of the relay network in Europe. The League could only make use of the medium waves assigned to the Swiss within the existing European frequency plans. By deciding to collaborate with the Swiss, the League chose to distinguish between Europe and other parts of the globe. Thereby it bound itself to European regulations and appointments for radio broadcasting that it could not influence. Its station, Radio Nations, went on air on February 2, 1932.

The plan for a global Catholic station encountered similar problems. Vatican Radio was established in response to political upheavals between the Papal States and the Italian government. The upheavals revealed the need to guarantee the voice of the Catholic Church in the future. Pope Pius IX had lived in exile in Vatican City for years until the signing of the Lateran Treaty in 1929. The Treaty made Vatican City an independent state with its own laws and its own currency, returning freedom of travel to the Pope. Shortly afterwards, tensions again arose between the Pope and Mussolini. The Church sought to strengthen the Vatican's power in order to avoid any repeat of defeat in modern world politics. According to Cardinal Eugenio Pacelli (later Pope Pius XII), the new medium of radio could be employed to spread the Catholic message. "Through radio, no pope could ever be driven

63 LoN doc. ser., A.23.1928.VIII, 6.

64 LoN, *Advisory and Technical Committee for Communications and Transit, Minutes of the Sixteenth Session, held at Geneva, May28th-June 2nd, 1931*, LoN doc. ser. C.417.M.173.1931.VIII (Geneva: LoN 1931), 75.

into isolation again; geographic and political borders had become virtually meaningless when confronted by the 'airwaves' of broadcast technology."[65] In 1930 the Church contacted Guglielmo Marconi, who not only was Italian by birth, but also a staunch Roman Catholic. A few months later, after a quick construction process, the station Vatican Radio went on air on February 12, 1931.[66]

Unlike the League, the Vatican could organize its broadcasting network far more quickly and easily. Marconi began by constructing a short wave station to secure the Vatican global reception of its own voice. After six months, the Vatican's programs could be broadcast in seven languages through seven non-directional shortwave transmitters. The Society of Jesus operated the station which had the call signal "Station HJV" referring to the Holy See, Jesus Christ, and the Vatican. According to the Vatican, the new station required a "certain knowledge, dedication and commitment to the Pope and the Catholic Church through the next millennium."[67] They envisioned a global Christian broadcasting network that despite transmitting from within the European region, did not pay any attention to Europe as a category.

Like the League of Nations, Vatican Radio experienced problems with reception. The short waves did not allow for the best quality of reception over short distances within Europe. In 1937 the Vatican gave its station a technological boost by adding a new 25kw shortwave station complemented by four 400kw shortwave antennas, and also by constructing a 5kw tower for medium wave transmission. These improvements greatly enhanced the Vatican station's reception in Europe and across the globe in new areas like Japan and South America. Broadcasting increased from seven languages to ten and then to twelve by 1939.[68] Contrary to the IBU and Radio Nations, Vatican Radio did not envision an audience based on the idea of nation-states. Like the League, by 1937 Vatican Radio technologically imagined a broadcasting space that distinguished between Europe over the medium waves and other parts of the world over the short waves. The medium waves had already been standardized in European institutions by 1931. Radio Nations and Vatican Radio could not avoid integration into the European broadcasting community if they wanted to have a say in their own network development.

65 Marilyn J. Matelski, *Vatican Radio: Propagation by the Airwaves*, Media and Society Series (Westport, Conn.: Praeger, 1995), xvi.

66 "Inauguration de la station du Vatican," Série 2386, 11 fév. 1931, CA Document Séries, IBU.

67 Matelski, *Vatican Radio*, xvi.

68 Ibid., xvi, 169. Radio Vatican added Polish and Ukrainian to its languages in 1938-39, and after the war it quickly added eleven new, mostly Eastern European languages, to its services: Hungarian, Rumanian, Czech, Russian, Slovene, Slovak, Latin, Bulgarian, Croatian, Belorussian, and Albanian.

The two stations related differently to the enduring European system. Despite no regulation of the short waves, the IBU and ITU together could still arrange the medium waves. Initially, Vatican Radio did not broadcast over the medium waves. Since the conventions designed under the auspices of the ITU allocated a small wave band to broadcasting, Vatican Radio joined the ITU and CCIR in 1931.[69] In this manner the Vatican could actively participate in international discussions on the allocation of waves between various forms of telecommunications as well as in technical discussions on the development of radio-electricity technology in broadcasting and wireless communications. Though Vatican City was largely interchangeable with the Roman Catholic Church and as such would not fit the idea of the genuine PTT administrations that joined the ITU, in legal terms it was a nation-state. This greatly facilitated its integration into the European community for radio communications. But Vatican Radio was not interested in IBU membership. The IBU dealt with European issues and could not influence the regulation of the short waves.

The League of Nations was a totally different matter. It wanted to join organizations like the ITU and the IBU, but did not meet their standards for adherence to the union. The developing European frequency standards assumed a nation-wide and public broadcasting network. The League not only sought to create global broadcasting, but was also an *intergovernmental* organization that at times operated a radio station of *another* broadcasting organization. The League's broadcasting activities caused considerable confusion. For years, the organization had joined in various international meetings of the IBU Technical Committee as well as the ITU consultative committees on telegraphy (CCIT) and radio-electricity (CCIR) as an observer with consultative voice. Now the League wanted to become a full IBU member in order to have some influence over the European part of its broadcasting activities with respect to Radio-Suisse. In 1930 the League requested IBU membership.[70] The IBU could simply not allow the League to join the Union, because the League station did not meet its standards.

Neither the IBU nor the League had any experience in this field. In particular the issues of ownership and transmission content caused complications. Regarding ownership, the IBU foresaw a problem with international rights, since the League had a special role in world affairs. The IBU Council asked its Special and Legal Committee to study the issue. The study should also explore the possibilities with respect to Vatican Radio, even though they had not approached the IBU with a

69 "Procès-verbal des réunions de la Commission Spéciale à Budapest, Octobre 1930," Série 2195, 6 nov. 1930, 2-3, P.V. Commission Mixte, Commission Relais, Commission Spéciales, IBU.

70 "Procès-verbal des réunions de la Commission Spéciale," Série 2195, 1, P.V. Commission Mixte, IBU.

request for adhesion.[71] The study results argued that the League could not be welcomed as full IBU member. First of all, the League used a station that was operated by another organization. That other organization could join the IBU, but the League could not.[72] Moreover, the League did not suit IBU standards for transmitting *broadcasting programs* to an *anonymous* audience either. The League rather sought to transmit *communiqués* to member governments and League Council members.[73] The IBU Council altogether sidelined the idea of making Vatican Radio a member. The organization did not intend joining the IBU anyway. The League would still be welcome as an observer to relevant IBU meetings.[74]

Only four months later, IBU director of legal affairs, Ladislav Sourek, and League wireless expert, Van Dissel, met each other half way. Sourek now recognized that the League wanted to broadcast genuine programs to a mass audience. Van Dissel approved a membership in which the League would have all the advantages of active membership without the right to vote. The League was an impartial international organization that represented many countries. A right to vote would undermine that position.[75] The IBU moreover thought that League membership would positively influence its own position during the Madrid Conference the following year. The IBU modified its statute.[76] With an eye to both Vatican Radio and Radio Nations, the IBU created the category of "special membership." Special members would have all the obligations of active members, but could only join with a consultative voice.[77] In spite of the fact that final decision power remained with Radio-Suisse, special membership would in any case give the League the right to participate in discussions on medium waves.

With international legal issues in place and broadcasting contents clarified, the international economic crisis brutally disturbed the formal bonding of the League and the IBU. In June 1932 Van Dissel informed IBU Secretary-General Burrows that he had to turn down, or at least postpone, League membership of the IBU. He

71 The Special Committee held a temporary nature. During the last year it had been examining the relationship between the IBU and other international organizations, questions very much in line with the League's request for adhesion. P.V. Conseil Administrative 1930, 229, IBU; P.V. Conseil Administrative 1930, 230, IBU.

72 "Procès-verbal de la Commission Spéciale tenue à Paris," Série 1996, 22 juillet 1930, 4-5, P.V. Commission Mixte, Commission Relais, Commission Spéciales, IBU.

73 "Procès-verbal de la Commission Spéciale," Série 1996, 4-5, P.V. Commission Mixte, IBU.

74 Ibid.

75 "Procès-verbal des réunions de la Commission Spéciale," Série 2195, 2, P.V. Commission Mixte, IBU.

76 The discussions about new forms of memberships would continue for almost a year. Ibid., 2; P.V. Conseil Administrative 1931, 28, IBU; Série 1886, 2156, 1995, 2389, 2616, 2647 révisé, 2653, CA Document Séries, IBU.

77 According to Van Dissel IBU membership would possibly concern the OCT, the League body that would be exploiting Radio Nations. P.V. Conseil Administrative 1930, 265-266, IBU; P.V. Conseil Administrative 1931, 28, 45, IBU.

remarked:

> My dear Burrows,…I am very sorry to inform you, considering the economic depression we all are passing through, I am obliged to adjourn the question of adhesion to the 'Union Internationale de Radiodiffusion' by 'Radio Nations' until next year. I hope the amelioration of the financial conditions will permit us then to offer you a formal demand for adhesion, as we are very interested, theoretically and practically, in the work of the Union.[78]

The plan did not die altogether. Having reconsidered its options, in 1935 the League decided that it wanted the same advantages as active members including the right to vote. A right to vote was indispensable for the League to be able to be "completely independent with regard to the Swiss Government," argued member of the League Secretariat, Metternich.[79] Since the League was an organization of international public rights, the League should be placed on equal footing with state organizations, on equal footing with a state administration charged with telecommunications. Moreover, the League would be represented by "a delegate acting as an expert for one of the interested bodies of the League of Nations" a similar construction as agreed to with CCIR and CCIT.[80] Both IBU Secretary-General Burrows as well as the by now Director of OCT Van Dissel, did not foresee any problems.[81]

In the end the matter was never resolved mainly because of the League's position as intergovernmental organization striving for international peace and collaboration. On the one hand, the main bottleneck would continue to be the right to vote relating to the League's legal position in world affairs. On the other, rapidly deteriorating international tensions due to ideological differences also curtailed more and more of the space left to the League to maneuver. Whilst Metternich advised the League to stay out of IBU discussions until the form and conditions for collaboration between the two organizations were resolved, Nazi Germany

78 "Letter from Van Dissel to Burrows," 2 june 1932, 1, registry file 9G, box R-2601, file 13698/2081, LoN.
79 "une indépendance complète vis-à-vis du Gouvernement suisse" "Report by Metternich," 9 Feb. 1935, 3, registry file 9G, box R-4321, file 663/663, jacket 1, XV, LoN.
80 "par un délégué agissant en qualité d'expert de l'organisme intéressé de la Société des Nations…" Though the archives reveal that the League desired to become an ITU member and to be able to join in the meetings of the ITU consultative committees, the documents do not give any indication that the League managed to acquire membership. The documents do indicate that the League wanted to give it a try during the 1932 Madrid Conference, and reveal that League representatives joined CCIR and CCIT meetings around the middle of the 1930s. "Report by Metternich," 4, registry file 9G, box R-4321, file 663/663, LoN.
81 Ibid., 5; "Rapport sur la session du Conseil de L'Union Internationale de Radiodiffusion tenue à Genève du 20 au 26 février 1935," 11 March 1935, 18, registry file 9G, box R-4321, file 15874/663, LoN.

decided to withdraw from the League.[82] When the League in Cairo requested placing the question of the re-transmission of League broadcasts on the agenda of the Montreux Conference, Germany and Italy objected.[83] The League even sent a delegate to Montreux the evening preceding the conference to calm the atmosphere and create favorable conditions for the discussion of "a particularly delicate question given the fact that certain organizations, those of Germany and Italy, oppose our request."[84] By the time of the outbreak of World War II, several League experts recognized that "[w]hatever little success is achieved by our radio service during its short career is due primarily to our contacts with members of the U.I.R."[85]

One year before its medium wave transmitter went on air, Vatican Radio for the first time showed some interest in "Europe" when it handed in a request for adhesion to the IBU in 1936.[86] By then the IBU had already amply discussed the issue of special membership. It immediately granted Vatican Radio access to the IBU broadcasting community as a special member. As a result, Vatican Radio could immediately join in all meetings relating to its technical and non-technical activities in medium waves. The Vatican, however, showed a broader interest in the IBU community. Its representatives actively participated in meetings of the Rapprochement Committees and assisted in the legal discussions of the Intercontinental conference for broadcasting in Paris in 1936.[87] Furthermore, Vatican Radio representatives also joined in mutual League and IBU activities with respect to broadcasting in the interest of peace (Chapter 5). Vatican Radio's clear religious orientation and its specific global interests deviated from established IBU standards. The station nonetheless shared an important ideological belief with the IBU and the League that made it an acceptable partner to the European broadcasting community. They shared a desire for the international peaceful use of the airwaves.

82 "Letter by Metternich to the League Secretary General," 30 janv. 1936, 1, registry file 9G, box R-4321, file 22092/663, II, LoN.

83 "Letter by Van Dissel to Metternich," 17 Feb. 1939, 1, registry file 9G, box R-4322, file 36772/663, jacket 1, XII, LoN.

84 "une question particulièrement délicate étant donné que certaines Compagnies, telles que la Compagnie allemande et la Compagnie italienne, s'opposent à nôtre requête." "Note by Vejerano to De Makay," 1, registry file 9G, box R-4322, file 36772/663, jacket 1, XV, LoN.

85 They agreed that in particular the role of IBU Secretary General Arthur Burrows had been essential to their success. Nevertheless, it is highly likely that the personal contacts between Burrows and Van Dissel formed the basis of whatever success they had. Metternich's advice did not stop Van Dissel from continuing to attend and participate in events organized by the IBU, like the 1936 Paris Meeting. "Dunning to Pelt," 25 April 1940, 2, registry file 9G, box R-4322, file 40118/663, III, LoN; IBU, *Documents of the Paris Meeting*, 12.

86 "Demande d'admission du service de radiodiffusion de la Cité du Vatican, (Rapporteur: A.R. Burrows)," Série 5661, 13 fév. 1936, CA Document Series, IBU.

87 IBU, *Documents of the Paris Meeting*, 9, 21.

Whereas both Vatican Radio and Radio Nations had initially envisioned a global broadcasting network to connect with their audiences, they could not but deal with Europe as a special category. The technicalities of short waves did not allow for high-quality reception of broadcasting within Europe. Consequently, they would have to transmit over medium waves as well, an activity that was already being standardized at European level. In the case of Radio Nations, contingencies with the Swiss made the organization decide to integrate Europe as a special category in its global network. Regarding audiences however, Radio Nations and Vatican Radio had different views. Radio Nations acknowledged and aimed at nation-states in its communiqués and broadcasts. Vatican Radio preferred Radio Moscow's plan, reaching for a global Catholic rather than socialist community. Similar to the case of Radio Moscow, over the years Europe became part of the Radio Nations and Vatican Radio's broadcasting projects for practical reasons.

Radio Luxembourg and Reluctant Experts

Radio Luxembourg was the sole station aimed explicitly at Europe. It was an initiative that deliberately desired to broadcast to a Europe-wide audience by means of a Europe-wide wireless network. Radio Luxembourg in fact would be the one station in the European Region that explicitly claimed to be European. Moreover, contrary to the developing European standards based on the idea of public broadcasting, Radio Luxembourg would be broadcasting on a commercial basis. Its case reveals how threatening the very idea of a European commercial broadcasting station appears to have been to the developing establishment of European broadcasters.

Before Radio Luxembourg there already was a commercial broadcasting tradition in Europe. When the French, in collaboration with Luxembourg PTT administration, established the Luxembourgeois Broadcasting Company, they were thus not doing anything highly unusual (Table 4.2).[88] By 1928, European stations like Hilversum (Netherlands), Radio Paris (France), Scheveningen (Netherlands), Tour Eiffel (France), and Radio Belgique (Belgium) at regular intervals issued commercial broadcasts that often specifically targeted British audiences.[89] French station Radio Normandy, which would become one of the most important commercial stations along with Radio Luxembourg, operated outside the IBU community. Other stations

88 Seán Street, *Crossing the Ether: British Public Service Radio and Commercial Competition, 1922-1945* (Eastleigh: John Libbey Publishing, 2006).
89 Street, *Crossing the Ether*, 147-148.

Table 4.2 – Commercial broadcasting stations in Europe in the interwar years

Name	Established	IBU membership
Radio Paris	1921	1925
Radio Normandy	1924	-
Poste Parisien	1924	1925
Radio-Lyon	1924	1926-1932, rejoined in 1943
Radio-Toulouse	1925	1926-1929, rejoined in 1943
Radio Méditerranée (Radio-Côte d'Azur/Nice Juan-les-Pins)	1927	1943
Radio Luxembourg	1931	-
Radio Athlone	1933*	1928 (in 1933 partially commercial)
EAQ Madrid	?	-

*The broadcasting service of the Irish PTT, member of the IBU, would commercialize part of its activities in 1933. It named that part of its activities Radio Athlone.
The stations in this table are an indication of the number of commercial broadcasting stations in Europe only. There have been other stations who over time incidentally or regularly sold airtime to commercial stations like the IBC.
Source: Seán Street, Crossing the Ether: British Public Service Radio and Commercial Competition, 1922-1945 (Eastleigh: John Libbey Publishing, 2006), 229-258; Arno Huth, La radiodiffusion: Puissance mondiale (Paris: Librairie Gallimard, 1937), 100-102; IBU, Twenty Years of Activity of the International Broadcasting Union (Geneva: International Broadcasting Union, 1945), 65-72.

like Radio Paris and Radio Toulouse, however, were respected members of the IBU. Several stations in Europe issued commercial broadcasts or advertizing.[90]

In the early years "commercial" stations mainly sought to cover costs via increased sales of receiver sets. The IBU whose 1926 statutes explicitly claimed that the "Union does not pursue any commercial scope" in reality consisted of many private broadcasters who created the IBU to stimulate the increase of set sales.[91] The IBU members mainly tried to achieve this goal via improvements in the quality of the technology and the broadcasts while they asked their listeners for a listening fee. Occasionally some members apparently had additional advertising activities. Along with several French stations, the German RRG for instance on occasion broadcast advertisements, in Germany "only if they promoted German-made products, and were placed through the German Reichspostreklame, a subsidiary of the German

90 The activity of commercial broadcasting in the north, east and south of Europe deserves further scrutinization.

91 UIR, "Statuts," Série 176, 23 mars 1926, 1, box 33 l'Histoire de l'UIR – statuts, IBU.

PTT..."[92] Consequently, in these early years there was no clear-cut view on commercial broadcasting in the IBU, or a shared view on what *no* "commercial scope" actually meant.

One of the most prolific personalities behind the commercial transmissions was Captain Leonard Plugge. At regular intervals stations like Radio Paris, Radio Toulouse, and the Spanish EAQ sold him airtime. Plugge formalized his commercial activities when he created the International Broadcasting Company (IBC). The IBC mainly purchased airtime from radio stations on the European continent to broadcast ads and sponsored programs from British advertisers into British territory. In this way, British and foreign industries could reach a large part of British society. The BBC had always pursued a policy of high quality and educative broadcasting that according to its management did not suit advertising activities. Commercial broadcasts rather consisted of low key and low-quality programs that attracted a large number of audiences. At some point, BBC's overseas director Atkinson stated: "...if we calmly allow this kind of thing to go on, sooner or later we shall be forced off the 'no advertising' standard, which, to my mind, would be disastrous."[93] Their protests and the efforts of the British government to lay down the no advertising standard for broadcasting into British law did not help. IBC continued to expand its activities and would also start to use stations other than the French ones, stations like Katowice, Poland.[94] By 1933 commercially sponsored Sunday programs had increased to some 12 hours over Radio Normandy, 2.5 hours over Radio Paris, 1.5 hours over Radio Toulouse, and 1 hour over Radio Côte d'Azur.[95]

The development of commercial broadcasting in Luxembourg followed naturally from the country's tradition of economic diversification. With an open market policy, the Grand Duchy had always tried to survive in a world of Great Powers. Not surprisingly, the first radiotelegraphic initiatives in Luxembourg in 1922 focused on economic interests and business support from the point of view of competitive advantage. The state rejected the idea. Unlike other countries, Luxembourg did not have a license system in place with which it could grant concessions to broadcasting stations. The Luxembourg government argued that the airwaves could not be seen as state property. Moreover, it was of the opinion that radio served as an extension of the press rather than other means of telecommunication. As a result, in July 1929 François Anen created the private National

92 Spohrer, "Ruling the Airwaves," 277.

93 Street, *Crossing the Ether*, 150.

94 Ibid.; Alan Davies, "The First Radio War: Broadcasting in the Spanish Civil War, 1936-1939," *Historical Journal of Film, Radio and Television* 19, no. 4 (1999): 473.

95 Street, *Crossing the Ether*, 151.

Luxembourg Broadcasting Company. The station used a medium wave frequency with a radius of about 70 kilometers which had been assigned to Luxembourg by the 1926 Geneva and 1929 Prague Plans. The station, Radio Letzeburg, was a quasi-amateur or at least low-capital operation aimed at a Luxembourg market but which also reached peoples and markets beyond the country's boundaries. In 1929 Anen requested IBU membership. The IBU refused on the basis that the station broadcasted without a concession granted by the state.

By December 1929 the Luxembourg government had finally put legislation in place and opened up bids for concessions. The state decided that Luxembourg would work with a monopoly concession as it would be highly unlikely that European frequency plans would grant the Duchy more than one frequency. The monopoly would be granted to a private party, the only option given the small size of the country. In the end, a French company the *Société luxembourgeoise des études radioélectriques* (SLER) won the concession. SLER would create the *Compagnie luxembourgeoise de radiodiffusion*, the Luxembourg Broadcasting Company (CLR) to operate the new station, Radio Luxembourg. Raymond Braillard, director of the IBU's Technical Committee, had been actively involved in the concession process in Luxembourg. He held an active stake in SLER, and had a lot to gain from SLER and Luxembourg PTT collaboration.[96] Braillard managed to discredit Anen's broadcasting initiative with the Luxembourg PTT on the basis of the station's lack of expertise.[97] Anen's station caused considerable disturbance to other European stations, he reported. As such, Braillard recommended to the Luxembourg PTT a company which had years of experience with the French station Radio Paris: SLER.

Before SLER had even built the station, the situation in Luxembourg almost immediately led to disagreements within the IBU. Braillard had favorably received a request from SLER to join the IBU. Nonetheless, when Burrows found out that SLER intended to establish a station for international commercial broadcasting, he immediately sought support from other IBU officials. German IBU delegate and IBU Vice President Heinrich Giesecke for instance shared Burrows antipathy towards low key entertainment programs, whereas Swiss delegate Maurice Rambert supported Burrows due to legal concerns. Braillard on the other hand gained support from IBU Vice President Robert Tabouis who also had a stake in SLER. Tabouis feverishly tried to influence discussions about a change in IBU statutes concerning special membership. Braillard and Tabouis ultimately failed to prevent

96 Braillard moreover had been working for the French CSF for years. CSF backed SLER and owned Radio Paris.

97 Spohrer, "Ruling the Airwaves," 58-59.

the addition of an important clause to the statutes.[98] The IBU would refuse the adhesion of any member that specifically aimed to broadcast across borders since it would only welcome all organizations "operating a broadcasting service in Europe with the main object of providing for national needs…"[99] Burrows got his way.

The commercial activities in Europe that considered the UK their main target area influenced decisively people's opinion on advertising sponsored, popular broadcasting. Since several IBU members were involved in commercial activities like advertising on occasion, the IBU used double standards for a while. It justified its decision to amend the statutes and thereby exclude a Luxembourg station from membership by referring to the educative high culture standard the organization pursued as well as the national basis for wireless broadcasting in Europe. In the end, the IBU judged SLER on the basis of a simple vision for international commercial broadcasting. They were judging the mere idea of a station that had not yet even been built.

Initially Radio Luxembourg combined commerce with internationalism. The station's management projected an international economic broadcasting network. The network would collaborate and align its broadcasting with the League of Nations' economic and ideological goals.

Radio Luxembourg's ideas were relatively progressive or at least different from dominant perceptions. According to managing director Jacques Lacour-Gayet and his colleagues, the private media should defend liberalism, conventional competition and private enterprise rather than promote cartel capitalism and protectionism. Furthermore, they argued that an economic liberalist use of the private media should become a proponent for European unity and growth.[100] With such perspectives, Radio Luxembourg was following in the footsteps of the first plans for broadcasting in 1922 when Luxembourg had sought to play an important

98 Other major IBU officials probably shared the opinion of Burrows. President Charles Carpendale for instance is known to have supported enthusiastically the BBC model. Moreover, it is highly likely that the Czech delegate and head of the IBU Legal Committee Ladislav Sourek supported this view as well, just like the Polish delegate and head of the International Relay Committee Zygmunt Chamiec, and the Austrian delegate and head of the Rapprochement Committee Oskar Czeija. Ibid., 277; Ryszard Miazek, *Przeminelo z radiem: opowiesc o Zygmuncie Chamcu – zalozycielu i pierwszym dyrektorze Polskiego Radia* (Rzeszów; Warszawa: Wydawnictwo Wyzszej Szkoly Informatyki i Zarzadzania; Polskie Radio, 2005); Reinhard Schlögl, *Oskar Czeija: Radio- und Fernsehpionier, Unternehmer, Abenteurer* (Wien: Böhlau, 2005); Eckhard Jirgens, *Der deutsche Rundfunk der 1. Tschechoslowakischen Republik: Eine Bestandsaufnahme* (Frankfurt am Main: Lang, 2005).
99 UIR, "Statutes of the International Broadcasting Union, Adopted by an Extraordinary General Assembly of the International Broadcasting Union, Budapest in October 1930," 123, box 33 l'Histoire de l'UIR – statuts, IBU; Spohrer, "Ruling the Airwaves," 273-280.
100 Ibid., 88-90.

information role in economic life. Such a role would allow the medium to become a vehicle for international peace and growth in economics and finance.

According to Spohrer, Lacour-Gayet "represented a new type of commercial professional."[101] He was not from a commercial family, but grew up in an intellectual milieu with cultural interests. He studied liberal arts at the French École normale supérieure but rather than following a career in academia or politics, he entered commerce. He became a retail industry lobbyist, and tried to free the way for the economic liberalization of European trade markets. In the years leading up to the establishment of Radio Luxembourg, Lacour-Gayet raised a large part of the capital for the new station. Spohrer shows how Lacour-Gayet personally sought to recruit like-minded individuals who desired to create a European market. He feverishly tried to give these men a position in companies and organizations in which he had vested interests.[102] These men often shared bourgeois backgrounds and ideas about commerce and liberalism. Their backgrounds led them to connect commerce and competition with international peace and progress.

When these men constructed Radio Luxembourg they connected their ideas by envisioning an international economic broadcasting network. Their station would adhere to the latest ideas expressed by the International Conference of Industry and Commerce. The conference had shown interest in the creation of a truly independent and international broadcasting station for industry and commerce. Such an international and independent station should above national broadcasts "create a larger space for international economic problems," for special international transmissions.[103] Radio Luxembourg considered it was qualified to do the job and could become part of the European broadcasting community within the institutional framework of the IBU, with a very specific niche function. It wanted to become an all-European or even global station that combined commerce with the internationalist ideals of the League and the IBU.

With a reluctant European community of broadcasters in the back of their minds, the founders of Radio Luxembourg took matters in their own hands. They made a *de facto* claim to the long wave frequency of 1190m (252 kc/sec) and decided to start experimental broadcasting in March 1932. The station agreed to start regular broadcasting only after the conclusion of the 1933 Luzern Plan. The claimed frequency fell outside the wavebands officially assigned to broadcasting

101 Ibid., 88.

102 Ibid., 88-89; Laurence Badel, *Un milieu libéral et européen: Le grand commerce français, 1925-1948* (Paris: Comité pour l'histoire économique et financière, 1999).

103 "une place plus large au problèmes économiques internationaux." BUIT, *Documents de la Conférence Européenne des Radiocommunications Lucerne, MAI/JUIN 1933* (Berne: BUIT, 1933), 123.

by the ITU.[104] In general, new stations in countries that had signed the 1927 Washington Convention had the right to claim unused frequencies as long as they did not cause any interference with existing stations. These *de facto* claims would then be taken into consideration the moment a new frequency plan was concluded.[105] Though Luxembourg had not signed the Washington Convention, the station hoped that the maneuver would secure its technical network for the future.

Meanwhile, the station promoted a "Europe" in accordance with the League's economic and commercial aims. The station portrayed itself as a "bon européen," and claimed to serve unity internationally.[106] In July 1932, several months before the Madrid Conference, Radio Luxembourg approached the League's International Co-operation Organization (ICO) that dealt with the diffusion of the League's aims in science and the arts. The station offered to try and put the station at the service of the ICO "for the scope that your committee pursues tirelessly."[107] With its high power transmitter and its central position in Europe, Radio Luxembourg might provide an interesting instrument for the aims of the League, Radio Luxembourg director Henri Étienne pointed out. In the absence of the League's wireless expert Van Dissel, the ICO favorably received the request, with "very lively appreciation and profound gratitude…"[108]

ICO invited Etienne to consider interesting options for collaboration within the League's ideological framework. In response Etienne offered the League the use of its microphone for an hour every day between 7 and 8 p.m. It would be better if the League itself thought of the kind of propaganda it desired to pursue first, in order to then find the most suitable broadcasting formula.[109] Etienne therefore encouraged the League to suggest its own broadcasting content available for an hour each day to all audiences in Europe. With its generous proposal to the League to broadcast an hour every evening at the very time most people were listening to the radio, the station was clearly trying to make its powerful medium available for the ideals and the needs of the League.

---

104 "Note à Monsieur de Montenach par Van Dissel," registry file 5B, box R-2259, file 38270/38270, 6.X.32, LoN; Spohrer, "Ruling the Airwaves," 147.

105 Ibid., 144.

106 Huth, *La radiodiffusion*, 19.

107 "vers le but que poursuit inlassablement votre commission." "Lettre du Compagnie Luxembourgeoise de Radiodiffusion au Président de la CICI," 20 juillet 1932, registry file 5B, box R-2259, file 38270/38270, LoN.

108 "la très vive appréciation et la profonde gratitude…" "Lettre de J.D. de Montenach à Etienne," 27 juillet 1932, registry file 5B, box R-2259, file 38270/38270, LoN; LoN, *International Committee on Intellectual Co-Operation, Minutes of the Seventh meeting held at Geneva, July 21st 1932*, LoN doc. ser. C.I.C.I./14th Session/ P.V.7, 11.

109 "Lettre de Etienne à J.D de Montenach," 19 août 1932, registry file 5B, box R-2259, file 38270/38270, LoN.

This generous offer did not stand a chance. Almost immediately, voices of protest arose within the League of Nations regarding possible collaboration with Radio Luxembourg. Though OCI was convinced it should make use "of the good-will of these people," doubts within the committee made them decide to inform the station that they could not undertake any definite action before consulting wireless expert Van Dissel.[110] In the meantime, the IBU found out about the request via its regular exchange of minutes with OCI. Burrows immediately mobilized his personal network to prevent such collaboration. Burrows contacted the chair of the OCI, British Professor Gilbert Murray.[111] He stressed that the station acted in contravention of the 1927 Washington Convention as much as with the national laws of several countries that forbade the broadcasting of advertizing. For more information Burrows referred Murray to the biggest opponents of Radio Luxembourg in Europe, IBU Vice-President Carpendale and BBC director John Reith.[112] A British offensive would have to undermine a League and Radio Luxembourg collaboration.

Van Dissel would come to the aid of the British. After ample deliberation with the British, Murray contacted the League Secretariat who assured him that the League had not made any promises to the station and could easily back out. When Van Dissel returned from his travels, he immediately put an end to the discussion. The "said station has been constructed at the margins and even in violation of the 1929 Prague Convention," he objected.[113] The League would have to act carefully and decided to "continue ad infinitum the preliminary measures."[114] It retreated silently without communicating its decision to Radio Luxembourg. Thus Radio Luxembourg had no idea that its idea to develop a specific niche role within the IBU and League of Nations collaboration had come to a halt almost the minute the proposal was made, having been boycotted by a tight, well-established and intimate personal network of stakeholders.

Seemingly unaware of these events, the Luxembourg PTT tried at the 1932 Madrid and 1933 Luzern conferences to turn its *de facto* claim for a long wave

---

110 "Letter by Montenach to Comert," registry file 5B, box R-2259, file 38270/38270, LoN; "Lettre de J.D. de Montenach à Etienne," 6 sept. 1932, registry file 5B, box R-2259, file 38270/38270, LoN.

111 Christopher Stray, ed., *Gilbert Murray Reassessed: Hellenism, Theatre, and International Politics* (Oxford: Oxford University Press, 2007), 297, 299.

112 Belgium, the Netherlands, Switzerland, Austria, Denmark, and Great Britain by law forbade broadcasting for advertizing purposes. "Letter from Burrows to Murray," 17 Sept. 1932, 1-2, registry file 5B, box R-2259, file 38270/38270, LoN.

113 "Letter from Van Dissel to J.D. de Montenach," registry file 5B, box R-2259, file 38270/38270, 6.X.32, LoN.

114 "de continuer ad infinitum ces mésures préliminaires." "Letter from International Co-Operation section to Van Dissel," 19 Oct. 1932, registry file 5B, box R-2259, file 38270/38270, LoN.

frequency into an officially accepted frequency for Luxembourg. The PTT explained that a long wave would suit their plan to build an international economic station. Such a station would have a special role in the European broadcasting community, serving the League's economic and financial aims. In addition to a political side there was also an economic side to internationalism. Radio Luxembourg wanted to provide this new kind of service complementary to, though without harming, the existing situation coordinated by the IBU, ITU, and national PTT administrations. Perhaps to show its sincerity, the PTT suggested that the organization of the station's programs could "be submitted to an international committee composed of the most qualified representatives from each country and that in keeping with the national broadcasting organizations."[115] This proposal meant the station was offering to subject its entire program schedule to an external international committee consisting of program experts and national broadcasting organizations. Radio Luxembourg thus wanted to create a Europe-wide supported and partially pan-national program schedule, constructing European economic culture by mutual consent.

The Luxembourg PTT argued that the station would not only cross national borders by choice with its material network, but would also respond to a dire need for a station that surpassed the national in its contents. Up until then national cultural products dominated broadcasting without having too much interest in the cultures of other countries. In particular this attitude existed amongst stations in the larger countries. PTT Administrator Jaaques argued that "maybe my small country could, without injuring other stations, serve as a conduit to relay the principal productions of the genius of the large neighboring countries."[116] Radio Luxembourg would fulfill an internationally oriented and impartial (from national interests) role. Its programs would feature pan-national economic broadcasts as well as programs promoting Europe's diverse national cultures. Such a vision of Europe would only be a partially radical departure from the IBU's efforts to create a Europe of bordering nation-states via its technical and public standards.

Radio Luxembourg strengthened its plea for full collaboration with the European community for broadcasters and its immediate stakeholders by referring to the practical cost-benefit advantages. The station was located close to the

115 "l'organisation des programmes sera soumise à une commission internationale composée des représentants les plus qualifiés de chaque pays, et cela en accord avec les organismes nationaux de radiodiffusion." BUIT, *Documents de Lucerne*, 123.

116 "Peut-être, mon petit pays pourra-t-il, sans nuire aux autres postes, servir de relais pour répandre les principales productions des génies de grands pays voisins." BUIT, *Documents de la Conférence Radiotélégraphique Internationale de Madrid (1932): Tome II: Travaux de la Conférence (Délibérations, Actes définitifs)* (Berne: BUIT, 1932), 862.

IBU Checking Center in Brussels. Therefore it could easily be employed to diffuse the ideological aims of the IBU and the European community of broadcasters at large. Moreover, from a technical point of view the station "could provide a crucial case study as part of the studies and experiments carried out by the international control center in the common interest of European listeners."[117] The station placed itself at the disposal of the Checking Center to be used for the IBU technical studies. These propositions suggest a technical and ideological participation of Radio Luxembourg from within the European IBU broadcasting community, directly matching the station's proposal to put its microphone at the disposal of the League of Nations.

The Madrid and Luzern conferences rejected Luxembourg PTT's proposal and refused to accept the *de facto* long wave frequency claimed by Radio Luxembourg. In particular the IBU and individual countries that pursued a policy of high culture and educative broadcasting objected strongly. Though the League did not interfere in the debate it silently supported the objections stating that it had not yet collaborated with Radio Luxembourg and had "simply taken note of the offer of collaboration made by the Luxembourg Administration, as it had taken note of other offers presented to the section by other organizations."[118] The Czech government supported the objections made by the British. One of the Czech delegates remarked that the broadcasting community in recent years had developed standards that substantially reduced chaos in the airwaves. What use would international regulations have if organizations did not adhere to them, he pondered. Immediate suppression of such behavior should prevent any future continuation of such irregularities.[119] They boycotted Radio Luxembourg on the basis of standards they had created in previous years. They boycotted the station seemingly without giving a second thought to the idea of a specific niche station in their network.

In the following years the experts worked around an unfortunate situation. Between 1933 and 1935, commercial broadcasting in France changed. Its focus shifted from an increase in sales of sets to the idea that the act of broadcasting could become a vehicle for the selling of airtime.[120] The shift in thinking evolved

117 "puisse constituer un élément de premier ordre pour les études et les expériences assurées, dans l'intérêt commun des auditeurs européens, par ce centre international de contrôle." BUIT, *Documents de Lucerne*, 123.

118 "...simplement pris acte de l'offre de collaboration faite par l'Administration luxembourgeoise, ainsi qu'elle a pris acte des autres offres présentées à la section part d'autres organismes." Ibid., 198.

119 BUIT, *Documents de Madrid Tome II*, 862-863.

120 Marc Martin, "Publicité et programmes radiophoniques dans la France d'avant-guerre," in *Histoire des programmes, histoire des jeux à la radio et à la télévision*, ed. Michèle de Bussierre, Caroline Mauriat, and Cécile Méadel (Paris: GEHRA, 1986), 121-123; Spohrer, "Ruling the Airwaves," 86. In the United States there

around the same time as Radio Luxembourg was rejected in Luzern. Initially Radio Luxembourg did not want to give up its idea to contribute to the League's cause. It created so-called national soirées that concentrated on one European country every evening of the week. Within a year, the station had already rescheduled the soirées to daytime hours when it did not manage to attract sufficient advertising revenues from all European countries.[121] In the meantime the station followed the French way of viewing commercial broadcasting. The station began selling airtime, that is, specific time slots, to advertisers who among other things could sponsor a program. Soon Radio Luxembourg would mostly produce sponsored advertizing and low-culture broadcast entertainment.[122] Its vision of becoming a specific niche for international economic affairs vanished.

Since Luxembourg is such a small country the station decided to proceed with its idea of becoming an all-European station. Not having been promised a long wave frequency in Madrid, the station used a long wave frequency which, to the annoyance of the Danish delegation, caused interference with the Danish station Kalundborg's officially assigned long wave of 1154m (260 kc/sec).[123] Moreover, when 24 hours after the 1933 Luzern Plan went into effect the Russian station in Minsk had not commenced broadcasting over its newly assigned long wave, Radio Luxembourg occupied this frequency instead. Usage of this frequency caused direct interference also with Danish stations. In the following year Radio Luxembourg would change frequencies at regular intervals leading to protests to the ITU by Denmark as well as the UK, and later by the Netherlands, Switzerland, Iceland, Italy, and Austria. An increasingly unworkable situation made the IBU decide to intervene by December 1934. The IBU arranged a frequency that Radio Luxembourg could use without too much disturbance. Even though the Danish station Kalundborg continued to dispute Radio Luxembourg, the station would continue to broadcast on this frequency of 232 kc/sec arranged by the IBU. This usage would be tolerated by the European community of broadcasters for the remainder of the 1930s.[124]

The reluctant toleration might seem odd given the initial harsh negotiations to exclude Radio Luxembourg from the European broadcasting community. Whereas

---

happened a similar shift although that preceded developments in Europe by a couple of years (see chapter 2).

121 Spohrer, "Ruling the Airwaves," 186-192, 219-220.

122 For an extensive elaboration on the programming activities of Radio Luxembourg see: chapter 3 of Spohrer, "Ruling the Airwaves."

123 This frequency fell within the band of 1250-1132m (240-265 kc/sec) which broadcasting had to share with other telecommunication services according to the Madrid Regulations. "Memorandum concernant Radio – Luxembourg, Annexe au Procès-verbaux de la 1ère séance de la Commission Technique," Série 4841, 17 fév. 1935, 1, registry file 9G, R-4321, file 15874/663, LoN.

124 "Memorandum concernant Radio – Luxembourg," Série 4841, 2-3, registry file 9G, R-4321, file 15874/663, LoN; Spohrer, "Ruling the Airwaves," 134.

on the one hand a desire for efficiency in the European airwaves formed the backbone for the final decision, intervention via personal networks played an important role as well. In 1933 IBU Vice President and proponent of Radio Luxembourg from its earliest days, Robert Tabouis, had left the IBU to join the executive board of the Luxembourg Broadcasting Company. Tabouis requested to speak before the IBU Council on behalf of the new station.[125] In the eyes of the IBU, Tabouis and his colleagues were "men of position, antecedents and high cultural standing."[126] Tabouis urged the IBU Council kindly but forcefully to bear Luxembourg in mind when making amendments to the Luzern Plan in the future. If "the IBU does not manage to resolve the problem, Radio Luxembourg will seek to conclude agreements between Administrations." The station would then conclude plural bilateral agreements and bypass the IBU.[127] The IBU took the "request" seriously. The IBU would tolerate the station's usage of a long wave outside the wave bands assigned to broadcasting in the 1932 Madrid convention.

Around 1936 Radio Luxembourg explored its options to turn its European network into a global one. The station thus made a deliberate connection between global broadcasting and commercial activities. By then, the international broadcasting community had already started exploring a global broadcasting space. The Luxembourg PTT, on the advice of its government, contacted the ITU requesting information about the technicalities of the construction of a short wave station. The ITU immediately referred the PTT to the IBU for the organization's expertise on the issue at hand.[128] Even though the IBU studied the technicalities of the short waves together with stations such as Radio Moscow and Radio Nations, the IBU refused to collaborate with Luxembourg. The IBU conservatively argued that it was not officially authorized to deal with short wave regulation. Moreover, this concerned broadcasting to areas beyond the borders of the European region, to areas beyond the influence of the IBU.[129] Luxembourg got stuck in a vicious circle in which the ITU and IBU continuously referred to one another on the basis of existing conventions and recommendations that the Luxembourg government had never accepted. Such

125 Ibid., 278; P.V. Conseil Administrative 1933, 164, IBU.

126 Richard Nichols, *Radio Luxembourg, the Station of the Stars: An Affectionate History of 50 Years of Broadcasting* (London: W.H. Allen, 1983).

127 "l'U.I.R. ne réussit pas à résoudre le problème, Radio Luxembourg essaiera de procéder à des ententes entre Administrations," P.V. Conseil Administrative 1933, 165, IBU.

128 BUIT, *Documents de Lucerne*, 123. "Demande de l'Administration Luxembourgeoise des P.T.T. en vue d'obtenir des renseignements sur les caractéristiques techniques d'une station de radiodiffusion à ondes courtes de rayon mondial (Rapporteur: le Secrétaire Général)," Série 6333, 26 janv. 1937, 1-3, CA Document Series, IBU.

129 "Demande de l'Administration Luxembourgeoise des P.T.T.," Série 6333, 3, CA Document Series IBU; P.V. Conseil Administrative 1937, 88, IBU.

behavior at least suggests that the ITU and the IBU boycotted Radio Luxembourg together.

In its attempt to prevent Radio Luxembourg from "going global" the European broadcasting community dealing with the material side of the medium suddenly used "Europe" (the European Region as registered in the international ITU Conventions) as a means of refusing data provision and collaboration regarding worldwide broadcasting. Thus the European Region became an excuse to hinder the establishment of a radio station with global reach on European soil. Since neither Radio Nations nor Vatican Radio, which also broadcast from within this European region, had ever experienced hindrance of this kind, the move can be considered as a deliberate attempt to exclude Radio Luxembourg.[130] Besides, the community participated with Radio Moscow on issues of global broadcasting even though European political society feared Russian socialist ideology. The entire broadcasting community dealing with the material side refused to support a station that wanted to link global broadcasting to commerce. The Grand Duchy of Luxembourg did not manage to realize its global aspirations in the interwar years.

Luxembourg continued to insist on building a veritable European network. On behalf of the government the Luxembourg PTT stressed the station's ability to "fill a very regrettable lacuna in European broadcasting, since no other broadcasting station could claim to be truly European."[131] Materially, Radio Luxembourg realized a European network with its long wave station that had a radius of at least 480 km covering the European continent and the British Isles. Nevertheless, the station would fall short of building the continental audience it promised its advertisers. Both its programming and its advertisements became increasingly west European, especially British and French, as well as urban oriented. The station failed to attract large audiences and advertisers for countries like Czechoslovakia and Italy, whereas programs for Luxembourg, Dutch, Flemish, and Swiss listeners continued in "abbreviated form and outside of prime-time hours."[132] Similar to its vision of a global station, Radio Luxembourg over time increasingly failed to realize its vision of an all-European station program-wise as well.

In the meantime, Radio Luxembourg's plans indirectly forced the IBU to redefine itself. Internally, IBU representatives clashed due to their different personal or national stakes in Radio Luxembourg. The IBU had to re-formulate its own scope and activities to ensure the support of all its members and representatives. Over the

130 P.V. Conseil Administrative 1937, 197-198, IBU.

131 Spohrer, "Ruling the Airwaves," 147; BUIT, *Documents de Lucerne*, 122.

132 Advertisers became big international enterprises which mostly targeted the British market. Spohrer, "Ruling the Airwaves," 193, 219-229.

years, the IBU projected itself more explicitly as a European organization aiming to create high culture educational standards for broadcasting. Its main objective, for national needs and acceptance of membership based on concessions granted by states, even caused statutory changes. The IBU nonetheless remained an expert in technology driven organization, reluctantly tolerating Radio Luxembourg on the long wave band to keep an efficient wireless broadcasting network in Europe.

Like the IBU, Radio Luxembourg had to regularly redefine its ideas about the kind of broadcasting network it wanted to create. Many of its ideas failed, being brutally rejected by the European community of the IBU, the ITU and League experts, including closely involved stakeholders. Although the expert community hid behind agreed standards and juridical arguments, sentiments of *fear* were probably an important reason for rejection. Before the establishment of Radio Luxembourg, commercial activities had already defined international ideas about the "do-nots" of commercial broadcasting whatever "commercial" meant. So why would the broadcasting community go to all the effort of rejecting a station that wanted to complement its ideological activities by becoming an economic niche rather than oppose the ideology of its community? And why would the same community actually tolerate the station broadcasting over the long waves after the station had let go of these ideals of peace and rapprochement and become an ordinary low-culture internationally broadcasting entertainment station? In spite of these rejections, Radio Luxembourg would have an enormous impact in Europe, with people listening in from all corners of the continent. Listeners had finally found a station that offered entertainment and relaxation.

Conclusion

Radio Moscow, Vatican Radio, Radio Nations, and Radio Luxembourg developed their systems based on their own ideas and interests. With the exception of Radio Luxembourg, which was the sole station that explicitly claimed to be European, these stations mostly had global plans for broadcasting. Radio Moscow for instance was initially not interested in participating in European standardization efforts. It changed its mind only when "Europe" could positively serve its new politico-economic strategy of collaboration with the West. In spite of most of these global outlooks, the notion of Europe did often enter these projects. Europe usually became important due to the need for system fine-tuning to prevent interference, and thus as a derivative of practical problems.

With the exception of Radio Moscow, most of these stations were constructed by the time the IBU and ITU frequencies were already well on their way to becoming the European standard. Whereas these standards envisioned a Europe of national wireless broadcasting networks, thus a Europe of nation-states, the plans of the promoters of the new stations did not always fit into this concept. These stations found out that they could not ignore the ITU and IBU's increasingly dominant standards. Nor could the IBU and ITU ignore these new stations either. They had to engage in a process of fine-tuning in order to save their standards.

Therefore the various parties had to renegotiate how they defined building a European network of national broadcasting stations. Fine-tuning standards with the structure of Radio Moscow, for instance, would redefine the boundaries of Europe in the east. Furthermore, Radio Nations and Radio Luxembourg threatened from within the very idea of a Europe of nation-states. They either sought a pan-national approach organizationally, as in the case of the intergovernmental Radio Nations, or materially, as in the case of Radio Luxembourg.

Fine-tuning happened via diffuse personal networks. Informal contacts between men like Burrows and Van Dissel for instance, affected the fine-tuning process between Radio Nations and the European standards defined by the IBU and ITU. Their relationship also played a decisive role in the deliberate boycott of Radio Luxembourg as a possible niche station in the IBU and the League of Nations' broadcasting efforts. The personal involvement of Raymond Braillard and Robert Tabouis with Radio Luxembourg played a dominant role in the ultimate creation of the station, including its unofficial tolerance in the European long wave band by the mid-1930s. At times, official routes would fall short or prolong procedures. The informal and personal networks then served as a route for immediate mediation or outright defense of technology related standards, including efforts of a more ideological nature.

The history of the fine-tuning process shows that this could be achieved without a shared ideology. The case of Radio Moscow shows that the station collaborated well with the IBU community as long as they managed to leave ideology out of their discussions. The case of Radio Nations reveals that a shared ideology would not be sufficient to make the station an IBU member in the end. Ideology did not seem to affect fine-tuning activities as long as international circumstances proved favorable for collaboration. Nonetheless, the medium of broadcasting was a culturally and politically sensitive instrument. In a situation when international relations deteriorated, the medium could easily become a bone of contention between nation-states. Ideological differences then re-entered the scene and could make effective fine-tuning of standards impossible. The following chapter focuses on the use of broadcasting for improving international relations.

Chapter 5
War and Peace in the Sky

Heute gehört uns Deutschland. Und morgen die ganze Welt.[1]

Thus reads a line from the well-known Nazi Party anthem. These anthems propagated a specific sense of belonging to the people of Germany. At the same time they communicated a message of German domination to all peoples outside Germany's borders. Hitler's Nazi Party found that this and similar propaganda activities were crucial for diffusing its message among the masses. Hitler's *Reichsminister für Volksaufklärung und Propaganda* Joseph Goebbels became the symbol of the propaganda machine that Nazi Germany established in the course of the 1930s. Already in 1933 he stated that broadcasting was the most influential and important intermediary between ideological movements and the nation-state as well as between ideology and people.[2] Broadcasting propaganda was a weapon that could effectively implement the ideas of the Nazi Party into the minds of the masses.[3]

Goebbels related infrastructure construction directly to program development, propaganda broadcasting in particular. He argued that effective and long-lasting propaganda should be accomplished by means of an efficient system that reached beyond the borders of Germany. Only a system that had a clear direction could create a sense of belonging.[4] Goebbels defined a kind of propaganda radically different from that of the past, referring to providing people with information, with facts. "Keep it simple," was his motto. "It is not a matter of doing the right thing; the people must understand that the right thing is the right thing. Propaganda includes everything that helps the people to realize this."[5] In his eyes, propaganda was not truth, but the act of communicating that the truth was the truth. This also

1 "Today Germany belongs to us. And tomorrow the whole world." "Es zittern die morschen Knocken" by Hans Baumann can be translated into "The frail bones are trembling." This was the official song of the Hitler Youth.

2 Joseph Goebbels, "Der Rundfunk als achte Groβmacht," in *Signale der neuen Zeit. 25 ausgewählte Reden von Dr. Joseph Goebbels* (Munich: Zentralverlag der NSDAP, 1938), 197-207.

3 Goebbels, "Der Rundfunk als achte Groβmacht," 197-207.

4 Joseph Goebbels, *Der Kongress zur Nüremberg 1934* (Munich: Zentralverlag der NSDAP; Franz Eher Nachfolger GmbH, 1934).

5 Goebbels, *Der Kongress zur Nüremberg 1934.*

Figure 5.1 – The Grand Hall of the Reichs Rundfunk Gesellschaft m.b.H. in 1936
Source: Box 58 Photographies, IBU. Used by the courtesy of the European Broadcasting Union.

applied to *political* propaganda, meaning state activities intended to create a feeling of belonging to a nation.[6] A broadcasting network that reached beyond the nation in combination with effective political propaganda broadcasts allowed the state to plant its ideals into minds at home and abroad.

In German propaganda activities, like the above Nazi anthem, *Deutschland* could easily have been replaced by *Europa*.[7] According to internationally active Norwegian broadcasting expert Arnold Raested, the Nazi Party discursively employed "Europe" as a stepping stone towards world domination. He argued that the Party could be linked directly with Napoleonic thinking, which claimed that "the master of Europe was the master of the universe."[8] Germany was (to be) the dominant leader in Europe, and therefore the one righteous master of the world. Such a German Europe would first of all have to combat Russian Bolshevism as a united front. One the one hand, this would serve Germany's *Drang nach Osten*. On the other, it would also serve the National Socialism ideals that did not tolerate

6 Goebbels, *Der Kongress zur Nüremberg*.
7 Arnold Ræsted, *Europe and the Atlantic World* (Oslo: I kommisjon hos Aschehoug, 1958), 43.
8 Ræsted, *Europe and the Atlantic World* , 42-43.

transnational communities like a global socialist society. Moreover, domestically it would have to disseminate the idea of *Blut und Boden nationalismus*, of a superior German Aryan race.[9] In Raested's view, it was clear that the propagation, via cross-border broadcasting, of the National-Socialist idea of Europe led by a Germany consisting of one pure race, had a substantial effect.

This chapter looks at the role of broadcasting in the light of interwar debates about war and peace. As the example of Goebbels' propaganda activities shows, broadcasting programs could be filled with any kind of meaning and therefore with all kinds of visions of "Europe."[10] Furthermore, Goebbels linked system construction to program construction, highlighting the spatiality of propaganda broadcasting. His contemporaries considered broadcasting a powerful instrument for shaping society. It could easily influence international developments, encouraging war or strengthening peace as desired. Consequently, decisions about the form and shape of propaganda broadcasts in the context of war and peace debates became crucial for broadcasting communities. The debate became the place where broadcasting communities envisioned and negotiated the formation of their systems and the direction of society at large.

This chapter concentrates on people's different notions regarding the potential roles for broadcasting as a society-building instrument. Immediately after World War I, a large number of people projected the reconstruction of European society. Above all the intellectual elites rejected the idea of employing technology as a means to such an end. In the meantime, the enlightenment-based ideals of the International Broadcasting Union (IBU) community clashed with propaganda efforts pursuing goals of international segregation or national unity at the cost of international peace. The IBU community sought to prevent "misuse" and push the "righteous" use of its medium in an international atmosphere of reluctance. These mixed feelings disappeared into the background when international relations improved by the late 1920s. The loosely coupled IBU agreements for propaganda broadcasting which had been established by then, would find their way into international government supported agreements in the first half of the 1930s. Rising nationalism by the mid-1930s however, caused international relations to quickly deteriorate, and pinned broadcasting to warfare. These changing relations impacted on the kind of society which was envisioned and constructed, in which Europe would only have a marginal or indirect role.

9 Peter Bugge, "The Nation Supreme: The Idea of Europe 1914-1945," in *The History of the Idea of Europe*, ed. Kevin Wilson and Jan van der Dussen (London: Routledge, 1993), 107-110.

10 Alexander Badenoch, *Voices in Ruins: West German Radio across the 1945 Divide* (Basingstoke: Palgrave Macmillan, 2008), 224.

Elites Doubting a Technology's Virtue

Elites doubted the virtue of broadcasting. Nazi Germany was not the first country to actively relate the diffusion of ideology via broadcasting to the creation of segregation, superiority, and even incitement to warfare. It was merely the first country that built a deliberately effective apparatus to achieve this. Broadcasting developed and crossed borders in a world and especially on a continent that had served as the main battleground for World War I and where people still felt anger, hate, sadness, and loss. Although people still felt the war intensely, they wanted to create peace and restore their daily lives. The political balance of power in Europe had shifted. The world created the League of Nations, and the political elites redrew many national boundaries in Europe, especially in the east. It was in this vulnerable and tense international atmosphere that stakeholders like governments, intellectual elites and broadcasters developed the first international standards in propaganda broadcasting. These stakeholders had different opinions about the role, if any, that broadcasting should have in the world.

Whereas the USSR immediately recognized the unifying prospects of broadcasting, other European governments and intellectual elites regarded the medium with reluctance. In the eyes of the Russian leader, Lenin, broadcasting was an instrument more than capable of creating global peace and democracy along socialist lines, while totally ignoring Europe and the nation-state (Chapter 4).[11] These Russian ideas contrasted sharply with the opinion of the majority of European stakeholders who tried to restore peace by designing special "projects" for Europe.[12]

Elites consciously related these projects to European civilization in a variety of ways. They felt this civilization should not be based on modernity and technological innovation. In fact, most elites rather blamed the destructive forces of modernity explained as materialism, technology, and mass society for the crisis in European civilization. In their eyes, technology in particular had characteristics of low, non culture.[13] Radio had even incited war when nationalists used the medium for their own ends. Radio technology should not be trusted as a society-building instrument.

Although many of these elites had zealously supported rising nationalism

---

11 Michael Nelson, *War of the Black Heavens: The Battles of Western Broadcasting in the Cold War* (London; Washington: Brassey's, 1997), 1; S.V. Kaftanov and Soviet Union. Gosudarstvennyi Komitet po radioveshchan i iu i televiden i iu, *Radio and Television in the USSR* (Washington: U.S. Joint Publications Research Service, 1961), 22.

12 For an extensive discussion of such European projects see: Kevin Wilson and Jan van der Dussen, eds., *The History of the Idea of Europe* (London: Routledge, 1993).

13 Jan Ifversen, "The Crisis of European Civilization after 1918," in *Ideas of Europe since 1914: The Legacy of the First World War*, ed. Menno Spiering and Michael Wintle (New York: Palgrave, 2002), 17; Count Richard Coudenhoven-Kalergi, *Paneuropa* (Wien: Paneuropa-Verlag, 1926), 16.

during the second half of the nineteenth century, after the war they blamed it for the outbreak of war. They blamed nationalism for the demise of European civilization. At the same time however, they had always rooted their nationalism deeply in a common awareness of a shared European destiny. Elites did not so much think that rising nationalism had destroyed European civilization, but rather destroyed Europe's sense of supremacy and confidence among other things.[14] In response, the international elites desired to rebuild Europe along similar lines as prior to World War I. This time they chose to appropriate their projects in order to avoid any repeat of war in the future.[15] Neither nineteenth century inspired ardent nationalism nor radio technology could find a place in these projects. It was in this atmosphere of reluctant politicians and elites that the new IBU took its first steps towards creating a cross-border broadcasting space in 1925.

Each in their own way, the USSR and the IBU argued that their medium could send out messages of unity and peace. Whereas the Russian government immediately adopted broadcasting, the IBU headed off as a private body, trying to construct a European broadcasting system without substantial government support. It was aware of the widespread reluctance with regard to its medium and the connection that elites made between radio, war, and nationalist sentiments. Meanwhile, the IBU also felt the threat of the Bolsheviks, who after the war "called for a class war in every country."[16] The Russian rejection of the idea of nation states as well as the idea of Europe as a site for building systems directly opposed the IBU's activities to create a European broadcasting system connecting nation-states. These attitudes hindered IBU members' pursuit of their tasks. With a reluctant government and elite community, they could only depend on their own collaborative efforts if they wanted to create efficient international broadcasting.

In 1925, the IBU devised a Gentleman's Agreement. Already at its first meeting, the IBU had discovered that there "was the need for some international method of dealing with what is known in French as *propagande inadmissible*."[17] Day to day broadcasting of illicit propaganda impeded IBU's efforts to create an efficient system.

14 Bo Stråth, "Introduction: Europe as a Discourse," in *Europe and the Other and Europe as the Other*, ed. Bo Stråth (Brussels; New York: P.I.E.-P. Lang, 2000), 15; Pim den Boer, "Europe to 1914: The Making of an Idea," in *The History of the Idea of Europe*, ed. Kevin Wilson and Jan van der Dussen (London: Routledge, 1993), 77.

15 Akira Iriye, *Cultural Internationalism and World Order* (Baltimore: Johns Hopkins University Press, 1997), 60, 70-78.

16 Bugge, "Essay 2: The Nation Supreme," 87.

17 "Address (made by Secretary-General A.R. Burrows) before the Conference on "Moral Disarmament" arranged by the International Students' Associations, Geneva, April 1st 1933," 3-4, box 74, file Propagande de l'Union ou des Membres, Général II, 1932/1936, IBU.

International litigation on controversial broadcasts could even impede collaboration between the broadcasters themselves. According to the IBU, illicit propaganda forwarded through broadcasting stations and wireless amateurs alike included all "... political, religious, economic, intellectual, or artistic matter liable to injure that spirit of international co-operation and understanding..."[18] Illicit or negative propaganda in the eyes of the IBU above all emanated in the form of pressing nationalistic propaganda by newly established states or the Russian Bolsheviks who aroused people to join the Socialist Revolution.[19] Using its Gentleman's Agreement as strategic tool, the IBU positioned itself politically and ideologically among the international community of politicians and international elites.

In fact, the Union did not choose the term "Gentleman's Agreement" arbitrarily. The 1926 Agreement stated that "all existing members of the *Union Internationale de Radiophonie* have agreed to assist mutually in promoting the ideal of international fellowship, by exercising the strictest possible control in the direction of avoiding the broadcasting of material calculated to give offence in other countries."[20] Since the IBU was a non-governmental organization, it could only make non-binding recommendations. The IBU's idea of society-building, civilization, and peaceful co-existence shared characteristics with what people generally understood as being a "gentleman." The idea of "civilization" implied a dubious correlation between mannerly conduct and ethical behavior. According to Terry Eagleton, the term gentleman similarly "equates manners and morals, and holds a cult of secular, progressive development."

> To be civilized or cultivated is to be blessed with refined feelings, well-tempered passions, agreeable manners and an open mind. It is to behave reasonably and moderately, with an innate sensitivity to others' interests, to exercise self-discipline, and to be prepared to sacrifice one's own selfish interests to the good of the whole.[21]

Because of the non-governmental character of the IBU, its members would have to behave in a gentlemanlike way. Other kinds of behavior would harm the success of its recommendations. The effectiveness of its activities was considerably improved

---

18 LoN, *Commission Consultative et Technique des Communications et du Transit, Procès-Verbal de la 9ième session, tenue à Genève, 12 au 17 juillet 1926,* LoN doc. ser., C.595.M.231.1926.VIII (Geneva: LoN, 1926), 28-30, box 92, file Société des Nations, Section des Communications et du Transit, Général, 1927-1930, IBU.

19 LoN doc. ser., C.595.M.231.1926.VIII, 28-30.

20 IBU, *International Union of Broadcasting Organisations – Union Internationale de Radiophonie* (Geneva: IBU, 1926), 15; IBU, *Twenty Years of Activity of the International Broadcasting Union* (Geneva: IBU, 1945), 28-29.

21 Terry Eagleton, *The Idea of Culture* (Oxford: Blackwell Publishing, 2000), 9, 18.

thanks to "the fact that delegates to the Union's meetings were always housed in the same hotel; they breakfasted, lunched and dined in the same restaurant and mixed together in the lounges in the evenings," as Secretary-General Arthur Burrows described. These delegates worked "in the same medium in their respective countries and developed common friendships. They became a little, unofficial, but very practical League of Nations."[22] The Gentleman's Agreement had the characteristics of rapprochement, cooperation, respect, and civilized behavior. This approach was the most effective way to achieve an international broadcasting network.

Figure 5.2 – A view to "Gentlemanness"
In the middle: Raymond Braillard, Director of the IBU Technical Committee and Brussels Checking Center with his colleagues at Ouchy, 1941.
Source: Box 58 Photographies, IBU. Used by the courtesy of the European Broadcasting Union.

22 "Address (made by Secretary-General A.R. Burrows)," 3-4, box 74, file Propagande, Général II, IBU.

The Gentleman's Agreement would above all focus on illicit propaganda problems resulting from zealous nationalism. As the war had redefined many national boundaries especially in Eastern Europe, their governments tried to create a sense of national belonging. Broadcasts from stations in border regions particularly hampered international relations. In these cases the agreements would have to include a code of conduct for civilized behavior. A few months after the signing of the Gentleman's Agreement, the Hungarian broadcasting station transmitted illicit propaganda, thus hindering Czechoslovakia. IBU Secretary-General Burrows volunteered to mediate.[23] The Czech delegate pointed out that the incident was part of a larger problem in central Europe. He urged the IBU to step up its efforts. In his eyes, transgressors should be sanctioned. A station that offended another station and continued to do so after repeated warnings should lose the protection of the Union if the offended country wished reprisal. The Belgian delegate considered the proposal rigorous and dangerous. He suggested contacting the League of Nations about adopting a resolution on propaganda broadcasting. This should in turn be passed on to the various governments. His idea found approval.[24]

In the meantime, the Gentlemen's Agreement continued. In August 1927, Burrows sent a confidential circular to all IBU members stating: "One of the best results, from a moral point of view, obtained by a mutual entente between the members of the International Broadcasting Union, has been the suppression of all transmissions of an offensive nature to another country."[25] Apart from the issues in the first half of 1926, the Gentlemen's Agreement had promising results. Program directors should continue their encouraging efforts. According to Burrows, there had been no new cases of illicit propaganda reported to the IBU as a result of ardent nationalism. The question remains why the IBU then felt the need to contact the League of Nations to make the Gentleman's Agreement legally binding and supported by government. This means the problem of illicit nationalistic propaganda still existed, and required firm legislative standards.

With a lack of legislation, the IBU sought to ensure adherence to its Agreement by dealing with the negative side of propaganda. It discursively defined illicit propaganda, the "misusage" of its medium as those efforts that opposed internationalist

23 "Commission juridique, Propagande inadmissible, Ladislav Sourek," box 17, file Désarmement moral et propagande inadmissible, Général I, jusqu'à 1928, IBU.

24 P.V. Administrative Council 1926, 59, IBU.

25 "L'un des meilleurs résultats, au point de vue moral, obtenu par une entente mutuelle entre les membres de l'Union Internationale de Radiophonie, a été la suppression de toute diffusion revêtant un caractère offensant pour un autre pays." "Circulaire confidentielle pour tous les membres de L'Union Internationale de Radiophonie," Série 641, 12 août 1927, box 17, file Désarmement moral et propagande inadmissible, Général I, jusqu'à 1928, IBU.

ideals of peace and rapprochement. The Agreement was a clear case of problem solving, dealing with negative experiences. Nonetheless, the IBU had nothing against nation-states or nationalism *per sé*. It promoted, however, any avoidance of broadcasting propaganda at the cost of other countries. According to the IBU, national and international interests did not have to be mutually exclusive. They could be pursued together. The Gentleman's Agreement communicated a vision of a very loosely coupled international world of sovereign nation states. In such a world people could cooperate peacefully along civilized lines of mutual trust. The IBU tried to create a firm basis for the national in its world views and in its international recommendations.

Reluctant intellectual elites hindered the efforts. The Gentleman's Agreement was a first attempt to set an informal standard focusing on the negative use of broadcasting. To solidify the Agreement, the IBU needed wide support. In 1926 already it asked its members to urge their governments to take legal and administrative action in line with the Agreement when they granted new broadcasting concessions.[26] Additionally, it also approached the League of Nations in July 1926 to discuss illicit propaganda. The IBU felt hindered by a continuing reluctance towards its medium by politicians and elites in Europe. Just like governments and the elite community, the League did not show any interest in propaganda broadcasting. In order to persuade the League, the IBU not only focused on propaganda problems but above all told the League "what we can do for you."

From day one, the IBU had demonstrated its desire to pursue the aims of the League by deciding to contribute to rapprochement, collaboration, and peace. As intergovernmental organization, the League represented nation states worldwide. At the time of the establishment of the League, people still regarded the young Soviet Russia as a possible model for social change.[27] The League provided an alternative. The IBU broadcasting community set its hopes on the League to create harmony internationally.

The IBU approached the League's Secretary-General, requesting effective and straightforward collaboration with the League's 6th Technical Committee: the International Committee on Intellectual Co-Operation (ICIC). IBU Secretary-General Arthur Burrows would be the main contact for the League on the matter of propaganda broadcasting. At all times he would be supported by the work of the IBU Legal Committee headed by the Czech Ladislav Sourek. Burrows was an

26 LoN doc. ser., C.595.M.231.1926.VIII, 84.
27 Stråth, "Introduction," 28.

expert in propaganda activities and wireless interception.[28] As such, he possessed the required skills to align national and international interests in war and peace related propaganda broadcasting.

Aiming at reducing the negative effects of propaganda broadcasting, the IBU above all advertized broadcasting as a cultural and technical medium beyond comparison, as a positive propaganda instrument. Broadcasting was a medium par excellence to achieve co-operation, mutual understanding, and peace, Burrows told the League's Secretary-General. The IBU proposed a regular exchange of documents with the ICIC, thus creating mutual documentation at the same time. Furthermore, it suggested that broadcasters could put their microphones at the disposal of the national affiliations of the ICIC.[29] In the eyes of the IBU, the ICIC was the place to be in order to gain widespread recognition for the positive nature of its medium. The IBU offered the Committee its full services.

In order to bolster international co-operation, dialogue, and peace, the League had established a number of technical committees. These committees had to resolve practical problems regarding disarmament, economy and finance, transport and communication, as well as intellectual life. In 1922 the League had created the ICIC to deal with the latter. The ICIC pursued quite literally the idea of international synergy via standardization in fields like academia, the arts, and the sciences.[30] The committee aimed to "be of direct service to intellectual life and intellectual workers, either by placing facilities at their disposal or by defending their rights."[31] One of the founding fathers of the ICIC, Henri La Fontaine, once stated that in "the work of promoting international co-operation it is important that the League, as a league of nations rather than as a league of governments, should not be lost sight of."[32]

---

28 Burrows had been employed at the Marconi Company during the war when it had operated as the British wireless interception service. Asa Briggs, *The History of Broadcasting in the United Kingdom, Volume I: The Birth of Broadcasting* (Oxford [u.a.]: Oxford University Press, 1961), 34, 73.

29 P.V. Conseil Administrative 1926, 59, IBU.

30 Until the present the scrutinization of the ICIC has been highly neglected. The one profound and well-founded study on the establishment, organization, and work of the ICIC is: Jean-Jaques Renoliet, *L'Unesco oubliée: La Société des nations et la coopération intellectuelle, 1919-1946* (Paris: Publications de la Sorbonne, 1999). Patricia Clavin only briefly touches upon the ICIC in: Patricia Clavin, "Europe and the League of Nations," in *Twisted Paths: Europe 1914-1945*, ed. Robert Gerwarth (Oxford: Oxford University Press, 2007), 325-354. Greaves describes its work in broad lines in: Harold R.G. Greaves, *The League Committees and World Order: A Study of the Permanent Expert Committees of the League of Nations as an Instrument of International Government* (London: Oxford University Press; H. Milford, 1931).

31 LoN doc. ser., A.23.1931.XII, 5.

32 The International Office of Bibliography in Brussels (1895) and the Union of International Associations (1910) preceded the establishment of the League of Nations. Via these organizations, the two prominent Belgian internationalists Henri La Fontaine and Prof. Paul Otlet had pioneered in promoting "literary and scientific co-operation between peoples." The origins of the League itself are intimately bound up with the ideas of the Union of International Associations, which before the war had already stated that

The ICIC created this international synergy by bringing together under one roof private and public organizations in the field of intellectual life.[33] ICIC experts and sub-committees formed the crème de la crème of the sciences, the humanities, and the arts, with people like Henri Bergson, Béla Bartók, Mme. Curie, Albert Einstein, Edouard Herriot, Salvador de Madariaga, Thomas Mann, and Paul Valéry, to name but a few.[34] Initially, these intellectuals mostly came from Europe. Only in the course of the 1930s would the ICIC reflect a more global representation.[35] Thus in the interwar years, European intellectual life substantially influenced the international standardization of the sciences, the arts, and humanities via the ICIC.

The backgrounds of these eminent intellectuals greatly influenced the ICIC's efforts to achieve intellectual collaboration. Its executive organ in Paris, the International Institute for Intellectual Co-Operation (IIIC), coordinated and stimulated studies in University Relations, Science and Bibliography, Letters and Arts, Intellectual Rights, the Instruction of Youth in the Aims of the League (1928), and the Interchange of Teaching Staff (1928).[36] The intellectuals in the ICIC did not relate broadcasting to their intellectual world or to culture. More importantly, they had not even decided to use music in general for their society-building activities. Broadcasting and music lay beyond their scope.

The ICIC did, however, try to pursue the League's goals in many areas. Both the wide diversity of topics and the inability to define its main spearheads were major obstacles for organizing the committee as a whole. Consequent lack of concrete action disturbed financers who decided to refrain from donating large sums of money.[37] The ICIC even felt that the League did not consider itself a truly official technical committee.[38] When the IBU approached the League in 1926 to collabo-

"'a League of Nations was the ultimate end of all international movements.'" According to Greaves, the "Brussels activities were not merely the precursors of the Committee on Intellectual Co-operation, they were responsible to a not inconsiderable degree for the materialization of the League itself." Greaves, *The League Committees and World Order*, 111-112, 129.

33 The American representatives G. Hale (1922), R. Millikan (1923-1932) and J. Shotwell (1933-1939) were funded by the Rockefeller Foundation and the Carnegie donation. Renoliet, *L'Unesco oubliée*, 184-185, 187, 197.

34 League of Nations, *Ten Years of World Co-Operation* (London: Secretariat of the League of Nations, 1930), 313-314.

35 At its inauguration in 1922 the ICIC had members from India (1922-1923), France, Norway, Brazil, France-Polish, Belgium, Germany, Great Britain, Switzerland, Italy, Spain and the USA, whereas in 1939 the members came from: Great Britain, Switzerland, Rumania, China, the USA, Portugal, France, Poland, Peru, the Netherlands, Hungary, Norway, Czechoslovakia, Egypt, Argentina, Brazil, Latvia and India. Renoliet, *L'Unesco oubliée*, 184-185.

36 Ibid., 187-191; League of Nations, *Ten Years of World Co-Operation*, 318-325.

37 Renoliet, *L'Unesco oubliée*, 184-185.

38 LoN doc. ser., A.23.1931.XII, 5.

rate with the ICIC on the issue of propaganda broadcasting, the Committee had little direction and achieved only very modest results.

In addition to the ICIC's lack of vigor, the question of whether the League should deal with broadcasting content at all, and if so, which technical committee would have to address this, led to internal discussion and delay. Since the Communications and Transit Committee was already dealing with the technical aspect of broadcasting, some time passed while the ICIC and the OCT debated their role. Whereas the OCT saw the technical advantages of broadcasting, the ICIC did not show any interest in the medium with respect to its cultural possibilities.[39] Only in 1928 would the League start to show an active interest in propaganda broadcasting. On the one hand, international relations improved with the signing of a number of international conventions. On the other, the League took its first steps towards employing electronic media for relaying its message.

In 1928 the international climate changed due to the signing of an important international treaty. Stalin had replaced Lenin in Russia. He pursued a seemingly more open tactic of European collaboration. Russia was one of forty-seven countries that in 1928 signed the Kellogg-Briand Pact, formally known as the International Treaty for the Renunciation of War as an Instrument of National Policy. The Pact was also known by its home base Paris. It renounced the use of force for resolving disputes. According to Zara Steiner, this Paris Pact "was symbolic of the coming together of different strands of internationalism."[40] In particular Akira Iriye views the Paris Pact with optimism. He claims that it "should be put in the context of the energetic efforts by men and women everywhere to construct a more interdependent world, to open up people's minds so that they would understand one another a little better…"[41] The Paris Pact signified an era of hope and optimism that the creation of a peaceful world would be possible.

With respect to broadcasting, Iriye's optimism was not misplaced. Around the time of the Paris Pact the League changed its mind about the cultural role of broadcasting. Rumanian League delegate Vespasian Pella argued that the question of illicit propaganda:

39 LoN, *International Committee on Intellectual Co-Operation, Minutes of the Eighth Session, held at Geneva, July 26th-29th, 1926*, LoN doc. ser., C.462.M.181.1926.XII (C.I.C.I/8th Session/P.V.) (Geneva: LoN, 1926), 21; LoN doc. ser., C.595.M.231.1926.VIII, 84; LoN, "Deuxieme Commission, Assemblée 1928, PV provisoire de la 6ème séance, tenue à Genève, 15 sept. 1928," A.II/P.V.6, 2, box 92, file Société des Nations, Section des Communications et du Transit, Général, 1927-1930, IBU.

40 Zara S. Steiner, *The Lights that Failed: European International History, 1919-1933* (Oxford; New York: Oxford University Press, 2005), 572-573.

41 Iriye, *Cultural Internationalism and World Order*, 88.

> is of particular interest, given the tendencies that manifest themselves in all states, pressing on moral foundations of the organization of peace, tendencies that have found their highest confirmation in the recently signed pact in Paris against warfare. Such a pact can have wholesome effects. And in spite of all international forces, it is absolutely necessary to consolidate the will of peace that inspires the nations today and to remove all misunderstandings and frictions that could still divide peoples.[42]

In the eyes of Pella, the Paris Pact reduced international tensions, and facilitated a spirit of working towards a peaceful world. Pella considered it time to act upon these changes. The League should safeguard broadcasting from propaganda. Broadcasting propaganda not only hindered the relations between states, but also hampered well-functioning communications.[43] According to Pella, broadcasting had proven to be one of the most efficient instruments to inform the people, the very basis for pacification between nations.

At the same time, the ICIC submitted a resolution to the League Assembly addressing the issue of illicit propaganda in relation to its new technology-based activity, cinematography. The Italian delegation proposed the establishment of an International Educational Cinematograph Institute in Rome (IECI).[44] The Italian proposal was in response to a discussion that had started outside the ICIC in 1926. During an International Cinematography Congress, participants had discussed the social and educational capabilities of the cinematograph.[45] Linking a technical medium with youth education neatly fitted in with the ICIC's recent creation of two new sub-committees, the Committee on Letters & Arts and the Committee on the Education of Youth in the Aims of the League. When the ICIC questioned the positive nature of the cinematograph, it included broadcasting in the discussion.

---

42 "présente…un intérêt tout à fait particulier, étant donné la tendance qui se manifeste dans tous les Etats de poser sur les assises morales l'organisation de la paix, tendance qui a trouvé sa plus haute consécration dans le Pacte puisse produire ses effets salutaires et à défaut de toute mesure de contrainte internationale, il est absolument nécessaire de consolider la volonté de paix qui anime aujourd'hui les nations et de faire disparaître tous les malentendus et les frictions pouvant encore diviser les peuples." LoN, "Deuxieme Commission, Assemblée 1928," A.II/P.V.6, 2.

43 Pella was professor in Law in Rumania and lectured in various countries including the Netherlands, France and Switzerland. Throughout the interwar years he occupied several positions within the League of Nations for Rumania. He was active in the Council (1936-1938) as well as in numerous expert groups such as *pour la repression du faux monnayage* (1926-1928) and the *Commission permanente du Régime des Eaux du Danube* (1934-1936). His work as the Rumanian delegate and reporter at the diplomatic conference that facilitated the international circulation of educational films (1933) had been of particular interest, as well as the Disarmament Conference 1932-1934. George Ripert, *Hommage au professeur V.V. Pella: Promoteur de l'unification du droit criminel, artisan insigne de l'organisation d'une justice pénale internationala.* (Paris: S.n., 1939), 43-44.

44 Zoë Druick, "The International Educational Cinematograhp Insitute, Reactionary Modernism, and the Formation of Film Studies," *Canadian Journal of Film Studies* 16, no. 1 (2007): 83-91.

45 League of Nations, *Ten Years of World Co-Operation*, 326.

The ICIC pointed out "the dangers of the cinematographic spectacles and the radiotelephony transmissions of a spirit *opposed* to that of the League of Nations. [italics – s.l.]"[46] The League finally acknowledged the problem of illicit propaganda. Improper use of broadcasting or the cinematograph went against the spirit of the League of Nations.

Shortly after its establishment, the IECI requested the ICIC to "extend its activities to questions related to broadcasting in the service of educational cinematography, to television, and to the recording of speech and sounds."[47] When the ICIC acknowledged the request, the IBU sought to secure its expertise in the area. Together with the BBC, the IBU asked to meet in person the head of the ICIC and supporter of broadcasting, Professor Gilbert Murray. Murray was a member of the British social elite, a professor of Ancient Greek and proponent of the restoration of civilization.[48] He fervently pursued his internationalism via the press and the cinema as well as broadcasting.[49] The IBU pleaded for an active role for broadcasting in the work of the ICIC.[50] To press for a breakthrough, the IBU also contacted the Secretary-General of the League, Albert Dufour-Mercier.[51] Although the League refused IBU a consultative voice in the Committee, the IECI soon thereafter concluded that almost all examined activities, including broadcasting, fell completely outside its scope.[52] The IBU thus managed to maintain its position as

46 "les dangers des spectacles cinématographiques et des émissions radiotelephoniques d'un esprit contraire à celui de la Société des Nations." LoN, *Travaux de la Commission et de l'Institut International de la Coopération Intellectuelle, Rapport de la sixième commission à l'Assemblée, Rapporteur: M.G. Gallavresi (délégué de l'Italie)*, LoN doc. ser., A.74.1928.XII (Genève: LoN, 1928), 3, box 92, file Société des Nations, Section de Coopération Intellectuelle, Général 1927, 1927-1932, IBU.

47 LoN, C.I.C.I./214, Geneva, July 20th 1929, 1; LoN, *International Committee on Intellectual Co-Operation, Minutes of the Eleventh Session, held at Geneva, July 22nd-26th, 1929*, LoN doc. ser., C.342.M.121.1929.XII (Geneva: LoN, 1929), 63.

48 Christopher Stray, ed., *Gilbert Murray Reassessed: Hellenism, Theatre, and International Politics* (Oxford: Oxford University Press, 2007), 233-234.

49 Already in 1926 he found that: "As one listened, and felt that some millions of one's fellow citizens, whatever their opinions, were listening at the same time and getting the same knowledge, one felt comfort in the thought of that living voice which could pass from end to end of the nation and amid the noises of discord make us one." Stray, *Gilbert Murray Reassessed*, 299. Quote from: "Broadcasting the News," *Radio Times*, 28 May 1926, 350.

50 P.V. Conseil Administrative 1929, 190, 202, IBU.

51 "Lettre d'Albert Dufour Mercier à Burrows," Genève 17 juillet 1929; "Lettre de Burrows à Albert Dufour Mercier," Genève 27 juillet 1929, box 92, file Société des Nations, III, Général 1929, 1930, IBU.

52 Television for instance, was still a "subject of scientific research pure and simple...the transmission of scenes, living images, etc., has not yet been completely achieved." LoN doc. ser., C.342.M.121.1929.XII, 63. Telecinematography, the broadcasting of moving pictures, of cinematographic films which could be picked up with special receiving fittings by anyone who was prepared to install this latest product of science and technical knowledge, indeed was possible and considered useful to the IECI. In broadcasting however, "enormous interests are involved, and for the complete study of which a huge and intricate organisation would be necessary." According to the IECI director the scope of its Institute could not be extended in such a way that these problems could be integrated in its activities and was beyond the possibilities of the

the main international expert organization in the area of broadcasting in Europe. Yet, above all it had finally drawn the attention of the League to the importance of combating the negative use of its medium.

In 1929 the ICIC decided to direct its focus more towards the aims of the League. It worked towards the creation of a new international state of mind. For years the ICIC had felt like the League's illegitimate child, not being taken seriously and uncertain about its spearheads. By 1930 its sub-committee of Letters & Arts even claimed that it "sometimes resembled a kitchen in which were prepared foods which did not appeal to the world's appetite."[53] The ICIC had always focused on the systemization of university life and the sciences. In 1929 the Committee decided to extend its activities "to render more direct service to the League,...to propagate the spirit of the League, to make it still better known, and to instill its ideals into the new generation."[54] On the basis of these new aims, the temporary *Committee of Enquiry to Study the Programme, Work and Organisation of the International Committee on Intellectual Co-operation and the Institute* conducted an inquiry. Its results led to a profound reorganization of the Committee and its agenda in 1931 (Figure 5.3).[55] The newly named Intellectual Co-Operation Organization (OCI) proved successful. The League received favorably the new agenda of the OCI and its new focus on building an international mind-set. It fully accepted the OCI as one of its technical committees.[56]

For the OCI, an international mind-set entailed more than just linking nation-states. It should also focus on communities beyond national confines. Sir Frank Heath, Chair of the Committee of Enquiry, had long term experience with the administration of academia and science in an international setting.[57] Heath argued

---

Institute. In the end IECI decided only to study "the relations between educational cinema and broadcasting...so far as tone – and speaking – films make use of broadcasting devices or so far as broadcasting may be used to supplement educational projections by means of lectures, lessons, etc..." LoN, C.I.C.I./214, Geneva, July 20th 1929, 2-4.

53 LoN doc. ser., C.428.M.129.1930.XII, 31.

54 LoN doc. ser., A.23.1931.XII, 5.

55 LoN doc. ser., C.428.M.129.1930.XII, 1-4. The ICIC replaced its sub-committees by so-called Committees of Experts; it established an executive and Director's Committee to follow and supervise closely the ever increasing progress in intellectual co-operation and provide for execution of decisions taken by the ICIC. Furthermore, the ICIC aimed to enhance closer relationships with its national committees to improve their engagement. Finally, the Paris Institute, the IIIC, was reorganized to facilitate alignment with the general scheme of ICIC's reorganization. Renoliet, *L'Unesco oubliée;* LoN, *International Committee on Intellectual Co-Operation, Report by the Committee on the Work of its Twelfth Session*, LoN doc. ser., A.21.1930.II (Geneva: LoN, 1930), 2-3, box 92, file Société des Nations, Section de Coopération Intellectuelle, Général, 1927-1932, IBU.

56 Renoliet, *L'Unesco oubliée*, 119.

57 Sir Frank Heath (1863-1946) had two university degrees, was a professor in English language and literature, and had wide knowledge and experience in education and research. His inquiries led to the

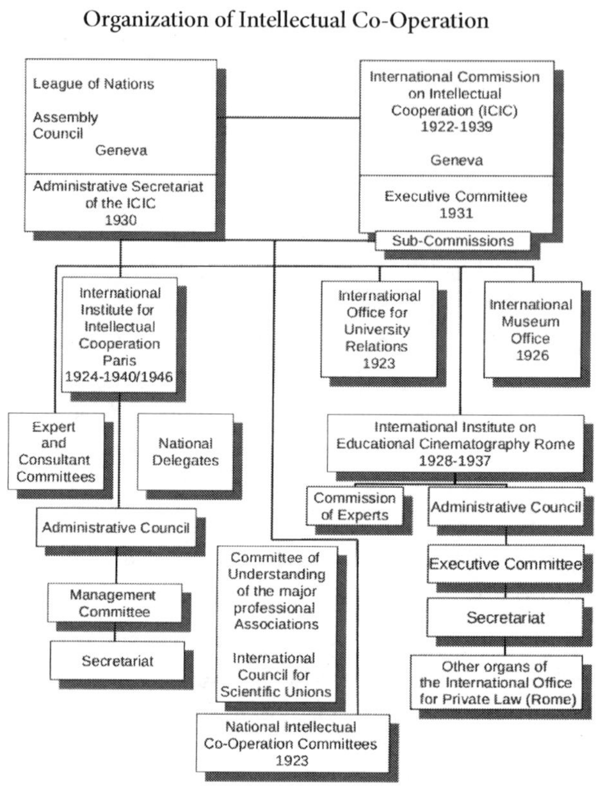

Figure 5.3 – The structure of the Intellectual Co-Operation Organization, 1931
Source: Jean-Jacques Renoliet, L'Unesco oubliée: La Société des nations et la coopération intellectuelle, 1919-1946 (Paris: Publications de la Sorbonne, 1999), 118.

that in the long run "progress in this matter [the development of an international mind] as in civilization in general would always be the result of individual thinking. Everything depended on how the individual mind developed and surrounded itself with the support of others."[58] As such, the OCI not only should employ the best intellectual minds for this cause, but also direct its attention to the worldwide masses. The war had brought together a large number of people whose aroused emotions had damaged stable international relations considerably. In order to create peace,

establishment of a Department of Scientific and Industrial Research in the British government. In the mid-1920s he performed similar studies in Australia and New Zealand. Afterwards similar departments were established in Canada, India and South Africa, all of them based on Heath's work. His knowledge of the organization of education and research must have formed an important input in the work of the ICIC regarding the organization of universities across the globe. Peter Gosden, "Sir (Henry) Frank Heath," *Oxford Dictionary of National Biographies*, 2008, http://www.oxforddnb.com/view/article/33791?docPos=14.

58 LoN doc. ser., C.428.M.129.1930.XII, 54.

one should prepare ordinary people throughout the entire world for more global and peaceful outlooks.[59] People would have to free themselves from nationalist sentiments fed by war. They would have to open their eyes to the world beyond their national borders, to feelings of international mutuality and peace.

Similarly they would have to open their eyes to all ethnic minorities living in their countries. To the OCI, the integration of ethnic and cultural diversity into visions of national communities and global interconnectedness of people and nations was fundamental to their concept of a true international mind. "No thoughtful person," Heath argued, "would deny that the progress of civilization depended more and more on the produce of countries whose riches could only be developed with the help of races which, up to present, had not participated to the same degree as some other races in the development of culture."[60] Up until then, not all ethnic groups had participated equally within civilization. It was vitally import that these people were heard and allowed to participate in society to a greater extent than they had done up until 1931.

Heath focused mainly on the European experience. In his view it was important to achieve acceptance through inclusion. In several regions "there were problems of tradition and economics which, if dealt with imprudently, would give rise to great dangers."[61] With a rapidly expanding economic depression and the unstable political and economic situation in eastern Europe since new boundaries were drawn in 1919, "great dangers" could lie just around the corner. The League named those who were not part of the dominant group in a country "minorities."[62] During its reorganization discussions, the OCI spoke of nation, race, and civilization rather than of national confines, ethnicities, and minorities. The OCI apparently used the discourse of the internationalist elite movement where race, culture, and civilization were closely related themes.[63] All in all, the OCI imagined a world in which all citizens would feel acknowledged: east, west, from every nation, from every local community, and every group. The OCI imagined an international mind with a veritable transnational character, based on European experience and European circumstances.

59 Iriye, *Cultural Internationalism and World Order*, 69.

60 LoN doc. ser., C.428.M.129.1930.XII, 25.

61 Ibid.

62 In the interwar years minority issues became fully integrated into the activities of the League of Nations. League of Nations, *Ten Years of World Co-Operation*, 354-378.

63 For extensive discussions of civilization, culture and races as well as nineteenth century cultural internationalism see: Wilson and Van der Dussen, *The History of the Idea of Europe*; Eagleton, *The Idea of Culture*; Iriye, *Cultural Internationalism and World Order*; Daniel Laqua, "Internationalisme ou affirmation de la nation? La coopération intellectuelle transnationale dans l'entre-deux-guerres," *Critique - Internationale, Une autre approche de la globalisation: socio-histoire des organisations internationales (1900-1940)* 52 (2011): 51-67.

The OCI decided that broadcasting would be the most important means of disseminating its idea of an international mind across the globe. It would educate and inform the masses concerning the new way of thinking about the free and peaceful co-existence of nations and peoples. A special sub-committee of the Committee of Enquiry foresaw a major role for broadcasting in education and museums. Mrs. Laura Dreyfus-Barney was a member of the expert committee for the Instruction of Youth in the aims of the League. Her main role was Vice-President of the peace section of the International Council of Women (ICW). She considered the cinematograph and wireless "…two of the most powerful weapons at the disposal of the teaching profession throughout the world."[64] The OCI now realized "that the League had actually in its hands an instrument of immense power which was scarcely being used."[65] When the IBU heard of these developments, Burrows wrote a letter to BBC director John Reith. For the first time the League officially showed a positive interest in the development of broadcasting by the IBU, and its relation to building social order.[66] The League not only wanted to deal with the problem of illicit propaganda. It also began focusing on broadcasting's positive use in the cause of peace, and as a propaganda instrument for diffusing an international mind-set across the globe.

Politicians Embrace Broadcasting

Soon politics firmly embraced broadcasting. Deteriorating international circumstances highlighted the importance and impact of the medium. Wall Street crashed in November 1929. What is more, the first traces of harshening political relations

64 LoN doc. ser., A.21.1930.II, 17; LoN, *International Committee on Intellectual Co-Operation, Minutes of the Thirteenth Session, held at Geneva, July 20th-25th, 1931*, LoN doc. ser., C.471.M.201.1931.XII (C.I.C.I./13th session/P.V.) (Geneva: LoN, 1931), 7. Laura Dreyfus-Barney was one of the leading ladies of the ICW, an organization that worked with the League of Nations in the 1920s in order to promote health, peace, equality and education. Leila Rupp, *Worlds of Women: The Making of an International Women's Movement* (Princeton, N.J.: Princeton University Press, 1997). She was American by birth and French by marriage. According to Burrows she was "much behind the scenes in international affairs." "Letter from Secretary General Arthur Burrows to BBC Director General Sir John Reith," 13 July 1931, 4, box 92, file Société des Nations, Section de Coopération Intellectuelle, Général, 1927-1932, IBU. The influence of women like Laura Dreyfus-Barney within the international elite community could be worth examination. Books like Rupp's are written from a largely feminist perspective. The case of broadcasting networks suggests that the influence of women must have reached a long way behind the scenes, on very practical issues as well.

65 "Letter from IBU's Secretary-General Burrows to Miss A. Hallstein-Kallis, co-worker of Gilbert Murray. Discussion of a letter from the President of the ICIC, Murray to the British Government," 29 July 1931, box 92, file Société des Nations, Section de Coopération Intellectuelle, Général, 1927-1932, IBU.

66 "Letter from Secretary-General Burrows to BBC's Director-General Sir John Reith," 1, box 92, file Société des Nations, IBU.

reared their head when new totalitarian regimes germinated in Germany, Italy, and Eastern Europe. Collaboration between the IBU and the ICIC took place in a setting that mirrored not only late 1920s optimism, but also gradually developing feelings of insecurity.

In these turbulent times, the IBU and the international cultural elites of the OCI tried to fine-tune their projects for Europe and the world by addressing legally binding standards for propaganda broadcasting. Similar to the frequency standard allocations, urgent practical problems had led to these negotiations. Propaganda broadcasts in the border regions of Poland and Germany caused stringent upheavals. As a result, broadcasters immediately refined the Gentleman's Agreement and reconsidered their plans for Europe. The Polish Government took the results of these discussions directly to the 1932 Disarmament Conference organized by the League of Nations. This act led to new international standardization attempts for propaganda broadcasting in which the OCI would play an important role. By the time the disarmament conference eventually failed, the OCI had firmly integrated the matter into its activities. With the help of the IBU, its efforts trickled down into the International Convention for the Use of Broadcasting in the Cause of Peace. The early 1930s thus saw a more formally shaping of their activities in propaganda broadcasting and their plans for Europe and the world.

In particular the 1931 German Polish Broadcasting Agreement was a watershed moment. Some four years after the signing of the Gentleman's Agreement, problems arose at Germany and Poland's borders in the Upper Silesian Plebiscite. These problems showed the vulnerability of the international situation there where peoples were separated by the newly drawn boundaries. Moreover, the Polish-German problems revealed the role that broadcasting could play in creating cohesion and segregation.[67] The international elite community would have to find ways to deal with these strongly related issues.

The repartition of the Silesia territory in 1919 had stirred local uprisings and diplomatic rivalries when the Allies assigned the major part of the area to Poland. Ever since then, the situation in the border region had remained delicate. Some Poles lived on the German side of the border. Some Germans now lived on the Polish side of the border. In 1925 and 1927 Germany had opened radio stations in Silesia that also radiated on Polish Silesian territory.[68] In 1927 the Poles opened

67 P.V. Conseil Administrative 1929, 202, IBU; IBU, *Twenty Years*, 29-30.

68 Its most well-known radio station was situated in Gleiwitz (Gliwicea in Polish). After WWII the station came into Polish hands and would be used for the broadcasting of Radio Free Europe programs across the Iron Curtain.

a radio station near their borders in Silesia as well, in Katowice. To celebrate the opening of this Katowice station, the Poles broadcast the opera *Halka*. This was one of the strongest expressions of Polish national music and Polish nationalism. According to Polish philosopher Dembowski, the opera promoted the cause of national realization and independence.[69] Works like *Halka* actually opened the opera season in Poland, "and the country's elites would gather at the event to celebrate the official national culture."[70] Near the Polish border, Germans could receive the Polish broadcast as well. They felt offended by the nationalist spirit and political claim of independence expressed by the opera. To avoid potential hostilities in the future, director of Polskie Radjo Zygmunt Chamiec travelled to Berlin twice. He stated that he would do everything he could to avoid harmful relations between the two countries.[71]

On air, German-Polish relations remained relatively stable until four years later, so the story goes, when the German Reichsrundfunkgesellschaft m.b.H. (RRG) decided to broadcast a German circus show. During the show "two clowns, joking with each other, made a few uncomplimentary remarks about the Poles, in a play of words."[72] In reaction to the broadcast, which the Poles had heard loud and clear, the Polish Government officially protested to its German colleagues. Shortly after, on the tenth anniversary of the Silesian area, several programs broadcast both in Poland and Germany contained less innocent remarks. These immediately led to considerable irritation on both sides of the border. In response to litigation, RRG and Polskie Radjo negotiated an agreement unique in its kind, designed to avoid future incidents.

The 1931 agreement was based largely on the 1926 Gentleman's Agreement though with some crucial improvements. Thus Germany and Poland now agreed they would:

> undertake in future to do everything in their power to ensure that matter – whether political, religious, economic, intellectual or artistic – broadcast from their stations shall not compromise in any way the spirit of co-operation and good understanding which is necessary if broadcasting is to fulfill its mission of drawing the nations together.[73]

69 Harry White and Michael Murphy, *Musical Constructions of Nationalism: Essays on the History and Ideology of European Musical Culture, 1800-1945* (Cork University Press, 2001), 167-173.
70 Philipp Ther, "The Genre of National Opera in a European Comparative Perspective," in *The Oxford Handbook to the New Cultural History of Music*, ed. Jane Fulcher (Oxford: Oxford University Press, 2011), 199.
71 Ryszard Miazek, *Przeminelo z radiem: opowiesc o Zygmuncie Chamcu – zalozycielu i pierwszym dyrektorze Polskiego Radia* (Rzeszów; Warszawa: Wydawnictwo Wyzszej Szkoly Informatyki i Zarzadzania; Polskie Radio, 2005), 137, 139-140.
72 John D. Tomlinson, "The International Control of Radiocommunications" (Ph.D. diss., Université de Genève, 1938), 226.
73 Extract from the Agreement of March 31, 1931, concluded between the German and Polish Broad-

The IBU strongly influenced these proceedings. The national departments of the German and Polish organizations allowed IBU Council members from their countries to come to terms.[74] These two men, Heinrich Giesecke and Zygmunt Chamiec, held key positions in the IBU outside the Council. Giesecke was one of IBU's Vice-Presidents. Chamiec was head of the International Relay Committee. They frequently collaborated in Geneva and had stood at the cradle of the IBU together, designing its goals from scratch.[75] Their IBU activities had been an important part of their daily lives for years. They knew each other well. Consequently, the entire agreement breathed the work of the IBU.

The two men also added some recent IBU work on propaganda broadcasting to the Gentleman's Agreement. Both organizations accepted certain amounts of positive propaganda with a view to national activities as long as "the matter broadcast does not in any way offend the national sentiment of listeners who are subjects of the other" country.[76] For the first time these two organizations officially distinguished between broadcasts intended for national audiences and those intended for foreign audiences (Fremdsprachsendungen). They agreed that in neither of the two cases was broadcasting to offend foreign listeners.[77] Secondly, because they were public broadcasters, their stations could on occasion be employed by their Governments. They added that all matters broadcast by their Governments over their stations, "in accordance with the concessions granted to the companies, the supervision of such matters is entirely in the hands of the Government authorities of each country."[78] As such, the organizations renounced any accountability for the latter.[79] These two appointments filled a lacuna in the internationally renowned 1926 Gentleman's Agreement.

casting Companies (Reichs-Rundfunk-Gesellschaft m.b.H. – Polskie Radio). LoN, *Moral Disarmament, Memorandum from the Polish Government*, LoN doc. ser., C.602.M.240.1931.IX [Conf. D. 16] (Geneva: LoN, Sept. 23rd, 1931), 4.

74 IBU, *Twenty Years*, 11.

75 Although Polskie Radjo, in contrast to the RRG which was one of the founding members of the IBU, only became an IBU Member in 1927, Zygmunt Chamiec had been working on the establishment of Polish broadcasting long before. In light of his interest and wish to establish broadcasting in Poland along the ideological lines of the international radio community regarding peace and mutual understanding, Chamiec had been present during the very first IBU conferences. In 1928 France awarded him the Order of the Legion of Honor for "his remarkable achievements in promoting peaceful cooperation across the airwaves." Miazek, *Przeminelo z radiem*, 122, 147, 311.

76 LoN doc. ser., C.602.M.240.1931.IX [Conf. D. 16], 4.

77 Heinz Pohle, *Der Rundfunk als Instrument der Politik: Zur Geschichte des deutschen Rundfunk von 1923/38* (Hamburg: Verlag Hans Bredow-Institut, 1955), 367.

78 LoN doc. ser., C.602.M.240.1931.IX [Conf. D. 16], 4.

79 Pohle, *Der Rundfunk als Instrument der Politik*, 367.

The real break with the past, however, was that both the German and Polish Ministries of Foreign Affairs supported the agreement.[80] In a more complex and extended form, the Gentleman's Agreement had found its way to the desks of two of the responsible government departments in Europe. Some would say that such bilateral agreements went against the grain of the IBU's non-governmental character. On the contrary, the IBU had always desired that its agreements would trickle down into national law. When the IBU agreed to the Gentleman's Agreement in 1926 it urged its members to accordingly bring the agreement to the attention of their respective governments.[81] The 1931 agreement responded to these wishes. Both countries, however, needed a pressing conflict to finally bring the problem to the attention of their governments. As the German and Polish governments fully backed the agreement on propaganda broadcasting, from the point of view of international law, this meant the arrangement did "not present any essential difference from the arrangements concluded between State Administrations such as those of Posts and Telegraphs."[82] The IBU's transnational work had filtered down into the national laws of two countries. Consequently, Poland and Germany acknowledged the importance of international activities for the smooth functioning of their nation-states.

The IBU immediately turned the bilateral agreement into a new recommendation. It added the improvements and promoted the German-Polish agreement as an example for similar initiatives. In October 1931 the IBU recommended all its members negotiate similar agreements if the occasion arose. Furthermore, in the hope uniformity on the matter would arise internationally, in May 1933 the IBU asked administrations who controlled stations outside the IBU to adhere to these principles as well.[83] Consensus should diffuse beyond the confines of the IBU's broadcasting space. Prior to the German-Polish agreement, several governments had already made arrangements with the USSR "that included more stringent anti-propaganda clauses."[84] Several others followed suit. In 1934 Poland and France agreed to add arrangements about regular program exchange to the

80 Tomlinson, "The International Control of Radiocommunications," 227; LoN, *Broadcasting and Peace: Studies and Projects in the Matter of International Agreements* (Paris: International Institute of Intellectual Co-Operation, 1933), 129-130.

81 LoN doc. ser., C.595.M.231.1926.VIII, 84.

82 LoN, *Broadcasting and Peace*, 30.

83 IBU, *Twenty Years*, 30.

84 Jennifer Spohrer, "Ruling the Airwaves: Radio Luxembourg and the Origins of European National Broadcasting, 1929-1950" (Ph.D. diss., Columbia University, 2008), 343; Stephen Duggan, "Radio: An Agency for International Understanding or Friction, In 'Radio: The Fifth Estate', Special Issue," *Annals of the American Academy of Political and Social Science* 177 (1935): 178; Vernon van Dycke, "The Responsibility of States for International Propaganda," *The American Journal of International Law* 34, no. 1 (1940): 60.

German-Polish decisions.[85] In 1936, the British and Italians would also follow the German-Polish example. The British requested that the Italians refrain from harmful propaganda broadcasts into its Middle East territories. In turn, the British would then recognize Italy's victory over Abyssinia.[86] The increasing number of bilateral agreements facilitated the practical realization of an efficient European broadcasting network.

These bilateral agreements along German-Polish lines reflected the importance of the work of international non-governmental experts in everyday life. Similar to the Gentleman's Agreement, the German-Polish agreement portrayed a Europe in which the Eurocentric international and national were being fine-tuned. Both co-existed in the international recommendations as well as the bilateral national agreements. Unlike the Gentleman's Agreement, the "national" within the German-Polish agreement now also consisted of levels below that of the nation state, levels which integrated all minorities. As such, the international became inclusive of the heterogeneity, of the transnationality, of the national. In turn, this refined internationalism infiltrated national laws. National laws were corrected for awareness and feelings of responsibility beyond their national confines. It was the first time transnational non-state efforts based on international elitist ideas of peace and rapprochement had found their way into legally binding national rules. The German-Polish agreement marked a watershed in broadcasting regulation.

The Agreement found its way to the Disarmament Conference. This event heralded the exploration of widely shared legally binding international standards for propaganda broadcasting. The Polish Government decided to use the agreement in its Memorandum on Moral Disarmament to the Preparatory Committee for the Disarmament Conference. The Polish Government used this to open a debate about establishing a widely supported international standard for propaganda broadcasting. If governments ratified these standards, they would trickle down into national laws on a far wider scale than the bilateral German-Polish agreement.

In his Memo of September 17, 1931, Polish Minister of Foreign Affairs, August Zaleski, requested placing the issue of *moral* disarmament on the agenda of the Conference for the Reduction and Limitation of Armaments. The conference would start in January 1932. It aimed to reduce the number of arms produced and owned by the various nation-states to avoid warfare in the future. According to Zaleski, factual disarmament could not be achieved without an equal disarmament

85 Miazek, *Przeminelo z radiem*, 157.
86 Spohrer, "Ruling the Airwaves," 343-344. There are indications that more such agreements were made across the European continent. Further research in national broadcasting archives could clarify this.

Figure 5.4 – War can start now!
Translation: "War can break out now! The American broadcaster is ready to fulfil its cultural task!" Press release in Dutch radio guide which announces that American broadcasters are going to bring the news from the frontlines in case war will break out between Italy and Abyssinia. The idea is to bring the news from both Italian and Abyssinian side.
Source: VARA, "De oorlog kan nu uitbreken!," De Radiogids, no. 47 (1935), 5. Used by the courtesy of the Netherlands Institute for Sound and Vision, Hilversum, the Netherlands.

of the mind.[87] The "dangerous agitation of certain elements which are endeavouring to poison relations between the peoples and to sow the seed of fresh conflicts in the minds of their supporters," increased the vulnerability of the world to yet another war.[88] Zaleski believed that at the basis of a negative mindset lay national laws that tied individuals to their national interests. What about the interests of a "higher society…outside the nation – namely the international community?" he asked.[89] Without a change in the direction of a moral international mind, practical

87 LoN doc. ser., C.602.M.240.1931.IX [Conf. D. 16], 2.
88 Ibid.
89 Ibid., 3.

disarmament efforts would not be effective. Zaleski pointed at the destructive forces of negative and nationalist propaganda. Positive propaganda, on the contrary, would disarm the mind. It would spread a message that looked beyond borders, to the community of international elites and their efforts to improve mutual benefits.

The idea of "moral disarmament" was nothing new. It already went back a long way. The idea had flourished widely amongst pacifists and internationalists ever since the International Peace Conferences held before the war and before governments had signed the Covenant of the League of Nations.[90] To pacifists, moral disarmament initially meant:

> 'the transformation of this [nationalist] aggressive mentality, vindictive and vengeful to a conciliatory mentality. It is...the sacrifice of national interests to interests no less real of the great human family. It is altruism substituted for egoism, it is reason and equity put in place of passion and injustice.'[91]

Pacifists and internationalists alike related moral disarmament to converting public opinion towards a mentality of peace. Pacifists considered it sufficient to relate this change of mind directly to international collaboration and to education. Internationalists rather linked this conversion of mind to international security, to treaties, to factual disarmament.[92] In pursuit of the latter, Aristide Briand had already addressed the first Disarmament Conference in Washington in 1921, arguing that "moral disarmament is as necessary as the material armament."[93] The 1928 Kellogg-Briand Pact also evolved around this internationalist view of moral disarmament. Though the entire pact was a factual measure of moral disarmament, it failed to integrate a paragraph on the route to go, on the technique that would realize such change of mind amongst the masses.[94]

In his memo Zaleski not only assumed an internationalist perspective, he also proposed a practical technique which could achieve disarmament of the

90 George A. Finch, "A Pact of Non-Aggression," *The American Journal of International Law* 27, no. 4 (1933): 725.

91 "'...c'est la transformation de cette mentalité aggressive, rancunière et vengeresse en une mentalité conciliante. C'est, quand le bien général l'exige, le sacrifice des intérêts nationaux aux intérêts non moins réels de la grande famille humaine. C'est l'altruisme substitué à l'égoïsme, c'est la raison et l'équité mises à la place de la passion et de l'injustice.'" Elly Hermon, "« Le désarmement moral en tant que facteur dans les relations internationales pendant l'entre-deux-guerres »" *Historical Papers/Communications historiques* 22, no. 1 (1987): 199.

92 Hermon, "« Le désarmement moral en tant que facteur dans les relations internationales pendant l'entre-deux-guerres »," 200.

93 International Conciliation, *Washington Conference on the Limitation of Armaments*, vol. 169 (Greenwich: American Association for International Concilliation, 1921), 516.

94 Julien Luchaire, *Le désarmement moral* (Paris: Librairie Valois, 1932), 83-85.

mind. He gave the example of the medium of broadcasting, and pointed to the German-Polish agreement as a guideline for possible international agreements. Broadcasting, according to Zaleski, was one of the crucial media for achieving moral disarmament. Moreover, the 1931 German-Polish Broadcasting Agreement was an excellent example of how such a focus on the international good could be realized practically. "This agreement may go to show that practical results in this field are in no way impossible," Zaleski argued.[95]

When the Preparatory Committee forwarded Zaleski's memo to the OCI, the pacifist and internationalist views on moral disarmament came together. Whereas the Disarmament Conference took an internationalist approach, the OCI worked on closely related themes from a pacifist point of view. With its reorganization from 1930 to 1931, the OCI and its community of intellectuals linked their peaceful international minds to education. According to former director of the IIIC, Julien Luchaire, moral disarmament should be seen as an intellectual movement that employed propaganda as a means of creating a mind of peace, as a common tactic, and an actual technique.[96] He foresaw a critical role for intellectuals who could create a rapprochement of minds, a unique and universal "soul," a universal civilization that presented the best of all existing civilizations.[97] Luchaire linked pacifism to internationalism when he spoke of moral disarmament.

These developments in the OCI went hand in hand with growing feelings amongst intellectuals outside the committee.[98] They argued, as recent isolationist economic developments showed, the pacifist approach had not sufficed. On January 18, 1931, roughly 180 French intellectuals followed by at least 188 German intellectuals signed a manifesto in which they promised to combat a spirit of war "at the moment when Europe must organize itself or perish, and cannot organize itself other than by *the pacification of minds* and the forgetting of hostilities..."[99] To these intellectuals, moral disarmament not only meant the birth of a mind of peace, but also the end of a mind of warfare; two things not altogether the same.[100] In response to these voices and Zaleski's Memo, OCI director Gilbert Murray emphasized "the need for a regular practice of international cooperation in the field

95 LoN doc. ser., C.602.M.240.1931.IX [Conf. D. 16], 3.

96 Luchaire had been the director of the IIIC in Paris before the 1930-1931 reorganization. He was a French writer and a member of ICIC's Committee on Letters and Arts.

97 Luchaire, *Le désarmement moral*, 39, 53.

98 For an exploration of the intellectual context in which the OCI operated see: Daniel Laqua, "Transnational Intellectual Cooperation, the League of Nations and the Problem of Order," *Journal of Global History* 6, no. 2 (2011): 223-247.

99 "'A l'heure où l'Europe doit s'organiser ou périr, et ne peut s'organiser que dans la *pacification des esprits* et l'oubli des ressentiments...'" Luchaire, *Le désarmement moral*, 8.

100 Ibid., 39, 53.

of science, arts and literature in order to achieve moral disarmament."[101] Murray especially referred to education, to youth, and to modern means of communication like broadcasting and cinema. The work on moral disarmament "could be regarded as the ultimate expression of all the work of the Intellectual Co-operation Organisation."[102] The OCI offered the Disarmament Conference its full cooperation.

The Political Committee of the Disarmament Conference decided to establish a Committee for Moral Disarmament in pursuit of the Polish memo. The Swiss Mr. Perrier chaired the committee. He was assisted by two Polish representatives Mr. Szumlakowski and Mr. Komarnicki. Their participation further strengthened Polish dominance regarding the question of moral disarmament.[103] The Committee for Moral Disarmament leaned heavily on the OCI's activities and results. Perrier joined in an OCI meeting in July 1932 when the OCI formulated its recommendations in line with the German-Polish broadcasting agreement. The OCI discussed the positive and negative side of propaganda. It would be wise to establish "agreements to eliminate from broadcast programmes elements prejudicial to good international relations." Moreover, one should pursue an international mind positively "by making the outstanding characteristics of the genius of different peoples known to each other, whether expressed by poetry, music, drama or in prose writings."[104] The OCI recommended that the Committee for Moral Disarmament take a positive rather than a negative approach to the issue at hand. Only then could one obtain a "fuller knowledge of each other's real characteristics."[105] Only by creating an international mind could one facilitate actual and mental disarmament.

The Moral Disarmament Committee addressed these recommendations in two of its studies featuring broadcasting as a technique to create disarmament of the mind. One study dealt with "questions concerning intellectual co-operation and technical means of spreading information, including the problems of education, utilization of cinematography and broadcasting." Another dealt with the

101 "la nécessité d'une pratique régulière de la coopération international dans l'ordre scientifique, artistique et littéraire pour atteindre le désarmement moral." Renoliet, *L'Unesco oubliée*, 303.

102 LoN, *International Committee on Intellectual Co-Operation, Minutes of the Fourteenth Session, Minutes of the Eight Meeting, held at Geneva, July 21st, 1932*, LoN doc. ser., C.I.C.I./14th Session/P.V.8 (Geneva: LoN, 1932), 8.

103 LoN, *Conference for the Reduction and Limitation of Armaments, Moral Disarmament, held at Geneva, July 25th, 1932*, LoN doc. ser., Conf. D. 138 (Geneva: LoN, 1932), 1.

104 LoN, *International Committee on Intellectual Co-Operation, Minutes of the Fourteenth Session, Minutes of the First Meeting, held at Geneva, July 18th, 1932*, LoN doc. ser., C.I.C.I./14th Session/P.V.1 (Geneva: LoN, 1932), 4-5.

105 LoN, *International Committee on Intellectual Co-Operation, Minutes of the Fourteenth Session, Minutes of the First Meeting, held at Geneva, July 18th, 1932*, LoN doc. ser., C.I.C.I./14th Session/P.V.2 (Geneva: LoN, 1932), 2.

legal aspects.[106] In light of the first study, Szumlakowski prepared a draft convention. This allowed for preventive measures to help avoid the diffusion of illicit propaganda.[107] The draft made four positive suggestions. These were with regard to the education of the younger generation, co-operation in the intellectual world, the use of technical means of spreading information, and the ways and means of giving effect to possible undertakings.[108] The study on legal aspects addressed and recommended negative measures. It addressed penal measures for diffusing illicit propaganda. The so-called Pella Report referred to its drafter Dr. Pella, the Rumanian delegate to the Disarmament Conference and member of the OCT. Pella had worked hard to place the issue of illicit propaganda broadcasting on the League's agenda in the second half of the 1920s. The two studies addressed all aspects of the disarmament of the mind via propaganda.

Throughout the discussion of these two reports, the Committee for Moral Disarmament strongly opposed any form of preventive control of programs close to censorship. The committee seemed "none too ready to include, in the convention…clauses destined to prevent certain wireless transmissions by the application of penal measures."[109] The committee asked its members to discuss the draft convention with their respective governments. Based on these suggestions a special drafting committee would prepare a new text to be discussed at a later meeting if necessary.[110] The committee thus designed a preliminary convention for the disarmament of the mind that provided ample space for propaganda activities contrary to warfare, in favor of peace. Though this draft focused on propaganda in its broadest sense, it was based on previous achievements, recommendations, and standards in broadcasting propaganda. It obtained a large space for broadcasting as a practical technique to pursue or not pursue propaganda.

The two studies opened up a discussion between the OCI and the Committee for Moral Disarmament about the role of the state if broadcasting were to be used to create moral disarmament. The discussion related the roles of the state and broadcasters to the extent of freedom of the individual in its broadest sense. It related broadcasting to the freedom to project societies. The OCI argued that moral disarmament should be pursued by furthering active interest among governments as well as co-operation with the IBU. The OCI acknowledged the broad expertise of the IBU and took note of its advice that too rigid an international convention would "preclude adaptation of the broadcasting programmes to the particular

106 LoN doc. ser., Conf. D. 138, 1.
107 LoN, *Broadcasting and Peace*, 156.
108 LoN doc. ser., Conf. D. 138, 2.
109 LoN, *Broadcasting and Peace*, 157.
110 LoN doc. ser., Conf. D. 138, 2.

needs of each country."[111] One should make room for national diversification. This could only happen if national laws and international conventions would leave some space for expert organizations like the IBU and its national members to maneuver.

Though Perrier thought it wise if OCI left the obligations of states to the Committee for Moral Disarmament, he agreed with its stance on government involvement. Perrier pleaded for an end to state supremacy. States would have to correct their legislation for international needs, and could not do everything themselves. Perrier urged "that in the cause of moral disarmament, great latitude must be allowed, as far as this was compatible with public order, to associations, organizations and even racial groups for the purpose of solving the difficult minority question."[112] Though the OCI agreed in principle, it argued that there should be some form of state intervention to recognize the importance of moral disarmament. According to OCI member Destree "it was a truism to some people that patriotism could be combined with international fellowship," but this was far from a general conception. He felt surprised to see that technology was much in advance of political theory which still focused narrowly on national sovereignty.[113] Real patriotism, in his opinion, consisted of recognizing interdependency between countries, though such "solidarity did not imply that individual liberty should in any way be restricted. Each country had its own sentiments, language, traditions. These must all be protected and guaranteed, as well as the rights of minorities."[114]

All discussants seemed to agree that there was a direct though inverse link between the degree of state intervention and the freedom of the individual. The higher the level of state intervention, the lower would be the freedom of broadcasters and their ability to correct for national differences. The lesser the ability to correct for diversification, the lower would be the degree of individual freedom in society. Freedom for broadcasters would secure the correction of such national and ethnic differences. In contrast, too little government interference would downplay any effective regulation of propaganda altogether. It would lead to complete anarchy of the airwaves. This would reduce individual freedom as well, giving way to propaganda which incited war. The parties agreed that the degree of government control was a measure by which they could influence war and peace propaganda efforts. The degree of state intervention and the closely related freedom for broadcasters defined the extent to which the international good could find its way into

111 LoN doc. ser., C.I.C.I./14th Session/P.V.2, 2.
112 LoN doc. ser., C.I.C.I./14th Session/P.V.8, 2.
113 Ibid., 3.
114 Ibid.

regulations and broadcast programs. It also defined the extent to which broadcasters could integrate the wellbeing of minorities within nations into these international agreements and broadcasts. The OCI favored a slightly larger control of the state than the Moral Disarmament Committee and the IBU, who both argued in favor of as limited state control and as much freedom for broadcasters as possible.

The Disarmament Conference created an international plenipotentiary hub for governments, the intellectual cultural elite, and lawyers, to discuss propaganda in the light of war and peace. The propaganda efforts of the European broadcasting community had been an important basis for discussion. Moreover, the medium of broadcasting became recognized as an obvious technique for pursuing propaganda. The Conference formally made the IBU's preventive and positive propaganda activities part of the international and political efforts to avoid war and create peace. These events freed the way for new international agreements on propaganda broadcasting. Such agreements would secure the legal implementation of the IBU's internationalist outlook among national governments across the globe. The question remained how the Committee for Moral Disarmament would present its conclusions. They could be in the form of an international convention on disarmament, or perhaps a separate convention, an additional protocol, or a relatively simple recommendation.

The League started on a broadcasting project for peace. This project was a new turning point in its efforts relating transnational broadcasting to projects for society. It built on the major results of previous studies tackling the problem of broadcasting within the broader context of moral disarmament. The OCI assembled a group of well-established experts to frame these results into a formal international convention. The experts could reflect on a number of parameters that would affect society such as the degree of state control over broadcasting content, the freedom of broadcasters to determine their radio schedules, and the freedom of the individual in its broadest sense. In this way, they would be able to influence substantially the place and role of broadcasting in times of war and peace. They could give direction to society-building.

In 1931 already the OCI asked the League Assembly to examine the role broadcasting could play in schools to educate young people about an international state of mind. First of all the Committee wanted to seek the opinion of qualified experts in the world of education. Furthermore it wanted to explore the best ways to employ broadcasting to develop such a spirit of understanding. The Assembly in particular valued the broadcasting part of the inquiry. The OCI should however extend its scope to "improvements of relations between peoples."[115] It should focus on the masses worldwide rather than only on young people. The OCI brought together a group of experts to study existing agreements on broadcasting for peace, notably the German-Polish agreement. The experts would question the purpose and underlying conditions of these agreements especially those governing the preparation of an international project.[116] Bearing in mind the previous studies, the experts aimed to make provisions for a possible *International World* Agreement or, in any case, a *European* Agreement.[117]

The Committee of Experts met in Geneva at the end of February 1933. The Norwegian Dr. Arnold Raested chaired the committee. His doctorate in law and his varied experience with governments and the broadcasting community made him the right man for the job. He was the former Norwegian Minister of Foreign Affairs, a current delegate to the League Assembly, the chair of the board of directors of the Norwegian Broadcasting Corporation, and part of IBU's legal team. In 1939 his work on propaganda broadcasting helped to mold him for the position of IBU Vice-President. Raested was assisted by Henri Bonnet, director of the IIIC, De Montenach, secretary of the ICIC, and a group of experts who held hybrid positions. These experts represented national governments, national broadcasting organizations, the IBU, the ITU, the International Broadcasting Committee, as well as the wireless listeners' associations via the participation of the *Association Générale des Auditeurs de T.S.F.*[118] ITU and IBU participation proved a valuable contribution as their experts had over eight years practical experience with propaganda broadcasting from a technical and political point of view.[119] All experts were European with the exception of the Mexican participant. They would have to translate the prevailing consensus on the problem of propaganda broadcasting into a workable and practical solution. The committee had a predominantly European though very diverse outlook. It dealt with global affairs, but mostly applied European experience.

The experts' General Report shaped the contours for agreements on the use of broadcasting to promote peace. Most of their five recommendations focused on

115 LoN doc. ser., C.I.C.I./14th Session/P.V.1, 4.

116 "Some notes on the U.I.R. Memorandum on the activities of the International Broadcasting Union for the use of broadcasting in the cause of peace and on the relations between the League of Nations and the U.I.R. in connection with this activity," 14.9.1936, 7, box 17, file Désarmement moral et propagande inadmissible, Général V, 1933-1939, IBU; Arthur R. Burrows, "La Convention Internationale sur l'Emploi de la Radiodiffusion dans l'Intérêt de la Paix," *Radiodiffusion*, no. 3 (October 1936): 32.

117 "Some notes on the U.I.R.," 19, box 17, file Désarmement moral, Général V, IBU.

118 Ibid., 14-15.

119 "Address (made by Secretary-General A.R. Burrows)," 2, box 74, file Propagande, Général II, IBU; LoN, *Broadcasting and Peace*, 19.

previously discussed topics. One of these looked at the use of positive and penal measures concerning broadcasting propaganda.[120] The experts recommended that preventive measures should be restricted "to the penalising of violations of manifest gravity." Such violations should be restricted to severe threatening or misleading news broadcasts "with a view to disturbing the interior or exterior security of another country."[121] Positive measures should rather be drafted in general terms for reasons "of the diversity of conditions and requirements" across the European continent and the world.[122] The experts reasoned that if the positive measures proved successful, in time the preventive measures could be applied less frequently. The experts' advice to draft the positive measures in general terms shows the importance they attached to drafting an agreement that could be sustained in the long run. The general terms would allow for the diversity in national, local, as well as regional circumstances, and any changes over time.

A similar intention came to the fore in another recommendation. This described the role that states and the IBU should play in ensuring that individuals, minorities, and nation-states were free to maneuver. The committee advised in favor of the importance of broadcasting related associations. Their work should be supported in any possible way. The experts specified this interdependency when they discussed "*what questions should be settled within each State* by means of instructions addressed by the government to the broadcasting companies."[123] Based on the IBU's long experience, the expert advice was varied and included: "preliminary examination (control) of messages and various matters likely to promote international understanding and also practical measures for the development of this kind of broadcast..."[124] Consequently, the experts advised governments to impose the measures on the organizations in their countries that these broadcasters had developed in the IBU. This advice implies a great amount of freedom for broadcasting organizations and limited government influence on program content. The experts projected the organization of propaganda broadcasting in relatively general and transnational terms.[125]

120 Burrows, "La Convention Internationale," 19, 29, 33, 49-51.
121 Ibid., 49-50.
122 Ibid., 29, 50.
123 Ibid., 19.
124 Ibid., 51.
125 The committee also made observations for discussion topics at a later date. One topic concerned "the distribution of authorised press information on events of international importance circulated by the broadcasting stations." Another was the possible development of a news service by the League in various languages. Burrows, "La Convention Internationale," 51. The relationship between the international press and the development of transnational broadcasting in the interwar years is worth studying. Whereas news broadcasting could be considered a threat to the established press, broadcasting could also be a business opportunity for the press. Furthermore, the international standards created for the regulation of the press across borders often served as example for the international regulation of broadcasting. See: the Archives

With this organization of broadcasting in mind, the experts advised the League on the most suitable type of agreement. According to IBU Secretary-General Burrows, an agreement should not be too rigid. Broadcasters had difficulty determining what issues or music harmed or did not harm listeners. “National feeling is much more aroused and wounded in some countries than in others. There are parts of the world which appear to be politically ‘volcanic’ in character; other parts where political life is much more tranquil.”[126] A rigid convention would not be able to correct for these differences, especially in the case of a worldwide convention. The experts recommended the League to create an international convention in general terms. Countries could then complement the convention with regional or individual agreements between governments or between broadcasting organizations. Such regional arrangements should be designed “to settle local difficulties and tighten the bonds between the neighbouring countries.”[127] Burrows pointed out that it would be impossible to imagine France not being capable of broadcasting the story of Joan of Arc, or the British being “so foolish as to protest against such broadcast.”[128] The experts proposed an international convention that settled the standardization of propaganda broadcasting in general terms complemented by regional, local, or bilateral agreements that filled out specifics as and when necessary.

The Committee for Moral Disarmament closely followed the findings of the group of experts. It integrated these recommendations in its draft convention for the high contracting parties of the Disarmament Conference in November 1933. At times, the committee found it hard to accept certain OCI formulations. The committee was an assembly of governments rather than a group of experts.[129] The draft convention stressed that governments should use broadcasting and cinematography to enhance a spirit of goodwill between nations. They should support any action of the OCI and other organizations with the same objective, and should avoid “the showing of films, the broadcasting of programmes and the organisation of performances obviously calculated to wound the legitimate sentiments of other nations.” Governments should monitor these activities “in accordance with the special system in force in their respective countries.”[130] Although the draft

of the League of Nations.

126 “Address (made by Secretary-General A. Burrows),” 2, box 74, file Propagande, Général II, IBU.

127 LoN, *Broadcasting and Peace*, 50.

128 “Address (made by Secretary-General A.R. Burrows),” 7, box 74, file Propagande, Général II, IBU.

129 The documents give no clear example of such difficulties. LoN, *International Committee on Intellectual Co-Operation, Minutes of the Fifteenth Session, held at Geneva, November 21st, 1933,* LoN doc. ser., C.I.C.I./15th Session/P.V.9 (Geneva: LoN, 1933), 10.

130 LoN, *Conference for the Reduction and Limitation of Armaments, Committee for Moral Disarmament, Text Adopted by the Committee*, LoN doc. ser., Conf. D/C.D.M./36 (Geneva: LoN, Nov. 21st 1933), 2.

convention focused on the disarmament of the mind in its broadest sense, it devoted an entire article to cross-border broadcasting and cinematography.

The committee meeting in October 1933 would be its last. In the course of two years their efforts had created international recognition among government plenipotentiaries on a number of issues related to broadcasting. Their final draft convention would pay tribute to the work of the OCI on questions of moral disarmament and the importance of technical media. The draft convention would also articulate that national and local diversity formed an inherent aspect of international projects. A member of the Moral Disarmament Committee, Mr. Peroso, expressed to the OCI his wish that the convention should be acceptable to non-League members as well.[131] A convention that was drafted in general terms would facilitate such development. For the first time, an assembly of governments had accepted broadcasting as a means of building peace. For the first time, an assembly of governments had drafted a convention using the expertise of the IBU, ITU, and OCI. Within months of the drafting of the convention, difficulties caused by changing international relations destroyed the hopes of the Disarmament Conference.[132]

The favorable international climate since the signing of the Kellogg-Briand Pact in 1928 had proved short-lived. The economic crisis had de-stabilized the world economy. Moreover, one of the two men greatly responsible for the relative entente in Europe died in 1929. In the post-war years the Frenchman Aristide Briand and the German Minister of Foreign Affairs Gustav Stresemann had managed to create relative stability in international relations between the largest enemies of the Great War.[133] Already during years of ill-health, Stresemann had lost domestic support for his international policy. When Hitler rose to power in 1933, Germany not only withdrew from the Disarmament Conference but also from the League at large. The disengagement by Germany meant the beginning of the end of the conference.[134]

Initial uncertainty about the future of the Disarmament Conference led the OCI to rework the draft into a convention for broadcasting in the cause of peace. Its commitment to work on such a convention was a desperate attempt to hold on to an internationalist ideology when international relations increasingly

131 LoN doc. ser., C.I.C.I./15th Session/P.V.9, 10.

132 LoN, *International Committee on Intellectual Co-Operation, Minutes of the Sixteenth Session, Provisional Minutes of the Eleventh Meeting, held at Geneva, July 21st, 1934,* LoN doc. ser., C.I.C.I./16th Session/P.V.11 (Geneva: LoN, 1934), 2.

133 Steiner, *The Lights that Failed*, 486.

134 Andrew Webster, "The Transnational Dream: Politicians, Diplomats and Soldiers in the League of Nations' Pursuit of International Disarmament, 1920-1938," *Contemporary European History* 14, no. 4 (2005): 516.

depended on dividing nationalistic ideals. The OCI relatively hopelessly pointed to the importance of the initial text of the Moral Disarmament Committee. The text had been drawn up with the consent of an assembly of governments. It was absolutely fundamental to secure "measures of execution in the field of moral disarmament."[135] The League Assembly appointed a committee of lawyers to draft a convention. Dr. Raested again chaired the committee, which drafted a *World International* Convention. In February 1934 as well as in April 1935, the IIIC communicated several drafts to the national governments for approval. Criticism by national governments, however, prevented the drafting of a final convention.[136] As a result, the League Assembly organized an international plenipotentiary conference. On September 23, 1936 the conference finally produced the International Convention for the Use of Broadcasting in the Cause of Peace.

The final convention integrated almost all the recommendations of the 1933 group of experts. Consequently, the key points of the 1926 IBU Gentleman's Agreement and the 1931 agreement on German-Polish Broadcasting had become the basis for an international convention for broadcasting in war and peace. Nevertheless, several countries objected to parts of the convention. Rumania, for instance, argued that the preventive and positive measures for broadcasting content were insufficient. It requested that a special clause be added. Each country should avoid the construction and operation of stations that due to their proximity to frontiers and their power "are capable of interfering appreciably with the reception of neighbouring countries of broadcasts of such countries."[137] Rumania had made the same request at the Madrid Conference in 1932, but was rejected for the political rather than technical nature of the request. The 1936 conference on propaganda also rejected the Rumanian request. The convention had already accounted extensively for the content of the broadcasts from these high power and frontier stations. These measures should suffice.

Not all expert advice found its way to the convention either. During discussions on the draft agreement, Poland questioned the experts' recommendation for supervising compliance with the convention. The experts advised that the IBU should fulfill that role. They argued that governments should act through the intermediary of their representative within the IBU. Poland disagreed. It argued that the OCI should be the one to act as intermediary. Article 7 of the convention

135 LoN doc. ser., C.I.C.I./16th Session/P.V.11, 3.

136 For some discussion of these drafts see: Registry file 5B, box R-3998, file 17704/1658 and 18271/1658, LoN.

137 LoN, "Intergovernmental Conference for the Conclusion of a Convention for the Use of Broadcasting in the Cause of Peace, Amendment Proposed by the Rumanian Government, Geneva, September 17th, 1936," Conf.E.R.P./13, registry file 5B, box R-3998, file 25664/1658, LoN.

complies with the Polish request. It states that involved parties would have to resolve the question via diplomatic channels. If these measures failed, one would have to bring the issue to the Permanent Court of International Justice. By common consent, however, the parties could also decide to appeal to the OCI. The OCI could then establish a special committee for this purpose. This solution could be an alternative to the diplomatic or international juridical path. The conference participants added Recommendation VII to the convention. As it was the first time the OCI would take on such a task, the special committee should preferably be chosen on the advice of and from the members of the IBU and the IECI. These advisors could possibly be complemented by recommended people from the ITU or the International Federation of Journalists. In the end, the convention did after all suggest involvement of the IBU in the supervision of the convention, although in a different form.[138]

Although the convention was drafted by European experts on the basis of mainly European experience with propaganda broadcasting, the final convention had a global scope. No fewer than 36 countries participated in the Paris Conference of 1936. In addition, Estonia, Latvia, Siam, the USA's International Section of the Federal Communications Commission, the IBU, and the Rockefeller Institute sent observers to the meetings. Ultimately a mere 28 out of 36 active participants signed the agreement in the course of 1936. Only Bulgaria, Chile, Cuba, Finland, Italy, Nicaragua, Portugal, and Sweden participated, but did not sign for a variety of reasons. Germany had refused to join the conference altogether.[139] At least two thirds of the signatories were European (see Table 5.1).[140] The list of countries that signed the convention suggests that the global convention had a largely European component both by reason of content and geographical territory.

138 LoN, "International Cooperation Organisation, Executive Committee, International Convention Concerning the Use of Broadcasting in the Cause of Peace, Draft Rules for Application, Explanatory Note, Geneva, December 12th, 1936," C.I.C.I./Com.Ex./79 (Geneva: LoN, 1936), registry file, 5B, box R-3999, file 27006/1658, LoN.

139 "Intellectual Cooperation Organisation, Executive Committee, Paris December 21st and 22nd, 1936," C.I.C.I./Com.Ex./79, 1-2, registry file 5B, R-3999, LoN.

140 It remains unclear how many countries signed the convention in between the end of 1936 and May 1, 1937, the official date of closure for the Signing of the Treaty. LoN, "Conference for the Conclusion of an International Convention for the Use of Broadcasting in the Cause of Peace, Action to be taken as a result of the Conference, Report by the French Representative, Geneva, October 9th, 1936," LoN doc. ser., C.440.1936.XII (Geneva: LoN, 1936), 2, registry file 5B, R-3999, file 25871/1658, LoN.

Table 5.1 – Signatories of the International Convention for the Use of Broadcasting in the Cause of Peace in the course of 1936

Albania	Czechoslovakia	Irish Free State	Rumania
Argentina	Denmark	Lithuania	Spain
Austria	Egypt	Luxembourg	Switzerland
Belgium	France	United States of Mexico	Turkey
United States of Brazil	Greece	Netherlands	USSR
United Kingdom	Hungary	New Zealand	Uruguay
Colombia	India	Norway	Yugoslavia

Source: LoN, Intellectual Cooperation Organisation, Executive Committee, Paris, December 1st and 22nd, 1936, C.I.C.I.Com.Ex./79 (Geneva: LoN, 1936), 1, registry file 5B, R-3999, file 27006/1658, LoN.

Table 5.2 – Ratification and accession to the International Convention for the Use of Broadcasting in the Cause of Peace by 1938

Ratified	Accessioned
Brazil	Australia
United Kingdom	Burma
Denmark	Southern Rhodesia
France	Union of South Africa
India	Guatemala
Luxembourg	
New Zealand	

Source: "International Convention Concerning the Use of Broadcasting in the Cause of Peace," The American Journal for International Law 32 (1938), 113.

The true reach of the convention did not deviate greatly from the number of initial signatories. In legal terms, the signing of a convention does not make its contents binding for national governments. A signature merely communicates an *intention* of the signatory country that it will deal further with the issue in its national parliament. Only when national parliaments implement the international convention into their national laws, does the country truly *ratify* it. In 1938 the Convention went into effect. By then seven countries that had participated in the conference had ratified the convention. Some five additional countries that had not participated accessioned it as well (see Table 5.2).[141] According to the OCI, by July 1939 some twenty-three countries had ratified the convention, which was "a striking result in the circumstances."[142] Furthermore, the members of the South American Broadcasting Union (USARD) agreed to the 1936 convention mostly indirectly by

141 "International Convention Concerning the Use of Broadcasting in the Cause of Peace," *The American Journal of International Law* 32, no. 3 (1938): 113-114; Spohrer, "Ruling the Airwaves," 349.
142 The documents do not specify further these twenty-three countries. LoN, *International Committee on Intellectual Co-Operation, Minutes of the Twenty-first Session, held at Geneva, July 20th, 1939,* LoN doc. ser., C.I.C.I./21st session/P.V.7, 53.

adopting a special clause in their own agreement that was based on previous drafts of the 1936 convention.[143] In spite of the large number of ratifications and countries that accessioned the agreement, it was an unfortunate state of affairs that the convention "could not be said to be thriving very vigorously yet," the OCI found in 1939.[144]

Over the following years the OCI stepped up its activities in day-to-day broadcasting. The committee increasingly worried about the further demise of civilization since the failure of the Disarmament Conference. Murray expressed it vividly: "The knowledge was there, the desire was there, but civilization, after three years of effort, seemed unable to save itself."[145] In case civilization was not lost forever it "at least merged in some unknown period of darkness."[146] In order to deal with this demise, the OCI above all tried to organize adherence to and ratification of the 1936 convention. Moreover, it tried to forge the convention by actively extending the positive measures for propaganda broadcasting in the cause of peace. In the years leading up to World War II, the OCI organized meetings of experts to discuss a large number of issues. These varied from options for unifying broadcasts to possible copyrights for music and authors' rights that until then had been arranged at a national rather than an international level.[147] The League frantically pursued its broadcasting project for peace within an increasingly deteriorating international climate.

Nationalism Shatters Peace Regulation

Rising nationalism shattered the IBU's options to regulate propaganda broadcasting. International relations changed drastically in 1933. Hitler rose to power in Germany and dictatorial regimes emerged in other parts of Europe such as Italy and Poland. Deteriorating national and international relations aroused slumbering nationalist sentiments. Incitement to warfare appeared to be just around the corner. These harsh sentiments immediately affected the drafting of propaganda broadcasting standards. On the surface, cooperation between the various parties

143 SdN, "Commission Internationale de Cooperation Intellectuelle, projet de résolution sur la radiodiffusion et la paix, Genève, le 18 juillet, 1935," C.I.C.I. 397 II, 1, registry file 5B, R-3998, file 18271/1658, LoN.
144 LoN doc. ser., C.I.C.I./21st session/P.V.7, 53.
145 LoN, *International Committee on Intellectual Co-Operation, Minutes of the Seventeenth Session, held at Geneva, July 15th, 1935,* LoN doc. ser., C.I.C.I./17th session/P.V.1, 8.
146 LoN, *International Committee on Intellectual Co-Operation, Minutes of the Seventeenth Session, held at Geneva, July 15th, 1935,* LoN doc. ser., C.I.C.I./17th session/P.V.1, 6.
147 See amongst others: LoN, *International Committee on Intellectual Co-Operation, Minutes of the Nineteenth Session, held at Geneva, July 15th-17th, 1937,* LoN doc. ser., C.I.C.I./19th session/P.V.1-8; LoN, *International Committee on Intellectual Co-Operation, Minutes of the Twentieth Session, held at Geneva, July 11th-15th, 1938,* LoN doc. ser., C.I.C.I./20th session/P.V1-10; registry file 5B, R-3999, file 27006/1658, file 27047/1658, file 27719/1658, file 30730/1658, file 32583/1658, 33585/1658, LoN.

continued without apparent difficulties. These sentiments nonetheless affected the international broadcasting community's working environment. Heated debates, suspicion, fear, and warfare over the airwaves typify the 1930s. The IBU, representing the major part of the European broadcasting community, had to obviate these contingencies.

Broadcasting's political implications severely tested the IBU's technical approach. Already in 1933, incidents forced the IBU Council to reconsider its officials' participation in external study groups on politically charged issues. The non-governmental IBU had always claimed to be what it called an "a-political" organization. Up to then this claim had allowed the IBU to work on politically charged issues in an efficient and technical manner. It never expected that its claim might be used against the Union. In contrast, the OCI study groups had an official political character. They were part of the intergovernmental League of Nations. For years the IBU could rely on its "a-political" approach and collaborate smoothly with the OCI on political issues. However by 1933, a number of incidents revealed that claiming an "a-political" policy could become a liability for the IBU if it wanted to pursue its internationalist ideals.

During discussions on the issue of broadcasting in the cause of peace, problems of a diplomatic nature arose. In 1933, the OCI invited several IBU officials to participate in their official capacity or as broadcasting experts in their own right. Official cooperation with the OCI in drafting agreements on propaganda issues contradicted the concept of an IBU that stood above politics. Nonetheless, no-one seemed to have thought about that. For the IBU it was a practical way of promoting the use of its medium for peace and rapprochement. It had no clear guidelines and none of its members questioned its participation. Then problems of a diplomatic nature arose between the Belgian IBU delegate and the OCI.

In 1932, IIIC director Henri Bonnet invited Baron van den Bosch to join the group of experts. The OCI regularly valued his opinion on various reports. When the OCI called the experts to meet in Paris, Van den Bosch had neither received an invitation nor a rejection. Van den Bosch explained to the IBU Council that an incident had led to an official protest at an IIIC meeting in Brussels.[148] As a result the Belgian Government had communicated the incident to the League of Nations. The OCI then decided not to invite Van den Bosch to the meeting of experts. The Belgian government in turn decided to refuse Van den Bosch participation in any study related to propaganda broadcasting in collaboration with OCI as well

148 P.V. Conseil Administrative 1933, 134, 137, IBU.

as possible voting on the topic.[149] The relationship between Belgium and the OCI became completely distorted.

Even though the nature of the incident remains unclear from the archives, it was important enough to provoke discussion in the IBU. Along with another related issue, it would eventually even change the IBU's international role as expert organization. Although in the end Van den Bosch was not part of the OCI study, other IBU officials like Burrows, Atkinson, Giesecke, Raested, and Sourek did participate. Sourek had performed an in-depth study on frontier stations and propaganda broadcasting. At the IBU Council meeting he clarified that the conclusions of the group of experts did not entirely match those of the individual broadcasting experts who had not been involved in the final editing process. Moreover, the OCI had consulted other experts outside European radio enterprises as well. The IBU Council decided in hindsight that all studies for the OCI on politically charged issues were and should be made in their own right. The OCI could not make any declaration in the name of the IBU. The IBU experts should warn the OCI that they had performed all their studies in their own right without any reference to the IBU.[150] The IBU officially distanced itself from any participation or advisory role in relation to politically charged issues. Cleverly, the decision to approve participation in their own right allowed these men to continue their activities on the very same politically charged issues.

The German-Austrian dispute that emerged in 1933 marked a turning point in transnational broadcasting. It can even be considered the first real broadcasting war over the airwaves. As Germany and Austria were both members of the IBU, the war affected the Union directly. The IBU's internationalist outlook turned out to be vulnerable to nationalistic thinking. Moreover, its claim of being "a-political" became an instrument that hindered rather than assisted its internationalist projects for society.

In 1934 the IBU Council decided that the study of political problems related to broadcasting was a matter for governments only.[151] For two years Austria and Germany had been at war over the airwaves. Their dispute led to litigations within the IBU between the German *Reichs Rundfunkgesellschaft* (RRG) and Austrian Ravag. Austrian Ravag protested strongly against German broadcasting politics. It also objected to one of Germany's most aggressive propaganda broadcasts under the leadership of Theo Habicht. He above all directed his propaganda broadcasts

149 P.V. Conseil Administrative 1935, 248, 267-268, IBU.

150 P.V. Conseil Administrative 1933, 146, IBU.

151 LoN, *Broadcasting and Peace*, 30.

towards Austria's dictatorial Dollfuß government. The Dollfuß government had been leading the Austrian Parliament with an almost non-existent majority. In March 1933 Dollfuß suspended Parliament indefinitely when expectations rose that the Austrian Nazi party, the SDAP, would gain a majority during the next election. Dollfuß now ruled by decree, to the dissatisfaction of the Germans.

In response, the Germans started serial broadcasts from the *Reichssender München* into Austria. For geographical reasons, these broadcasts could be received in most parts of Austria. Sometimes the reception was even better than that of Austrian stations. Habicht issued a series of broadcasts designed to rouse the Austrian population to revolt against their government. The broadcasts not only contained speeches by Habicht and by Frauenfeld, leader of the Austrian National-Socialist Party. They also included brief notes on the situation in Austria, describing the distress of the population and the curtailing of their freedom. Habicht warned that their government would further restrict their freedom in February 1934 by banning the Austrian SDAP altogether. Dollfuß was not pleased.

The subsequent dispute touched the very heart of IBU standards on propaganda broadcasting. Austria accused its German counterpart of deliberately broadcasting harmful propaganda to audiences across its borders. Germany claimed that it was merely informing its domestic audiences about Austrian political developments. In line with IBU recommendations, Austria requested rectification and sight of the scripts beforehand. Germany refused. RRG considered it impossible to control every single presenter in German broadcasting. This was a rather weak excuse. For years the European broadcasting community had been working on this particular propaganda issue. When Austrian protests had no effect, other countries like Italy, France, and Britain tried to interfere. The broadcasting war had become a diplomatic dispute.

Italian, French, and British ambassadors took turns in pressing Germany to stop nationalistic broadcasting activities against Austria. They urged Germany to comply with the Lausanne Protocol of 1932. This protocol obliged Austria to refrain from economic or political cooperation with the German Reich. More importantly, they urged Germany to comply with the Four Power Pact signed in July 1933, the Locarno Treaty of 1925, and the Kellogg-Briand Pact of 1928. The Four Power Pact between Germany, Great Britain, France, and Italy was designed to introduce a new era of cooperation in Europe under the guidance of the League. The pact should guarantee peace for the next ten years, but turned out to be relatively weak. In spite of diplomatic efforts, Germany refused to cooperate, claiming that it was adhering to international standards for propaganda broadcasting.

The German-Austrian broadcasting war was the main topic of a turbulent IBU Council meeting in 1934. Austrian Ravag accused the Nazi controlled RRG of poor broadcasting politics. Both organizations threatened their withdrawal from the IBU. Ravag would withdraw if the IBU could not prevent further German propaganda broadcasts. RRG would withdraw if it was accused officially of transmitting illicit propaganda.[152] To avoid such a development, RRG tried to detract attention from the subject.

Together with other IBU members, RRG rallied against the idea to discuss a draft of the Convention on the Use of Broadcasting in the Cause of Peace.[153] It argued that the issue did not belong on the Council's agenda as it had a political nature and addressed governments only. The official IBU contact for the League of Nations Maurice Rambert responded that he understood the critique. He did not want to ignore the Union's official statement that it did not interfere in political discussions, but rather tried to keep the microphone away from government influence. In particular, articles 4 and 5 of the convention would allow immediate and direct government participation in broadcasting programs. Rambert argued that IBU members should examine the consequences for their organizations and for the development of broadcasting at large. The IBU statutes pointed out that the consequences for the development of broadcasting did fall within the direct jurisdiction of the IBU.[154] Rambert put considerable effort into directing the microphone away from government influence via his politically charged negotiations with the League of Nations. The National-Socialist RRG did not want to collaborate with the League and had no interest in keeping the microphone away from governments. Its National-Socialist ideology clashed with the internationalist ideals of the IBU.

When the situation threatened to escalate, IBU President Carpendale made it his personal task to find a solution. A presidential delegation consisting of the Czech, Polish, and Swiss representatives considered all complaints and issues, including those of Austrian Ravag and German RRG.[155] In the end, the Council accepted a resolution stating that the IBU officially refrained from participation in political discussions. The decision was in compliance with the wishes of RRG. Furthermore, the Council advised its members to contact their governments about

152 Werner Schwipps, "Deutschland im Weltrundfunkverein 1933-1938," *Rundfunk und Fernsehen* 19, no. 4 (1971): 435.

153 Heinrich Giesecke, "Der Weltrundfunkverein: Seine Enststehung und sein Wirken bis zur Luzerner Konferenz," in *Weltfunkverein: Studien zum Weltrundfunk und Fernsehrundfunk* (Heidelberg; Berlin; Magdeburg: Kurt Vowinckel Verlag, 1941), 48.

154 P.V. Conseil Administrative 1934, 187-188, IBU.

155 Schwipps, "Deutschland im Weltrundfunkverein," 434.

the draft convention if they deemed it necessary. They could then take measures accordingly if desired.[156]

The Council decision suggests a profound change in IBU's approach to politically charged problems. The IBU would no longer participate in the discussions on the Convention for the Use of Broadcasting in the Cause of Peace. Nevertheless it could attend the 1936 conference as observer. In previous years the IBU had deliberately functioned as an intermediary in quarrels on propaganda broadcasts. Now the Union officially refused to interfere in such affairs. Furthermore, the IBU decided to halt all its previous efforts in the OCI relating to broadcasting propaganda disputes. The increased influence of nationalist ideologies and state intervention in broadcasting affairs forced the IBU to give way to the OCI and national governments. They reduced its ability to realize an international society characterized by friendship between nations and inclusion of all peoples. These decisions foregrounded the technical expertise of the organization over any possible form of techno-political expertise. The IBU now possessed fewer tools to pursue its internationalist aims for society.

The Spanish Civil War showed how much space to maneuver the IBU actually lost. Sören Brinkmann terms the Spanish Civil War "a kind of dress rehearsal for the catastrophe of World War II."[157] According to Alan Davies, it was also the first radio war where all broadcasting techniques were put to the test over a period of four years.[158] It was a war that quickly acquired a substantial international dimension, even though that would only become clear with hindsight some years later. It was a war between National-Socialists and Fascists on the one hand, and Russian Bolshevists on the other. Yet they never confronted each other directly. It was also a war that left international organizations like the League of Nations and the IBU completely helpless. The war opposed their principles as the Gentleman's Agreement already reflected in 1926. The broadcasting battles of the Spanish Civil War reveal the firmness of the shift from internationalist ideology towards nationalist ideologies in the European broadcasting community.

Initially the Spanish War was a domestic affair. It began with a military coup supported by poor land workers who threw off the ruling Republican socialist

156 P.V. Conseil Administrative 1934, 188, IBU.

157 "ein Art Generalprobe für die Katastrophe des Zweiten Weltkrieges." Sören Brinkmann, "Bilder eines Krieges: Europa und der Bürgerkrieg in Spanien," in *Europäische Öffentlichkeit: Transnationale Kommunikation seit dem 18. Jahrhundert*, ed. Jörg Requate and Martin Schulze Wessel (Frankfurt am Main; New York: Campus, 2002), 250.

158 Alan Davies, "The First Radio War: Broadcasting in the Spanish Civil War, 1936-1939," *Historical Journal of Film, Radio and Television* 19, no. 4 (1999): 494.

government. Compared to European averages, modern economic development in Spain had spread very unevenly across the country and across society. The coup had its origin in "the cumulative political, social and cultural anxieties provoked by a rapid, uneven and accelerating modernization." An unequal diffusion of modernization and fear of the progressive technical means of achieving ends had caused fear about "where change was leading."[159] The coup resulted in a civil war between the Nationalists who had started it and the Republicans who opposed it. The Republicans had control over the major transmitters covering the urban areas. The Nationalists used recently invented mobile transmitters. Both sides set up a center from which they organized their propaganda activities.[160]

The war quickly acquired a European character when Germany, Italy, and Russia started meddling in Spain's affairs. COMINTERN mobilized its entire logistical apparatus to support the Republicans in Spain. On a Europe-wide scale, all communist press offices recruited volunteers to join the war. No fewer than an estimated 59,000 mostly European volunteers joined. In the eyes of Brinkmann, this action set the tone for a more open and more connected Europe via communication structures like broadcasting.[161] Though COMINTERN supported the Republicans, its overall foreign policy had become less radical during the 1930s. Stalin favored a soft approach and did not want to antagonize western powers or Germany out of fear of repercussions in the east of Europe. Consequently, Minister of Foreign Affairs Litvinov applied some sort of crisis-management in Spain rather than a frontal attack on National Socialism.[162]

In response to Hitler's overt attack on Bolshevism, Russia redefined its role in Europe. It now promoted itself as the "stabilizer of the European order."[163] Russia became a member of the League of Nations in 1934 and signed the 1936 Convention on the Use of Broadcasting in the Cause of Peace. Beneath this surface it still aimed to separate itself from the post-war international order in Europe. Doubts amongst Soviet leaders, western distrust, as well as hostility by authoritarian states in Eastern Europe fed this ongoing desire for separation.[164] Via its approach in Spain, COMINTERN thus not only created an open European socialist communication structure. It created a great feeling of brotherhood among the

159 Helen Graham, *The Spanish Republic at War, 1936-1939* (Cambridge: Cambridge University Press, 2002), 1.
160 Davies, "The First Radio War," 484.
161 Brinkmann, "Bilder eines Krieges," 257.
162 Silvio Pons, *Stalin and the Inevitable War, 1936-1941* (London: Frank Cass, 2002), xi.
163 Pons, *Staling and the Inevitable War*, x.
164 Ibid., x-xi.

immense group of European volunteers and communist society as well.[165]

Germany and Italy in contrast came to the aid of the Nationalists. They provided them with support on all fronts. At the heart of German activities lay the work of Anti-Komintern established by Goebbels in 1933. Anti-Komintern sought to undermine Russian power in front of a global audience by means of anti-Bolshevist propaganda.[166] As such, it supported the Nationalists to degrees beyond comparison. The Nationalists acquired valuable information, technical equipment like a 20kw transmitter, knowledge on propaganda broadcasting and the like. Moreover, Anti-Komintern introduced the so-called ghost broadcasts. Such broadcasts appeared to be Republican, but were Nationalist instead.[167] The Italians supported these activities by covering broadcasting in the east of Spain. [168] Together Germany and Italy established a very effective support for the broadcasting war in Spain.

In a similar way to Russia, Germany also promoted itself as the defender of European hegemony. With hindsight Arnold Raested argues that in the eyes of Germany, Europe was a "field of power politics in which Germany would have to try to gain a hegemony of some kind."[169] Raested even recalls Hitler's use of "Europe" to achieve anti-Bolshevist hegemony. Hitler said that "the bolshevizing of Spain both menaced the peace of Europe and threatened to upset the balance of power in Europe." Furthermore, in 1939 Hitler added that Germany had intervened in Spain to "protect the European culture, and the real culture, against its Bolshevist enemies."[170]

The use of pro-European propaganda in a war like the Spanish Civil War became a means of gaining international support for personal ideologies. Russia used positive propaganda to promote Bolshevism as the stabilizer of European order. It avoided destructive propaganda against nationalist ideologies. On the contrary, nationalist ideology tried to weaken a global socialist ideology by claiming to defend European hegemony. It used a considerable amount of negative destructive Bolshevist propaganda. By then Hitler's attack on Soviet Bolshevism had, however, lost credibility. Countries in Europe seem not to have feared Socialism anymore. Even in Britain, where Hitler's propaganda had been effective for long, Russians were no longer feared.[171] Thus, even though the Nationalists won in Spain, Hitler's

165 Lorna L. Waddington, "The Anti-Komintern and Nazi Anti-Bolshevik Propaganda in the 1930s," *Journal of Contemporary History* 42, no. 4 (2007): 586-587; Brinkmann, "Bilder eines Krieges," 258-259.

166 Waddington, "The Anti-Komintern and Nazi Anti-Bolshevik Propaganda in the 1930s," 575.

167 Ibid., 582; Davies, "The First Radio War."

168 Davies, "The First Radio War," 484.

169 Ræsted, *Europe and the Atlantic World*, 42.

170 Ibid., 42.

171 Waddington, "The Anti-Komintern and Nazi Anti-Bolshevik Propaganda in the 1930s," 591.

anti-Bolshevist attack failed. It was National-Socialism and its *Lebensraum* politics that created widespread fear by 1939.

In the meantime, the League of Nations and the IBU saw a government that had supported their ideals of peace and understanding, collapsing in disarray.[172] They could not intervene. When war broke out in Spain, the Spanish broadcasters withdrew from the IBU. Their act rendered the resolutions and agreements of the IBU obsolete. Moreover, when Spain signed the 1936 Convention on the Use of Broadcasting in the Cause of Peace, it added a special clause. Spain retained the right "to put a stop by all possible means to propaganda liable adversely to affect internal order in Spain and involving a breach of the Convention..."[173] At the League Assembly, administrations discussed an OCI resolution about non-intervention in the conflict in Spain. They differed in opinion whether intervention or non-intervention would assure that the conflict did not cross Spain's borders.[174] The OCI foresaw a danger to European conditions. The League might degenerate into "a mere alliance of one group of States against another group." The Irish OCI participant De Valera even stated that it "would be the end of our hopes for a real League."[175] The League decided not to interfere in what it considered to be a domestic affair. In retrospect, the League decision made room for Germany and Russia to propagate and realize their ideologies at the expense of the internationalist ideals of the League and the IBU.

Rising nationalism would even divide the IBU from within.[176] In the second half of the 1930s, nationalist tendencies were spreading across Europe. They changed regimes in the east and gained support across countries in the south and west. Especially the smaller European states became main National Socialist targets. Whereas in Poland an authoritarian government came to power, Germany occupied smaller countries like Czechoslovakia and the larger Austria. These nationalist

172 The ruling government had supported the 1931 constitution based on democratic, liberal, and reformist principles.

173 LoN, "Intergovernmental Conference for the Conclusion of a Convention for the Use of Broadcasting in the Cause of Peace, Declaration by the Spanish Government, Geneva, September 23rd, 1936," Conf. E.R.P./26, 1, registry file 5B, R-3998, file 25664/1658 XXVI, LoN.

174 LoN, *The Eighteenth Ordinary Session of the Assembly, Verbatim Report of the Eleventh Meeting, held at Geneva, October 2nd, 1937*, box 92, file Société des Nations, Général VI, de 1936 à 1940, IBU.

175 LoN, *The Eighteenth Ordinary Session of the Assembly, October 2nd, 1937*, box 92, file SdN, Général VI, IBU.

176 This section is an exploratory attempt to delineate the influence of the individual in international non-governmental organizations at the hand of biographies, archive material from international organizations, and available secondary literature. In order to get a fuller and more reliable grasp these results should be complemented by the consultation of private archives or the archives of national organizations where those individuals have been employed.

ideologies imagined society in an essentially different way from the ruling international elites of the League of Nations and the IBU. Nationalist ideals entered such international organizations, creating ongoing internal tensions. A technical committee like the OCI dealt with these European changes by attracting a larger number of committee members from outside of Europe. The OCI maintained a global rather than a European outlook.[177] Such an organization as the IBU could not resolve the problem that easily. The IBU represented most of the European broadcasting community and considered Europe its core area of activity. In spite of its non-governmental character and its own claim of being an "a-political" Union, internally the IBU encountered changes and firm clashes between ideologies.

Individuals would play a key role in the fine-tuning of these ideologies inside the IBU. With the rise of new regimes, several members of the European broadcasting community were replaced by people who were loyal to these new regimes. In Germany, for instance, Hitler immediately reorganized the infrastructure of German broadcasting. He fired the heads of broadcasting, flattened the organization structure, and placed both organization and control over radio schedules under the direct supervision of Goebbels.[178] Consequently in 1933, IBU Vice-President Heinrich Giesecke had to leave RRG and was replaced by someone who represented the new regime. Shortly after, Polish and Austrian broadcasting faced a similar fate. Zygmunt Chamiec from Polskie Radjo had to leave the IBU in 1936, followed by Oskar Czeija of Austrian Ravag in 1938. In just a few years, the IBU lost its long-term vice-president, its director of the International Relay Committee, as well as its Rapprochement Committee director. These three men had been involved in the founding of the IBU, designing from scratch their idea of an internationalist society via broadcasting. New people who supported the totalitarian regimes replaced these men domestically and in the IBU. The IBU could only partially downplay the repercussions of these changes.

In Poland the economic recession of 1932 led to riots that replaced the Polish government with the Piłsudski regime in 1933. This regime wanted to have full control over Polskie Radjo. Such control nonetheless was impossible without the majority consent of the stakeholders.[179] Zygmunt Chamiec was not only the station's managing-director but also one of the major stakeholders. Always a former government supporter, Chamiec soon became isolated internally. The number of employees supporting the Piłsudski regime increased steadily. When the station

177 Renoliet, *L'Unesco oubliée*, 184-185.

178 Goebbels, "Der Rundfunk als achte Groβmacht."

179 The Polskie Radjo financial structure from 1926 onwards: 40% government; 40% stock market (specific banks and individual stakeholders like Chamiec); 20% stock market. Miazek, *Przeminelo z radiem*, 93.

appointed Piłsudski supporter Roman Starzynski as Vice Managing Director, Chamiec faced increasing difficulties. He now had a hard time keeping the microphone away from the government, which had been one of the leading principles of Polskie Radjo since its foundation. After fighting a losing battle, he was discharged as major stakeholder in August 1935.[180] Government now fully controlled Polskie Radjo and desired it be "led by completely loyal people, Zygmunt Chamiec was not one of them."[181] By preventing the government from employing broadcasting to achieve its own ends, Chamiec had become an enemy of the state.

The IBU managed to cope with the regime shift in Poland relatively easily. With the departure of Chamiec, the IBU had lost one of its most notable figures and most ardent supporters of its internationalist ideals. Poland lost its important role in the IBU. Chamiec had been IBU Vice-President since 1935 and head of the International Relay Committee since 1928. The new president of Polskie Radjo, Roman Starzynski, replaced Chamiec at the IBU. Starzynski did not get the chance to continue Chamiec's work. Instead he became head of the Budget Committee in 1937.[182] The IBU ejected Polskie Radjo from major IBU roles, placing Poland in a "leading" finance position where its nationalistic visions could do no harm. Moreover, the IBU merged the International Relay Committee over which Chamiec had presided with the Rapprochement Committee. The IBU appointed the Dutch Antoine Dubois as director of the new Program Committee.[183] The IBU relocated Chamiec's posts to men from small west European broadcasting

180 In the end the shareholders voted in favor of the change in organization structure, because, according to what Chamiec told his family, the Polish PPT delayed the listener payments to Polskie Radjo (license fees were paid to the PPT and were then forwarded to Polskie Radjo). As a consequence of these PPT actions, dividends to the stakeholders were delayed. Ibid., 187-215.

181 Ibid., 312.

182 The Budget Committee seems to have been relatively small and unimportant in respect to the other committees. The IBU did not have a large budget. The committee merely seems to have handled admission fees, and day-to-day administration costs such as meetings, conferences, committee budgets, and the like. Charles E. Sherman, "The International Broadcasting Union: A Study in Practical Internationalism," *EBU-Review* 25, no. 3 (1974): 32-36.

183 Dubois (1997-1956) was an electro-technical engineer and radio-expert. He had a background in wireless telephony for which he had travelled to the Dutch East-Indies and China. He had led and established the Radiotelegraphic service of the Dutch Ground forces, and became director of the Nederlandse Seintoestellenfabriek (NSF) after the war. The NSF built navigation apparatus for ships and aircraft, and constructed transmission and reception apparatus for wireless telegraphy and telephony. The NSF broadcasted from a transmitter in Hilversum. To finance transmissions, Dubois created the broadcasting organization, the Hilversumse Draadloze Omroep, in 1923. From 1928 onwards Dubois chaired the Dutch Comité van Omroepverenigingen, a consultative body that represented the interests of all Dutch broadcasting organizations. After having been the head of the IBU Program Committee for two years, he would become the new president of IBU in 1937. The Swiss C.A. Dümling would succeed him as new head of the Program Committee. J.H.J. van den Heuvel, "Dubois, Antoine (1887-1956)," *Biografisch Woordenboek van Nederland. Deel 1* ('s Gravenhage, 1979), 153-154, URL: http://www.inghist.nl/Onderzoek/Projecten/BWN/lemmata/bwn1/duboisa. Accessed on March 13, 2008.

organizations, men who had a history with the IBU and its internationalist ideals.

The merger of the International Relay Committee and the Rapprochement Committee also cushioned the problems resulting from the regime shift in Austria. Ravag managing-director Oskar Czeija had been the director of the Rapprochement Committee for ten years. To the dismay of the Nazi Party, Czeija had always received the full support of the Dollfuß government. According to the Party, Czeija was the personification of all Austrian evil. He was *politisch unzuverlässig*. After the Germans invaded Austria, they dismissed him in March 1938.[184] The Germans gave Ravag the name *Reichssender Wien*. The station fell under full National-Socialist control and was integrated into the German RRG.

The German IBU representative now represented the former Ravag as well. With the creation of the Program Committee, the IBU gave Czeija's former position to someone in the West. Consequently, the German IBU representative could not claim the one position in the IBU which allowed the holder to shape the image of a cultural society via music and plays. In Austria, Hitler did not replace all the main broadcasting staff as he had done in Germany. Several Ravag employees had been active in the IBU for a long time and could continue their work. For years E. Künsti had represented Czeija when he was unable to attend IBU meetings. He was well known at the IBU. Moreover, chief engineer Schwaiger had been a respected member of the Technical Committee from the days of the foundation of the IBU.[185] Schwaiger continued his work at the IBU and would become head of the committee when Raymond Braillard left in 1940. Both Künsti and Schwaiger had a long history of working along internationalist lines.[186] These developments suggest that the IBU managed to downplay several options that the National-Socialist invaders might have utilized to increase their influence over the Union.

Notwithstanding, the IBU could not refuse the new German representative an important IBU position. Germany was one of the largest and most important broadcasting nations in Europe. Its membership fee accounted for a large part of IBU revenues. German IBU representative Heinrich Giesecke had been known for his support of internationalist projects. The Nazi Party named him a "*Träger des alten System-Rundfunks*," a "barrow boy" of the former ideology. Goebbels replaced him with Kurt von Boeckmann. He had been forced to join the Nazi Party

184 Reinhard Schlögl, *Oskar Czeija: Radio- und Fernsehpionier, Unternehmer, Abenteurer* (Wien: Böhlau, 2005), 129.

185 "Conférence européenne d'ingénieurs en radiophonie (Salle E.) Palais des Nations, Genève, 6 & 7 juillet 1925, Disposition des places," Série no. 19, 6 juillet 1925, registry file 14, box R-1139, file 43391/28231, LoN.

186 See: file P.V. et documents complémentaires Commission Technique, 1925-1930, IBU; file P.V. Commission des Programmes [anciennement commissions de Rapprochement et Relais], 1926-1939, IBU.

and stood under clear watch of Goebbels himself. Von Boeckmann would replace Giesecke in the IBU.

Von Boeckmann was educated in history and philosophy and had a doctorate in law. He was fluent in Italian, English, and French, and had travelled the world from Italy and Greece to the United States. He had been director of the Research Institute for Cultural Morphology in Munich for five years when he started working at the *Deutsche Stunde in Bayern GmbH* in May 1925. Within months he was promoted from the position of program director to artistic director and business leader. In early 1928 he again got promoted to the position of managing-director.[187] In 1933 he would take over Giesecke's position in RRG and became managing-director of *Deutsche Welle*, the German short wave station near Berlin.[188] Von Boeckmann had been involved in broadcasting developments from the start when Giesecke and his companions managed RRG. He had a largely internationally oriented background, a love of music and a love of the medium of broadcasting.

Von Boeckmann seems to have struggled with his internationalist background and his work for the Nazi Party. According to Wilhem Schwipps, who has examined Von Boeckmann's role in German broadcasting during the Nazi regime, he opposed an ideology that he had to support, and supported an ideology that he had to reject. Consequently, Von Boeckmann was an important cog in the organization where the internationalist ideals of the IBU were fine-tuned with the nationalistic ideals of Nazi-Germany. According to Schwipps, his pre-1933 humanistic ideals and international orientation towards cooperation did not disappear when he joined the Nazi-Party.[189] In addition to his position as IBU Council member and vice-president from 1935 to 1936 and 1938 to 1939, he devoted much of his time in the IBU to his favorite topic: music. Between 1935 and 1939 he was on the International Relay Committee where he negotiated television's rightful place with the IBU rather than with the IECI in Rome.[190] He also took the international program *Youth sings across the frontier* (1935) into his brief.[191] Schwipps argues that in his heart, Von Boeckmann was an internationalist. According to his wife,

187 Schwipps, "Deutschland im Weltrundfunkverein," 430.

188 According to Wilhelm Schwipps the Party had to use soft pressure to make Von Boeckmann accept Giesecke's former position. Ibid., 430; Kurt von Boeckmann, *Von der Arbeit des Deutschen Kurzwellensenders* (S.l.: S.n., 1939).

189 Unfortunately, Schwipps neither reflects on the form and shape of these humanistic ideals nor the nature of the orientation towards international cooperation. Furthermore, though he clearly has consulted the archive of Deutsche Welle and the personal archive of Von Boeckmann, his references at times appear careless and sometimes even arbitrary.

190 P.V. Conseil Administrative 1935, IBU; IBU, "Procès-Verbaux des Réunions de la Commission de Rapprochement, tenue à Varsovie, juin 1935," Série 5273, 2 août 1935, 4-5, file P.V. Commission des Programmes [anciennement commissions de Rapprochement et Relais], 1926-1939, IBU.

191 P.V. Conseil Administrative 1936, 6, 22-23; IBU, *Twenty Years*, 63-64.

he resigned from the Nazi Party when war broke out, but the Party refused to accept his resignation. Von Boeckmann escaped into illness, an excuse that enabled him not to return to his desk but rather to spend his days in Bayern.[192]

Germany used the IBU and its internationalist ideals strategically as instruments to pursue its own targets. It was Von Boeckmann who brought such a strategy to the attention of Goebbels. In line with Germany's withdrawal from the League, Goebbels had also wanted to leave the IBU. Von Boeckmann argued that strategically such a move would be undesirable. Germany would lose its advantage of participating in important frequency debates. He pointed out that the IBU was also a key player in program exchange. The IBU could be used to propagate German culture and politics abroad. Nazi-culture could then reach areas like Scandinavia, the Baltic States, Hungary, Italy, and maybe even Turkey. Goebbels agreed that such a strategy could only benefit the propagation of Nazi-ideology.[193] In his eyes, such a move was an effective way of keeping up appearances. He positioned National-Socialism within an internationalist ideology that embraced the idea of sovereign nation-states. In reality he would rather use this position to gain support for Nazi-ideology and the destruction of internationalism from within.

Goebbels managed to greatly frustrate international activities even though Von Boeckmann was able to limit the damage. He ordered Von Boeckmann to create confusion in some parts, thwart resolutions unfavorable to the Nazi regime, and pursue Nazi ideology via the IBU's international programs. Von Boeckmann mostly impeded processes around IBU-League collaboration in propaganda broadcasting. These activities disturbed German transmissions of its ideology across borders. Sometimes Von Boeckmann disrupted negotiations on authors' rights and gramophones. These discussions were part of IBU and League collaborative efforts within the OCI to improve peace via broadcasting.[194] Von Boeckmann knew the IBU would merely appoint a sub-committee to study his questions further then continue its activities as planned. Other than mere irritations, these disturbances were not too damaging.

---

192 Schwipps, "Deutschland im Weltrundfunkverein," 440.

193 Ibid., 432-433.

194 See for extensive discussions on the topic: Box 86, file Relations internationales, Collaboration avec la S.D.N. & l'I.I.C.I. X, 1938, IBU; File P.V. Commission des Programmes [anciennement commissions de Rapprochement et Relais], 1926-1939, IBU; LoN, *International Committee on Intellectual Co-Operation, Minutes of the Nineteenth Session, held at Paris, July 16th 1937*, C.I.C.I./19th Session/P.V.6&8 (Geneva: LoN, 1937), 60-61,74; LoN, *International Committee on Intellectual Co-Operation, Minutes of the Twentieth Session, held in Geneva, July 14th, 1938*, C.I.C.I./20th Session/P.V.7&9 (Geneva: LoN, 1938), 57-60, 85-89; C.I.C.I./C.E.E.R.1-C.I.C.I./C.E.E.R.6, LoN; C.I.C.I./C.E.E.R./P.V.1- C.I.C.I./C.E.E.R./P.V.3, LoN; Registry file 5B, box R-3999, file 33583/1658, file 34534/1658, LoN.

The IBU did not passively watch RRG destroy all its propaganda activities. Von Boeckmann could easily refuse to conclude bilateral agreements with other countries on the illicit use of propaganda.[195] He simply never followed up on any of their requests.[196] He could less easily ruin the IBU's work in relation to international conventions. In 1936 Von Boeckmann tried to exclude the IBU from intervening in propaganda affairs even further. Vatican Radio had reopened the discussion. It proposed integrating the 1936 convention into the IBU Statutes.[197] This would lead to RRG's isolation in the IBU and from international relations at large. The question could best be dealt with in private meetings with the IBU president. Von Boeckmann objected to the idea on the basis of the 1934 Agreement that stated the IBU would not deal with politically charged questions. Although Vatican Radio withdrew its proposal in response, several IBU members regretted accepting the 1934 agreement. The BBC requested a revision. The Council reaffirmed that the IBU did not deal with politics, but decided that Burrows could attend the 1936 Conference in Paris as observer.[198]

This decision might be viewed as a first attempt to maneuver away from nationalist domination of discussions related to propaganda broadcasting. On behalf of the IBU, Burrows and legal director Sourek continued to attend expert meetings on broadcasting organized by the OCI. Burrows even took part in an advisory role, something the 1934 resolution had prohibited.[199] Moreover, Raested and Sourek remained involved in the drafting of the final 1936 convention. The IBU Council had encouraged its members to infiltrate their governmental delegations which were to draft the convention.[200] The IBU stepped up its activities to counter internally flourishing nationalist ideals via its normal and alternative routes.

Over the years, the Nazi Party not only used the IBU for its own ends. Schwipps suggests that the same thing happened the other way around. Von Boeckmann managed to counter some of these National-Socialist attempts to disturb the IBU's

195 Alexander J. Mackenzie, *Propaganda Boom* (London: J. Gifford, 1938).

196 Both Sourek of the Czech Radiojournal authorized by his Czech Minister of Foreign Affairs Edouard Beñez and Oscar Czeija of Austria Ravag made a proposal for a bilateral agreement with RRG along lines similar to the 1931 German-Polish Agreement. Schwipps, "Deutschland im Weltrundfunkverein," 435-436.

197 IBU, "Conférence mondiale sur la radiodiffusion et la paix à Genève – 1936, Lettre reçue de la station de radiodiffusion de la Cité du Vatican," Série 5931, 17 juin 1936, CA Documents Series, IBU.

198 Schwipps, "Deutschland im Weltrundfunkverein," 437-438. These quarrels seem to have remained outside the minutes of the IBU Council and Rapprochement Committee. The IBU does examine its possibilities for activities at the meeting in Paris. See: P.V. Conseil Administrative 1936, 39, 64.

199 SdN, *Organisation de Coopération Intellectuelle, Comité d'Experts chargé d'élaborer des suggestions concernant l'utilisation de la radiodiffusion dans l'intérêt de la paix, Procès-verbal provisoire de la deuxième séance, tenue à Genève, 17 juin 1938,* C.I.C.I./C.E.E.R./P.V.1-C.I.C.I./C.E.E.R./P.V.2, box 86, file Relations Innternationales, Collaboration avec la S.D.N. & l'I.I.C.I., X, 1938, IBU.

200 IBU, P.V. Conseil Administrative 1936, 58-59.

Figure 5.5 – IBU's visit to RRG in 1936
Source: Box 58 Photographies, IBU. Used by the courtesy of the European Broadcasting Union.

international peace-related activities. Von Boeckmann addressed relatively unimportant issues and managed to promote international peace-related efforts to Goebbels as National Socialist activities.[201] Though the IBU initially kept a close eye on Von Boekmann, Schwippes concludes that over the years, the European broadcasting community accepted him as one of their own. After the war, IBU general director H.W. Glogg apparently referred to him as a man who guarded

201 Schwipps suggests this, and the IBU archives seem to support this view. The personal archives of Von Boeckmann should confirm this more strongly. They can be found at the Bayerischer Rundfunk Historisches Archiv, Bestand NL/20 Nachlass Kurt von Boeckmann 1885-1950.

ideals of rapprochement and peace, and "suffered from…a world view with which he could not identify."[202] Von Boeckmann could not however stop National-Socialism attacking the internationalist ideals of the IBU from within.

In the course of the 1930s, the IBU became less effective in realizing its internationalist ideals. The IBU was not a static entity. As an "a-political" organization, the IBU could not refuse membership to the National-Socialist RRG, nor interfere in broadcasting wars like the German-Austrian or Spanish ones. It could not even ignore Germany as a venue for a meeting in 1937 (Figure 5.2). Moreover, the role of individual experts in reacting to changes in war and peace-related activities appears to have been incredibly important. IBU's claim of being an organization that did not meddle in political affairs and its open organizational structure made it an increasingly vulnerable organization. In a world of strong international tensions, the IBU proved less effective than it would have liked.

Conclusion

This chapter has explored broadcasting construction in the context of war and peace. In an internationally unstable situation, broadcasters tried to cement their first steps towards a transnational system. They envisioned and constructed a medium that in the eyes of the dominant elites was exemplary for low culture and segregation between people, and could easily be used for illicit propaganda across borders. The opinion of the elites and state officials was important to the broadcasting community, because these people influenced nation-state broadcasting policies, and tried to reconstruct Europe after the war. IBU promoters needed their collaboration in order to transform recommendations into legally binding agreements.

The IBU community rejected the use of broadcasting for illicit propaganda. In their eyes, illicit propaganda was the broadcasting of national programs directed towards the construction of national unification, causing political and ideological tensions across the border. These programs disturbed the IBU's efforts to build a European broadcasting network of national stations. Over the years, the IBU pursued preventive measures to counteract illicit propaganda. It also pushed for correct use of the medium which would help to build peace in Europe. Propaganda was not necessarily bad, as long as it was used for the benefit of all. To pursue its aim, the IBU sought integration into the European intellectual elite community

202 "…der persönlich unter dem zunehmenden Druck der Nationalsozialismus litt, mit dessem Weltanschauung er sich nicht indentifizieren konnte." Schwipps, "Deutschland im Weltrundfunkverein," 440-441.

as represented by the OCI. However, by the time these two organizations had established a good understanding, strained international relations interrupted their efforts.

In this context, the IBU increasingly became a victim of its own policy of depoliticization. Pressed by some of its members, the IBU decided to step down its efforts to regulate propaganda. It increasingly placed the regulation of propaganda broadcasting in the hands of the League's OCI, who pursued and designed the efforts amid quickly growing international tension. Consequently, the IBU had fewer tools to pursue its broader internationalist ideals, in particular fostering peace.

Over the years, fear of war and hope for peace affected the IBU's broadcasting efforts. Discussions on the standardization of propaganda broadcasting show that most of these projects were in line with plans for a global society. The IBU community did not concentrate on building a European society. During the various design processes they rather spoke the language of the internationalist movements focusing on international, national, local, civilization, and ethnic questions. Such discourse served the integration of the national into the international and vice versa. As a result, nationalist sentiments would become detached from warfare and coupled to ideas of peace and prosperity. Although Europe as a project did not feature in these discussions, broader visions of European civilization and culture formed the backdrop for the broadcasting expert community's global visions.

The change in who regulated propaganda might have reinforced the global dimension of the transnational discussions on standardizing propaganda broadcasting. Just like the League, the OCI pursued a global rather than a European program. The internationality of the program became even more pronounced when the OCI, anticipating increasing European tensions, hired more and more non-European members for its committee. The idea of propagating peace via broadcasting was a global aim.

Contrary to the IBU, both National Socialism and Russian Bolshevism began to exploit the European theme more prominently in their broadcasting projects in the 1930s. To the Nazi Party and to Stalinist Russia, the idea of a peaceful and united Europe served as a kind of Trojan Horse that would have to win over as many souls as possible for their true ideals. Although they employed similar rhetoric to the European broadcasting community, their ideas for European peace and prosperity rather served as a stepping stone towards an integrated third *Reich* and a Socialist world without borders.

Governments' role in propaganda broadcasting was considerable and they were inclined to mainly use the medium for domestic nationalist purposes. Consequently, broadcasting propaganda systems in Europe increasingly projected a Europe of national differences, stressing the dominance of some nations and the inferiority of others. At the same time, regulations on propaganda broadcasting increasingly found their way into international and national laws. These created an international world of nation states that peacefully co-existed and understood one another based on civilized European behavior. The intellectual elites in the OCI, the IBU, and governments worldwide tried to balance their interests and plans against a backdrop of greatly changing international relations. The outbreak of war brought all this international regulation of propaganda broadcasting to a halt.

To Goebbels the IBU was a potentially interesting tool for his propaganda efforts. He desired to "integrate" the National-Socialist RRG into the European broadcasting community because these were the people who discussed the day-to-day content. Over the years, broadcasters and related stakeholders became increasingly aware of the fact that the medium could diffuse and change ideas, culture, and opinion via international programs and program exchange. The following chapter focuses on the programming side of broadcasting where visions of European culture were negotiated.

Chapter 6
Broadcasting a Musical Culture

My own idea…– of which I have been fully conscious since I found myself as a composer – is brotherhood of peoples, brotherhood in spite of all wars and conflict. I try – to the best of my ability – to serve this idea in my music; therefore I don't reject any influence, be it Slovak, Romanian, Arabic, or from any other source. The source must only be clean, fresh and healthy![1]

Béla Bartók (1881-1945) wrote these words on January 10, 1931 in a letter to the Rumanian diplomat and music historian Octavian Beu. Bartók was an internationally renowned Hungarian composer, pianist, and ethnomusicologist. He was a man with a vision. He had an idea of what society should look like and thought this could be created via cultural expression like music. In his eyes, culture lay at the heart of society. Bartók recalled how his visions were born in a bar in 1904 when he heard a Transylvanian-born barmaid singing. He immediately committed the song to paper. "Now I have a new plan: to collect the finest Hungarian folksongs and raise them, adding the best possible piano accompaniments, to the level of art-song."[2] In his letter to Beu, Bartók explicitly promoted the integration of original, and therefore "clean, fresh, and healthy" influences of ethnic groups, be they "Slovak, Romanian, Arabic, or from any other source" into musical compositions. In his view, ethno-music and folk music in particular, should be part and parcel of the national canon, the national musical heritage.

With such modernist views on music, Bartók distanced himself from the majority of composers in small countries who had created a single style national musical canon.[3] Ideals and visions influenced music. As soon as a market developed

1 János Demény, ed., "Bartók to Octavian Beu, January 10, 1931," in *Béla Bartók: Letters* (London; Budapest: Faber and Faber Ltd.; Corvina Press, 1971), 201.

2 Malcolm Gillies, "Bartók, Béla," *Grove Music Online. Oxford Music Online*, n.d., http://www.oxfordmusiconline.com/subscriber/article/grove/music/40686pg3. Accessed July 1, 2009; "Notice sur le programme du concert européen hongrois, Budapest, le 4 novembre 1933," Série 3871, oct. 1933, 5, box 85, Concerts européens [subdivisé par pays], IBU.

3 Modernism in music refers to the development of new experimental forms of music which developed during the first half of the twentieth century. In between 1910-1945 there developed several styles within modernism like futurism, expressionism and impressionism. Composers like Stravinsky and Bartók formed a category all by themselves since they had gained inspiration from almost all periods in western

Figure 6.1 – Yugoslavian folk dance music
Source: "Ein glückliges neues Jahr 3.1.1937," box 58 Photographies, IBU. Used by the courtesy of the European Broadcasting Union.

for music in Europe in the course of the nineteenth century, these visions diffused further and faster. Composers in several countries, as in Bartok's motherland Hungary, had aimed to produce specific styles of music that best reflected their nation. As such, they developed very particular cultures of listening to and experiencing music. Single style musical canons formed the ultimate expression of national culture. They were a means of "catching up, of being among the most 'civilized' nations, and of acquiring power in fields other than opera or music."[4] Modernist compositions such as the works of Bartók, Leoš Janaček in the Czech lands or Karol Szymanowski in Poland actually destroyed such national canons from within since they progressively took account of ethnic diversity.[5] The visions on musical culture varied so much between modernist composers like Bartók and the majority of the composers in their countries, that a man like Bartók temporarily withdrew from public musical life in Hungary in 1912.

music history, including folk music. For a discussion of modernism in music see: Leon Botstein, "Modernism," *Grove Music Online. Oxford Music Online*, n.d., http://www.oxfordmusiconline.com:80/subscriber/article/grove/music/40625. Accessed September 5, 2011.

4 Philipp Ther, "The Genre of National Opera in a European Comparative Perspective," in *The Oxford Handbook to the New Cultural History of Music*, ed. Jane Fulcher (Oxford: Oxford University Press, 2011), 198.

5 Ther, "The Genre of National Opera," 197; Gillies, "Bartók, Béla."

Internationally, his visions and his music found more resonance. Bartók traveled Europe and the United States where he gave concerts and participated in live broadcasts.[6] Moreover, Bartók wrote his words to Beu exactly three days before he started work as a member of the Committee of Letters & Arts, one of the subcommittees of the League's International Institute on Intellectual Co-Operation (OCI).[7] In his letter, he positioned his interests beyond the nation-state into the heart of the ideals of the international cultural community and that of the League of Nations. He told Beu, that to the best of his abilities, he would try to serve the "brotherhood of peoples, brotherhood in spite of all wars and conflict." Bartók could now pursue his plans via his work on the Committee of Letters & Arts. Bartók not only radically changed music itself, he also performed live before the microphone in many countries, and participated actively in defining international cultural standards for broadcasting. Bartók therefore changed Europe's musical culture in many ways: through his music and his performance, as well as his international activities in international organizations.

The example of Bartók shows that there were people who actively changed culture and related visions of society via musical compositions or international cultural standards. Broadcasting developed at a time when a culture for music already existed in Europe, but where a tradition of listening to broadcasting was absent. By the turn of the century, listening to music had become a *social experience* that happened in *one place*, in a theater or at home, where people listened *attentively* to live performances or gramophone records *together*. As soon as broadcasters started to make programs, they found out that their medium did not fit these existing "standards," these existing cultures of listening. In these early years, listening to the radio was not a social event that people could experience together for an evening in a place like a concert hall. As Bartók had done when integrating folk rhymes into what he called "art songs," broadcasting experts tried to redefine music itself, adapting the music, the message, to their medium.[8] By resolving the problem of discontinuity in listening traditions, broadcasting experts specifically aimed to renegotiate culture, the heart of civilization, of society, in a way they considered appropriate.

6 In 1923 and 1925 Bartók traveled from Czechoslovakia to the Netherlands, Italy, and Switzerland. In 1929 he made a grand tour of three weeks in the Soviet Union, followed by a series of concerts in Switzerland, Denmark, Britain, the Netherlands, Germany, France, Italy, Austria, and Hungary. Gillies, "Bartók, Béla"; Malcolm Gillies, *Bartók Remembered* (London: Faber and Faber Ltd, 1990), 102-103, 225.

7 Though he would continue to hold this position for about five years, it is barely discussed in existing literature on his life and work. Gillies, "Bartók, Béla"; Gillies, *Bartók Remembered*; Kenneth Chalmers, *Béla Bartók* (London: Phaidon Press Ltd, 1995).

8 Brian Currid, *A National Acoustics: Music and Mass Publicity in Weimar and Nazi Germany* (Minneapolis: University of Minnesota Press, 2006), 119-120.

This chapter looks at the day-to-day programming side of broadcasting where visions of European culture were negotiated. In practice it was a continual process of trial and error. The question "*What was a good program?*" dominated policy negotiations until the outbreak of World War II and would remain unresolved. This chapter begins by showing how the IBU, the OCI, and the international cultural elite envisioned a different role for broadcasting in the creation of a European musical culture. They established a loosely coupled collaboration, but differed in opinion as to who was responsible for the content of these programs. They desired to create international programs and exchange. To make these programs a reality, experts had to resolve a number of issues. These included a lack of music supply for broadcasting, the technical and organizational construction of a social experience via live programs, as well as a suitable repertoire for broadcasting music. Through these efforts, they envisioned and shaped a European society, and with that a European identity.

A Cultural Policy for a New Medium

One pre-requisite was a suitable framework for programming. A cultural policy for broadcasting would establish an international program tradition that was absent. The IBU, OCI and international cultural elite built on their experience with the theater and the gramophone record. In the course of the nineteenth century, musical elites, artistic affiliations, traveling musical theater, and the opera had shaped listening cultures across borders in Europe.[9] A core characteristic of these cultures was that listening to music had become a social experience. Broadcasting experts and intellectual elite alike considered such social experience a bench mark for broadcasting programs. They nevertheless had diverging opinions about the role that broadcasting should have in society-building. Moreover, they did not agree on what "quality" actually meant in this respect. International cultural elites in general considerably valued the impact of what they called "art music" on identity building. Yet, it would be 1931 before the OCI evaluated the influence of music and broadcasting on unification efforts in a positive light.[10]

9 Heinz-Gerhard Haupt and Philipp Ther, *Europe and Beyond: Transfers, Networks and Markets for Musical Theatre in Modern Europe, 1740-1960*, Project proposal (Florence: European University Institute, 2008). The ongoing research program "Europe and Beyond: Transfers, Networks and Markets for Musical Theatre in Modern Europe, 1740-1960" is based at the European University Institute in Florence, Italy. The research is supervised by Prof. Heinz-Gerhard Haupt and coordinated by Prof. PhilippTher. The program links a large number of cultural research institutes from all over Europe.

10 "Radiodiffusion éducative: Collaboration avec la S.D.N., Rapporteur A.R. Burrows," Série 2796, 25

Broadcasting experts did not hesitate to take up the challenge. They felt they could create a social experience among listeners through broadcasting programs. Forming a cultural policy accordingly proved no easy task. Broadcasting did not automatically create the kind of musical experience with which people were familiar. Its technical peculiarities required a different approach. This envisioned social experience would immediately impact on the kind of society that was communicated and shaped via programs. In creating their cultural policy, broadcasters in fact actively and "hidden" from official government-related activities, worked on what could be defined as a "hidden" cultural unification of society.

IBU Council members discussed the problem at their first meetings already. They agreed that on admission, all members would have to agree to the "principle of cultural collaboration."[11] This was a commitment to exchange knowledge and technical information on programming. Moreover, members committed to helping forward IBU's ideal of international fellowship in the cultural area of programming, suppressing all possible causes for friction.[12] As such, the IBU directly related its programming activities to internationalist ideas of peace, understanding, and collaboration as pursued by the League of Nations. However, an IBU request to the League to collaborate on the issue did not find any resonance. Notwithstanding, in 1926 the IBU established the Committee on Intellectual, Social, and Artistic Rapprochement (from now on referred to as the Rapprochement Committee) to deal with this issue. The principle and the Rapprochement Committee provided the IBU with a practical means of giving cultural expression as it saw fit to these ideals.

The committee's core activity was to bring these visions into peoples' homes and educate their minds. To do so, the Rapprochement Committee constructed an extensive cross-border exchange of programs and an extensive supply of what it called international programs. People would have to have a social experience when they listened to broadcasting. They would have to feel part of a society built on internationalist values by getting acquainted with the music of other countries. The realization of such a sense of belonging and mutual interdependency was not

---

sept. 1931, 4, box 92, file Société des Nations, Section de Coopération Intellectuelle, Général, 1927-1932, IBU. For an exposé on German identity formation via music see: Celia Applegate and Pamela Potter, eds., *Music and German National Identity* (Chicago: University of Chicago Press, 2002). For an exposé on German identity formation and broadcasting programs see: Currid, *A National Acoustics*.

11 "Procès-verbal de la Commission de rapprochement, intellectuel, artistique et social tenue à Ouchy," Série 567, 17 juin 1927, 3, file P.V. Commission des Programmes [anciennement commissions de Rapprochement et Relais], 1926-1939, IBU.

12 Antoine Dubois, "L'Activité de la radiodiffusion internationale dans le domaine du rapprochement," *Radiodiffusion*, no. 1 (1935): 42; IBU, *Twenty Years of Activity of the International Broadcasting Union* (Geneva: IBU, 1945), 32.

easy to accomplish via a medium like broadcasting. Listeners did not gather in public places like a theater to take part in a communal experience using all their senses. People listened in the privacy of their homes, initially via headsets, later via loudspeakers. It was far more difficult to create an international feeling of unity in an area as large as Europe and where such huge numbers of ethnicities lived alongside each other. An extensive international exchange and supply of live cross-border programs would increase the amount of foreign music in broadcasts. In the eyes of the IBU, these activities would allow people to have the necessary communal experience via their radio sets.

The committee members did not agree about the kind of communal experience programs should express. Whereas some stressed the importance of actuality or entertainment, others stressed the need for educating the public and cultivating their tastes and visions of society. The committee for instance discussed writing radio drama and radio theater in 1929.[13] Several members felt that such plays should not focus on somber topics. Radio plays should lift listeners' tastes and educate their minds. Apparently, these discussants considered "the uplifting of tastes" and "avoidance of somber aspects of life" in radio plays to be ingredients for building *joie de vivre* among listeners. Others nevertheless argued that these plays should deal with all aspects of life.[14] Only by providing a full picture would people's minds be truly educated, they found. Although their ideas for a "suitable" program format differed, the committee members generally agreed that broadcasting served the education of society. Tastes would be raised, creating an international society based on peace and understanding.

Regarding international programs, the IBU particularly valued the idea of airing "live" events. It built on previous developments in the United States where broadcasters found airing so-called "live" programs an excellent way of giving listeners the impression of participating in a communal live experience. The transmission of live talent offered listeners a unique listening experience that they could not have elsewhere, neither by listening to records nor tuning into the channels of their competitors. Airing live talent in the United States had quickly become a quality standard for broadcasters. According to Michelle Hilmes, this standard would continue well into the 1960s.[15] Although the IBU certainly valued the competitive advantages and unique character of live events, it considered the cross-border bonding effects of particular importance. Participating in a live event

---

13 Since radio broadcasting by character differed from the printed arts, there was a need to rewrite existing plays or design new ones in order to come across clearly to audiences.

14 "Procès-verbaux de la réunion de la Commission de Rapprochement tenue à Londres, décembre 1928," Série 1171, 16 janvier 1929, 3-4, file P.V. Commissions des Programmes [anciennement commissions de Rapprochement et Relais], 1926-1939, IBU.

15 Michele Hilmes, *Only Connect: A Cultural History of Broadcasting in the United States* (Belmont, CA: Wadsworth/Thomson Learning, 2002), 43.

Table 6.1 - Series of programs arranged by the IBU, 1926-1939

Program	Technology and liveness	Geographical participation	Organization and liveness	Announcement	Total nr. of broadcasts	Success / failure*
National Nights (1926-1931)	Initially no cross-border relays and usage of national wireless stations only; from 1929 where possible usage of international long-distance relay network	European zone before Madrid Convention: European continent stopping at the borders of the USSR	Same time, same date, same program script; Monthly basis	Bilingual: [or trilingual?]	34	Relative success
European Concerts (1931-1939)	International long-distance relay network, incl. usage of short wave stations	European zone as defined by Madrid Conference & relays to USA and European colonies	Same time, same date, same program; Monthly basis	Trilingual: English, French, language of honored country	48-58; 30 of which between 1931-1934	Failure (1931-1934) Success (1935-1939)
National Concerts (1934)	Parts of the international long distance relay network	European zone as defined by Madrid Conference	Arbitrary time & date Requested foreign program	Bilingual: Languages of the two participating countries	unclear	Failure – Too many offers of too low quality
International Concerts (1932-1939)	Parts of the international long distance relay network	European zone as defined by Madrid Conference	Same time, same date, same program	Bilingual or multilingual**	1932: 123 1933: 367 1934: 762 1935: 779 1936: 881	Success
World Concerts (1936-1939)	International long-distance relay network and short wave connections to wireless stations or relay networks outside European borders	European zone as defined by Madrid Conference, USA, Japan, and European Colonies	Same time, same date, same program	Trilingual: See European concerts format	1936-1939: 6. Nr. 7 on Brazil Spring 1940 happened?	Success

* The success/failure as according to the IBU and its Rapprochement Committee: success = satisfactory; relative success = ok, could be better; failure = failure.

** Either bilingual in language originary country and French or in language of both relaying countries. In case of more than three participants the latter choice could not proceed. If so, than the languages of all participants needed be included.

Source: P.V. Conseil Administrative 1925-1939, IBU; Box 80ter Rapports Commission de Rapprochement et Rapports Commission de Relais/Programmes, 1927-1950, IBU; IBU, Broadcasting: An Instrument for the Creation of a Better Understanding Between the Peoples (Geneva, 1937), 91-93.

Table 6.2 – Unique events arranged by the IBU, 1934-1937

Program	Technology and liveness	Geographical participation	Organization and liveness	Announcement	Description
Christmas program (1934 -1935)	National broadcasting networks	European zone as defined by Madrid Conference	Exchange of gramophone records – no centralized broadcast		Christmas Hymns, carols and other musical expressions of the Christmas season enjoying popularity in Europe
Youth sings over the Frontiers (1935)	International long-distance relay network and intercontinental short wave connections to wireless stations or relay networks outside European borders	European zone as defined by Madrid Conference & relays to North and South America, Japan and European colonies	Live coordinated from Berlin – switching connections live from various countries		Choirs of children and young people in thirty-one countries situated in five continents singing sang to each other, and to the world at large, the song they love.
A bright and happy new year to all the World (1937)	The international long distance relay network	European zone as defined by Madrid Conference	Live from Berlin – 1 minute pre-recorded contributions of national organizations.	French, then announcement of the receiving country, 15 sec., then announcement + contribution of the honored country, 45 sec.	New Year whishes from European IBU members. Contributions are chosen freely per country, being of great variety and as typical as possible (bells, chansons, dance, choirs, etc).
Travelling Microphone	With the aid of portable transmitters and recording apparatus	European zone as defined by Madrid Conference			Employees of the IBU members visited historical places, festivals, or industries, building a panorama and sound of events.

Source: P.V. Conseil Administrative 1934-1937, IBU; Box 80ter Rapports Commission de Rapprochement et Rapports Commission de Relais/Programmes, 1927-1950, IBU; IBU, Broadcasting: An Instrument for the Creation of a Better Understanding Between the Peoples (Geneva, 1937), 92.

became associated with high-quality broadcasting as well as communal participation in a "unique" experience.

Over the years the IBU experts designed a wide variety of programs. It started by coordinating series of international live programs to be organized in rotation by its member organizations (Table 6.1). In the course of the 1930s it would add a list of unique and one-off programs to the repertoire (Table 6.2). Daniel Dayan and Elihu Katz call such live performances "media events."[16] They state that if "festive viewing is to ordinary viewing what holidays are to the everyday, these events are the high holidays of mass communication."[17] Events that were broadcast only once could have a considerable impact because of their unique character. The series of events could impact because of their repetitiveness. This "routinization" of media rituals would make such events an integrated part of the lives of the listeners.[18] According to Nick Couldry, serial events each in their own way simulate mass participation and cement the idea that broadcasting has ritual authority. Given the technical and organizational limits of the time, the programs which the IBU designed more or less fulfilled the quality ideal of society-building.

The expertise to build such programs and resolve any problems rested with the program directors of the IBU member organizations. In 1927 the Rapprochement Committee for the first time expressed a desire to organize regular meetings between these program directors, chaired by the IBU. However, the committee immediately rejected the idea due the lack of technical and organizational networks. Programming was mostly a national affair. The committee decided to consult its members when necessary, inviting the program directors to its meetings.[19] There they exchanged knowledge, dealt with international problems resulting from differences in national approach, and discussed cross-border program exchange as well

16 Andreas Fickers and Suzanne Lommers, "Eventing Europe: Broadcasting and the Mediated Performances of Europe," in *Materializing Europe: Transnational Infrastructures and the Project of Europe*, ed. Andreas Fickers and Alexander Badenoch (London: Palgrave, 2010), 225-251.

17 Daniel Dayan and Elihu Katz, *Media Events: The Live Broadcasting of History* (Cambridge, Mass.: Harvard University Press, 1992), 1. The definition of media events by Dayan and Katz is generally criticized for being of too rigid a nature for its stress on "pre-planned and highly stereotyped forms of mediated social interaction." Nick Couldry, instead wonders whether such events rather than telling something about social order tell us something about medial constructions of collective identities. Fickers and Lommers, "Eventing Europe"; Nick Couldry, *Media Rituals: A Critical Approach* (London; New York: Routledge, 2003), 56.

18 Couldry, *Media Rituals*, 55-74.

19 Over the years these directors regularly attended committee meetings. In 1935 the IBU decided to step up collaboration and organized its first meeting of program directors. "Procès-verbaux de la Commission de Rapprochement: Séance du 26 septembre 1927," Série 723, 13 oct. 1927, 7; "Procès-verbaux de la Réunion de la Commission de Rapprochement," Série 2671, 7 juillet 1931, 4; "Procès-verbaux des Réunions de la Commission de Rapprochement," Série 2924, 25 nov. 1931, 1-2, "Procès-verbaux des Réunions de la Commission de Rapprochement tenue à Varsovie," Série 5237, 2 août 1935, file P.V. Commissions des Programmes [anciennement commissions de Rapprochement et Relais], 1926-1939, IBU.

as new international program formats. Consequently, forthright international collaboration and exchange of ideas developed between national broadcasting organizations through the intermediary of the IBU. Working in a reluctant international climate, they shaped a cultural policy for broadcasting outside government legislation.

Differences in national legislation, vocabulary, and technological and organizational problems especially hindered their efforts. Several issues could not be dealt with beyond the realm of governmental regulation. Differences in national law with respect to authors' rights, artists' rights, and copyrights prevented an easy exchange of music across borders or international relay of live events. Moreover, radio reporters who had to travel for such programs experienced delays and problems at border control.[20] In order to resolve these problems, the broadcasting expert community required the support of governments, international governmental organizations, the music industry, and the international cultural elite.

The first signs that the cultural policy being formed would find resonance within a broader international community appeared in 1929. In February and November respectively, IIIC member Mr. Weiss and Director of the Information Section of the League Secretariat Mr. Pelt attended a Rapprochement Committee meeting.[21] Shortly afterwards, with its reorganization in 1930, the League OCI put the medium of broadcasting and the activity of music on its agenda.

Collaboration did not come about easily. Unlike the IBU, the OCI initially considered broadcasting an instrument for low culture expressions. Culture related questions were delicate since they directly expressed ideas of civilization, and touched at the heart of sentiments of identification. In their eyes, only expressions of high culture would be good enough to restore civilization. In response, the IBU invested years in building programs "with the aim of spreading the notion of radio's usefulness and making the public understand that there exist certain manifestations of art that can unite all European listeners."[22] When the IBU convinced the OCI and

20 LoN, *Committee of Experts to draw up suggestions on the use of broadcasting. Provisional minutes, First meeting held on June 17th 1938*, LoN doc. ser., C.I.C.I./C.E.E.R./P.V.1, 4-5, box 86, file Relations internationales, Collaboration avec la S.D.N. & l'I.I.C.I., X, 1938, IBU.

21 "Procès-verbaux de la Commission de Rapprochement tenue à Ouchy-Lausanne," Série 1411, 5 juin 1929, 1; "Procès-verbaux des réunions de la Commission de Rapprochement tenue à Barcelone," Série 1672, 16 déc. 1929, 1, file P.V. Commissions des Programmes [anciennement commissions de Rapprochement et Relais], 1926-1939, IBU.

22 "dans le but de répandre l'idée de l'utilité de la radiodiffusion et de faire comprendre au public qu'il existe certaines manifestations de l'art qui doivent unir tous les écouteurs européens." "Concerts EUROPEENS pour la saison 1931-1932, letter from Secretary General A.R. Burrows to IBU Members," Série 2694, 22 July 1931, box 85, file Relais, Concerts Européens, Général 1931 à 1939, IBU.

international musical elites to focus on broadcasting as a society-building instrument, they were critical of IBU's cultural policy.

When the IBU first contacted the OCI in 1926, the OCI had little or no interest in focusing on music as an art form that might promote society-building practices. It had no committee members with a background in music. Only the sub-committee of Letters & Arts could draw on the expertise of one associate partner, Austrian composer, conductor, and author, Felix von Weingartner.[23] His background was philosophy and he had forged an international career in the classical and opera scene in Germany and Austria. Weingarter also frequently traveled Europe, the United States, and Latin America as a guest composer.[24] Two years later, Letters & Arts could count on another associate participant with a background in music, the British Edward Joseph Dent. Dent was a professor of music at Cambridge who translated many libretti, wrote on Handel, Mozart, and several English operas.[25] If the occasion required, the committees representing other fields like literature and the sciences could consult Von Weingartner and Dent.

Applegate and Potter show that those who wrote about and debated music – musicians like Weingartner and Dent, but also writers, thinkers, statesmen and educators – were able to shape identity through music. Discussions about unity in taste and the effects of literary matters would "eventually spill over into musical matters."[26] This is exactly what happened in the OCI. Internationally renowned writers, thinkers, scientists, and educators joined hands in developing international consciousness, peace, and understanding. In light of reorganization, the Committee of Letters & Arts successfully advised that the OCI "should endeavour to extend its activities in the sphere of artistic relations by dealing with music as it has hitherto done with the plastic arts."[27] The issue of music now followed previous and ongoing international discussions on literature and the plastic arts.

It was at this point on January 13, 1931 that Béla Bartók joined the now permanent Committee of Letters & Arts for about five years. His joining coincided with the moment music and radio broadcasting were placed on the OCI agenda.

23 Associate partners were not an official committee member, but could be consulted if the occasion required. LoN, *International Committee on Intellectual Co-Operation. Minutes of the Eighth Session, held at Geneva, July 26th-29th, 1926*, LoN doc. ser., C.462.M.181.1926.XII (Geneva: LoN, Aug. 1926), 5.

24 Edler von Münzberg, "Weingartner, (Paul) Felix," *Grove Music Online. Oxford Music Online*, n.d., http://www.oxfordmusiconline.com:80/subscriber/article/grove/music/30042. Accessed April 21, 2011.

25 LoN, *International Committee on Intellectual Co-Operation. Minutes of the Tenth Session, held at Geneva, July 25th-30th, 1928*, LoN doc. ser., C.533.M.160.1928.XII (Geneva: LoN, Oct. 1928), 7.

26 Celia Applegate and Pamela Potter, "Germans as the 'People of Music': Genealogy of an Identity," in *Music and German National Identity*, ed. Celia Applegate and Pamela Potter (Chicago; London: University of Chicago Press, 2002), 3.

27 LoN, *International Committee on Intellectual Co-Operation. Minutes of the Eleventh Session, held at Geneva, July 2nd-26th, 1929*, LoN doc. ser., C.342.M.121.1929.XII (Geneva: LoN, Sept. 1929), 99.

The OCI, and in particular its Committee of Letters & Arts, started to value music as an important means of achieving the aims of the League. An interesting point to note is that Bartók had a clearly more modernist and progressive view of music than the associate advisor of the initial ICIC, Weingartner. Whereas Bartók was despised by the Hungarian elites for his progressive and inclusive compositions, Weingartner has only been scarcely associated with progressive schools of music. He is remembered as "one of the most eminent classical conductors of his day," and known "for the lack of exaggeration in his interpretations."[28] Bartók's progressive and modernist view of the use of music and broadcasting as instruments of society-building gave a new impetus to international efforts in the field. During his years with Letters & Arts, this committee would define how to influence the IBU's cultural policy efforts, in particular its programming activities.

Immediately after the reorganization of the OCI, well-known Czech author Karel Čapek proposed putting the creation of international broadcasting programs on the committee agenda. Čapek valued the possibilities of the medium for achieving mutual understanding. It could acquaint the public with the musical choices of different peoples. In his view, the OCI should study "the means to establish international radio broadcasting programs responding to this intention" in collaboration with the IBU and other interested organizations.[29] The committee decided to send the IBU a questionnaire to get a feel for its efforts and plans for musical repertoires and international programs. Part of the questionnaire focused on the kind of musical repertoire, while another part dealt with the practical organization and distribution of music. The questionnaire, subsequent exploratory studies, and discussions on the use of broadcasting for illicit propaganda (as discussed in Chapter 5), informed the OCI on the state-of-the-art of broadcasting policies. The OCI seriously studied the relationship between broadcasting programs and shaping society.

Shortly after a first in-depth study in 1933, the OCI and members of the musical elite reasoned that previous and ongoing international program efforts fell short in communicating the aims of the League. In 1934, head of the committee, Gilbert Murray, pleaded for intellectuals to have more influence over those broadcasting experts involved in designing broadcasting programs.[30] Moreover, a 1935 OCI study, in

28 Weingartner seems to have had a relatively traditional view on music and therefore seems to have contributed to single style musical canons. It remains unclear what his visions on the relation between music and broadcasting technology was. In any case he had an interest in recording his music. Von Münzberg, "Weingartner, (Paul) Felix." Accessed April 21, 2011.

29 "les moyens de faire établir des programmes internationaux de radiodiffusion répondant à ce dessein." "Radiodiffusion éducative," 5, Série 2796, box 92, file Société des Nations, Coopération Intellectuelle, IBU.

30 LoN, *International Committee on Intellectual Co-Operation, Minutes of the Sixteenth Session, held at Geneva, July 17th, 1934*, LoN doc. ser., C.I.C.I./16th session/P.V.4, 9.

collaboration with external experts, concluded negatively on ongoing efforts. These efforts fell short in educating and raising the public's cultural and civilized world view. One external expert was Franz Wilhelm Beidler-Wagner, doctor of letters and grandson of the master Richard Wagner. He had extensive knowledge of musicology. In his view, the current transmission of music via broadcasting was "already to some extent, condemned to remain deficient."[31] He recognized that broadcasting due to its technical nature could not transmit music in exactly the same way as people had known before. The ultimate specifics of the adaptations, he concluded mercilessly, had made the transmitted music of an inferior and popular quality.

Rather than rejecting the medium altogether, Beidler-Wagner suggested that musicians and the international cultural elite should still embrace the medium. They could employ it to promote their music, and open up their high art music to the masses. This would not only overcome their artistic isolation, but also bring the medium to a higher level. As such, they would resolve the existing problems with the medium.[32] Beidler-Wagner thus argued in favor of a growing influence of the musical elites in order to raise the content-side of program formats for broadcasting to a higher level.

Director of the OCI Institute Henri Bonnet, shared Beidler-Wagner's opinion. He felt that the IBU's ongoing efforts were too shallow. They required additional input of intellectual material, and the organization was inadequate. Their preparation times were too long and should be replaced by new mechanisms. Only then could one start to think about suitable types of programs. These could be commemorations of exceptional events, speeches that gave national perspectives on international matters, daily radio reports, as well as the regular exchange of programs on peoples' lives. Bonnet particularly valued the concept of broadcasting as a medium that aired live and current affairs, foreseeing that these would have an important impact on identity formation.[33] Such broadcasts would educate the public by uplifting their musical tastes and their awareness of international events. Bonnet argued that the broadcasting community should be aided by the intellectual elite to resolve the issue of low quality programs.

Bonnet envisioned a new central program committee. This could streamline the international coordination of programs, and improve their intellectual content more in line with the League's aims. In order to be effective, the committee would

31 "se trouve déjà, dans une certain mésure, condamné à démeurer deficiente." F.W. Beidler-Wagner et al., *Le rôle intellectuel et éducatif de la radiodiffusion*, Cahiers, 2 (Paris: Institut International de Coopération Intellectuelle, 1935), 215; Dieter Borchmeyer, *Drama and the World of Richard Wagner* (Princeton: Princeton University Press, 2003), 329.

32 Beidler-Wagner et al., *Le rôle intellectuel et éducatif de la radiodiffusion*, 210-215.

33 Ibid., 273, 275-276, 283-287.

require a firm national basis. Where bilateral agreements were already in place, these could form the basis for international collaboration, Bonnet believed.[34] Nonetheless, each country should have its own program committee consisting of qualified representatives outside broadcasting like the social services, intellectual and artistic institutions, the press, and user organizations. Together with technicians, these program groups would have to unite internationally and construct international programs that crossed borders. Although technical knowledge was necessary for their success, the intellectual elites should be responsible for content. Countries could speak to one another directly by means of programs in order to get to know, to understand, and to respect each other. What better route to pursue than music, which was a source of infinite joy, and provided the occasion for communication because of its universal language?[35]

Bonnet and Beidler-Wagner's plans reveal a shift in thought among the international intellectual elite. Gradually a sense of awareness emerged of the role that music and the medium of broadcasting could play in shaping society across borders. Although Beidler-Wagner was more reserved than Bonnet, they both understood that they could influence content and turn broadcasting into a medium that propagated their visions of society. What is more, Bonnet envisioned the use of such a medium and art to be an interaction between transnational developments without interfering with the sovereignty of national cultures.

The OCI's plan to streamline international programs immediately caused friction with the IBU. Bonnet had wondered whether an international committee would be a suitable foundation for giving people a feeling of brotherhood, of being part of a thriving civilization based on a common culture of nations. Bonnet questioned whether the IBU would be capable of developing the international organization of such programs. Previous experience had shown the multiple aspects one had to deal with internationally before an effective organization of international live programs could be put in place.[36] The IBU felt offended. It had invested years in creating a type of program that corresponded to its ideals and those of international society in a proper manner. Their content had even been completed in collaboration with institutions like national PTTs and national artistic affiliations.[37] The IBU had the feeling

34 Ibid.

35 Ibid., 275, 283-287.

36 Ibid., 285.

37 IBU, P.V. Conseil Adminstrative 1926, 58; "Nuits Nationales, Rapporteur A.R. Burrows, Commission de Rapprochement, no. 6 de l'Ordre du Jour," Série 1152, 4 déc. 1928, box 85, file Relais, Programmes Nationaux (Nuits Nationales), Général, 1926-1931, IBU; "Conseils pour l'organisation de nuit nationale allemande prevue pour le 18 décembre 1929," Série 1530, 10 oct. 1929, 1, box 85, file Nuits Nationales [subdivisé par pays], 1926-1931, IBU; Fickers and Lommers, "Eventing Europe," 231, 233.

that the OCI was trying to take over its role as the main developer of international broadcasting programs after all that it had done to make the OCI understand the importance of the medium.[38]

In the following months the IBU and the OCI managed to come to a kind of collaboration with which the two parties were comfortable. The issue was delicate. After a firm lobby, the OCI recalled that it had "never wanted to create an organism that is a kind of super state."[39] The IBU was capable of solving international program problems.[40] The IBU and the OCI developed a loosely coupled collaboration that reached beyond their institutional boundaries, including experts from international musical and intellectual society and PTT Administrations, as well as national artistic affiliations. The question remains how frequent and straightforward these national and local efforts were.[41] Via discussion and the exchange of knowledge, this widely diffused transnational collaboration below, at, and also beyond the national level impacted on the way quality standards for broadcasting have evolved.

One of the main challenges during discussions was that the IBU and intellectual elites had different definitions of high quality standards in international programs. By the early 1930s they valued equally their importance for society-building, in particular *live* broadcasting of *high culture* music. However they defined terms like "high quality" and "high culture" or "high art" music in a different way. What exactly did "high quality" mean when it came to educating the public with suitable visions of society via live program events? What kind of live broadcasts should be created to communicate feelings of belonging? And what caliber of music was of high enough cultural quality to truly reflect a feeling of international brotherhood? Then there was still the question whether the technology would be able to realize this sense of communal participation, and these quality standards. The discussions returned regularly when broadcasting experts worked on specific problems related to programming.

38 LoN, *International Committee on Intellectual Co-Operation, Report of the Committee on the Work of its Seventeenth Plenary Session*, LoN doc. ser., C.290.M.154.1935.XII (Geneva: LoN, 1935), 43.

39 "...celui-ci n'ayant jamais voulu créer un organe qui soit une sorte de super-état." IBU, P.V. Conseil Administrative 1936, 9.

40 Arthur R. Burrows, "Les moyens modernes de diffusion utilisés dans l'intérêt de la paix," *Radiodiffusion*, no. 5 (November 1937): 88.

41 The whereabouts of the practical collaboration between these institutions happened nationally and locally. They therefore do not surface in the archives of international organizations like the IBU or the League of Nations. These collaborative efforts deserve further examination.

Solutions to "Discothèque" and Copyright Problems

A proper music supply was lacking. It was virtually impossible for broadcasting organizations to provide round-the-clock live programs. The IBU sought a solution with the help of the OCI. The IBU tried to get artists to perform before the microphone and examined the possibility of an audio library especially for broadcasting. Both efforts immediately raised problems with the record industry. The diffusion of music across borders also caused legal problems. The OCI recognized it needed to help resolve those problems because they obstructed the airing of desirable high quality live events. Nevertheless it felt that for expressing European culture, other affiliations than broadcasters would be better equipped to envision the "just" musical choices that best expressed society.

In order to meet the demand for music, broadcasting organizations in particular tried to relay live performances of concerts, artists, and events. Over the years, they encountered considerable difficulty in attracting artists of worldwide fame to perform live. Many artists were accustomed to working with gramophone reproducers and hesitated to participate in live broadcasts.[42] In 1926 Austrian Ravag for instance complained to the IBU. The organization had tried to relay a live opera from the Viennese Opera house to its audiences, but failed. The main opera singer contracted by the Viennese Opera for the evening, Mrs. Jeritza, refused to sing before the microphone that Austrian Ravag had installed in the theater. It transpired she had an American gramophone company contract that contained a clause stating that she could not participate in radiotelephony transmissions.[43] The record industry feared that listeners at home would record such live performances of artists whom they had contracted. Such home recordings would reduce their profits. They preferred their artists to abstain from broadcasting performances.

With an unparalleled growth in demand for music and a reluctance among musical artists to participate in broadcasting, the IBU envisioned a central musical library under the auspices of the IBU in Geneva. It imagined becoming a clearing house for music suitable for broadcasting. Such a central library should contain as complete a collection as possible of all the finest national music of its members. This should also include recordings of live concerts, of specific musical tunes as well as of entire broadcasting programs. The IBU would make and then distribute complete lists of the musical collection to its members, giving them a clear indication of the readily

42 IBU, *Union Internationale de Radiophonie* (Geneva: IBU, 1926), 16.

43 "Commission juridique et Commission intellectuelle, artistique & social. Opposition des companies des gramophones aux radiodiffusions. Exposé," 3 juillet 1926, 1, box 32, file Gramophones. Le dossier général I, à 1933, IBU.

Figure 6.2 – Audio archives of the RRG, Germany, 1936
Source: Box 58 Photographies, IBU. Used by the courtesy of the European Broadcasting Union.

available musical repertoire. In February 1929 the IBU Council decided the idea would be too comprehensive and too costly.[44] High customs-duties that were levied on records imported to Geneva as well as their re-dispatch at a later date would not only incur high expenses but also cause considerable delay. An international library containing a collection that was the sum total of the finest of national musical heritages in Europe was a practical impossibility.

An exchange of music organized between broadcasting organizations would be more beneficial.[45] Each would have to create its own library consisting of as complete a collection of national music as possible. It should contain "songs, melodies, national and popular dance music in the form of sheet music, phonograph cylinders, and mechanical reproduction of vocal and instrumental performances from their own respective countries."[46] Members would then have to keep adequate lists of the recordings and send these to the IBU on a regular monthly basis.[47] The IBU

44 IBU, P.V. Conseil Administrative 1929, 180.
45 LoN doc. ser., C.I.C.I./C.E.E.R./P.V.1, 19.
46 "chants, melodies et danses populaires et nationales sous forme de partitions, cylindres phonographiques, reproduction mécaniques d'exécutions instrumentales ou vocales propres à leurs pays respectifs." "Procès-verbaux de la réunion de la Commission de Rapprochement," Série 1171, 10, file P.V. Commissions des Programmes, IBU.
47 "First list of records which members have ready to place at the disposal of other members of the

in turn would redistribute the lists to all its members. In 1932 the IBU added that the actual exchange of records based on the distributed lists could take place bilaterally between members. In addition members should maintain and continue to complete their library collections. Over the years, these libraries would then provide an invaluable amount of added value to broadcasting.[48] In spite of the decentralized organization of these audio libraries, the central distribution of the lists resulted in an intense cross-border exchange of recordings.

Both the employment of artists in front of the microphone as well as this intense cross-border exchange forced broadcasters to deal with the established record industry. Rather than buying the gramophone records, people everywhere in Europe could now listen to the recordings for free. As a consequence, manufacturers, merchants, and hagglers (in particular composers and editors) tried to hamper the easy transmission of art music over the radio, and as such hampered the exchange of music or musical programs between broadcasters. Moreover, the record industry's national legislation was not necessarily the same in every country. Whereas these activities had caused legal problems within borders already, the cross border exchange and transmission of live events further complicated matters. Both the IBU and the international musical elites considered such a state of affairs problematic and disadvantageous to their interests. Musicians and artists were hindered in using the medium to promote their work, and the record industry regularly accused broadcasters of illegally using recorded music. On several occasions the industry even took legal steps to secure its economic interests.[49]

Initially the broadcasting organizations established relatively positive contacts with the record industry. They agreed to specify in their program guides the particulars of the recorded music by showing the name and reference number of the gramophone record. This meant listeners could easily trace the music and know where to buy it. Initially, the increase in sales of recordings went hand in hand with the broadcasting of these recordings. According to the IBU, matters were arranged to the satisfaction of all parties in spite of occasional accusations. By the early 1930s nonetheless, relations with the record industry deteriorated rapidly. In the eyes of IBU legal director Sourek, the general economic crisis had caused a devastating drop in record sales. The record industry however, argued differently. It claimed that the frequent broadcasting of their recordings had made the public stop buying the records.[50]

Union," Série 7476, July 1932, CA Document Series, IBU.

48 IBU, P.V. Conseil Administrative 1938, 228.

49 Beidler-Wagner et al., *Le rôle intellectuel et éducatif de la radiodiffusion*, 211.

50 "Revendications des sociétés de gramophone en europe centrale, Rapporteur: Mr. le Dr. Sourek," Série 3193, 26 mai 1932, 1, box, 32, file La radio et l'Industrie du disque, IBU.

The IBU tried to resolve the friction with the music industry and with artists. It stimulated the fine-tuning of national legislation across borders for copyright, authors' rights, artists' rights, and performers' rights. In this way it was trying to resolve border problems and at the same time protect artists and authors against the illicit reproduction of their work. The IBU wanted to show artists that despite their agreements with editors, they could freely accept radio work. Their participation, whether live or recorded, should be protected internationally "against the commercial use of broadcasts transmission of recordings and the right to broadcast recorded works."[51] To this end, the IBU prohibited its members from using recordings for other than the principle of cultural collaboration, assuring the record industry that their records would not be used for commercial purposes.[52] Its moral codes could prevent irregularities between the record industry, editors, and broadcasting organizations. Over the years, international companies demanded large sums of money from the broadcasting organizations in their own countries. Already in 1932 a French company started to print on its records "radio broadcasting prohibited."[53]

A Committee of Letters & Arts proposal put forward by Bartók and Čapek created unexpected opportunities for standardizing the international exchange and relaying of music. In 1931, they proposed studying international collaboration in collecting registered music, *musique enrégistrée*.[54] Bartók and Čapek stressed the need for an extensive listing of existing national and international music as well as extensive documentation of these recordings. Bartók had a long history of collecting recordings of "genuine" music. He had created a private library consisting of music recordings as well as specific folk and dance tunes, which he had recorded with the phonograph. In his view, a listing like this would complete what he then called the "actual existing collection," nowadays known as "musical heritage."[55] Such lists and collections would "modernize" music and inspire composers by provoking new musical compositions. This part of the proposal leaned heavily on the life and work of Bartók. However the two men also contended that such lists and collections would make a wide diversity of music more readily available

51 IBU, *Twenty Years*, 41.
52 "Procès-verbal de la Commission de rapprochement," Série 567, 3, file P.V. Commission des Programmes, IBU.
53 "Radiodiffusion interdite." "La radio et l'industrie de disque," Série 8793, 9 juin 1944, 2, box 32, file La radio et l'Industrie du disque, IBU.
54 "Radiodiffusion éducative," Série 2796, 4-5, box 92, file Société des Nations, Coopération Intellectuelle, IBU.
55 Gillies, "Bartók, Béla."

to artistic institutions and broadcasting organizations alike. They envisioned an international "discothèque."[56]

Both the problem and the charm of the resolution was that the proposal related to music beyond its use for broadcasting. The charm was that an incredible number of interested institutions and artistic affiliations would collaborate internationally in establishing an extensive and highly diversified collection of European music. Similar to the IBU's audio library effort, this international "discothèque" should be decentralized within countries and connected via distributed lists. Local, national, and international artistic institutions would collaborate closely. The problem with this collection was that broadcasters would not be able to influence the process and its success would depend on whether others were willing to let them use the recordings. Such an international "discothèque" would only really benefit broadcasters if they themselves were able to influence and coordinate its collections.

Over the years the League would study in depth the plan for an international convention related to the recording, documentation, and cross-border exchange of music. The convention was a complicated idea. It meant organizing an international "discothèque," as well as dealing with authors' rights, artists' rights, copyrights, and custom duties. The League formed a study group to examine the possible links between institutions in charge of sound and gramophone music along with broadcasting organizations. Furthermore, the League's Technical Committee on Economic and Financial questions would examine the aspect of custom duties on cross-border record exchange.[57] The League's efforts to create some kind of international convention in collaboration with a large number of stakeholders gave a new impetus to IBU's initial dealings with the record industry. With the League directly involved, the chances of dealing successfully with differences in national legislation increased, establishing a sound basis for international live broadcasting events.

Whereas the IBU was pleased to have governmental parties on board, it desired the control and security of a certain supply of music for broadcasting. The question of broadcasting organizations' participation was discussed. In 1938 the study group envisioned how to set up national documentation of recorded music. OCI member Laura Dreyfus-Barney, posed the question whether the national sound and gramophone institutions were willing to lend their material to broadcasting organizations in their own countries. Burrows, who had been invited to the meeting, argued that years ago IBU members had already started developing their own

56 "Bibliothèque de disques reproduissant des danses et chansons populaires (Rapporteur: A.R. Burrows)," Série 3154, 19 mai 1932, 2, CA Document Series, IBU.

57 LoN, *Intellectual Co-Operation Organisation, Modern Means of Spreading Information Utilised in the Cause of Peace*, LoN doc. ser., A.19.1938.XII (Geneva: LoN, 1938), 6, box 86, file Relations internationales, Collaboration avec la S.D.N. & l'I.I.C.I., X, 1938, IBU.

sound archives and so these were already being created. He gained the support of Murrow, European director of the American CBS. Murrow argued that broadcasting organizations would be best suited to collect musical repertoires. Already "through the system of exchange that had been organized, each broadcasting company could quite easily procure any records which it needed, but the establishment of richer and richer collections ought to be encouraged."[58] Their arguments did not help them.

The final resolution favored the establishment of national sound and gramophone repositories "with a view to international loans and exchange, which may prove to be invaluable auxiliaries to broadcasting."[59] National musical collections should be established separately from the broadcasting organizations' existing sound archives, but be open to use for broadcasting. The League clearly desired that artistic affiliations and the international intellectual elite would have control over the creation of national musical repositories. These experts were better able to create and look after Europe's musical repertoire, its musical heritage.

The plan for an international convention did not make it through the 1930s. The convention had envisioned a European musical repertoire consisting of decentralized national collections, registrations and exchanged musical recordings. Heated debates on all aspects of this international exchange proceeded slowly towards an agreement. They culminated in the organization of a conference to conclude the Berne Convention in 1939. Discussions were fruitful though difficult.[60] By the time of the outbreak of World War II, one of the IBU study group stated that there was resistance from the authors' societies. The main bottleneck was their traditional approach to music, including the principles attached to copyright and their lack of knowledge of the medium. The plan for an international convention was halted by the outbreak of World War II.[61] After the war, the international partners would take up the discussion again.[62]

A lack of clear agreements did not prevent an extensive cross-border flow of broadcasting programs and recorded music. While the international cultural elite discussed an international convention, the IBU continued coordinating the practical construction of decentralized sound libraries for broadcasting. Their collections of recorded music played an increasingly important role in daily broadcasting.

58 LoN doc. ser., C.I.C.I./C.E.E.R./P.V.1, 20-21, 24, 25.

59 LoN doc. ser., A.19.1938.XII, 6.

60 For in-depth study see: Box 32, gramophones, IBU; Box 86, Relations Internationales, IBU; C.I.C.I./C.E.E.R.1-C.I.C.I./C.E.E.R.6, LoN; C.I.C.I./C.E.E.R/P.V.1 – C.I.C.I./C.E.E.R/P.V.3, LoN; registry file 5B, box R-2239 music, LoN; registry file 5B, box R-2258 Library of Sound films on folklore, LoN; registry file 5B, box R-2259 Broadcasting and Peace, LoN; registry file 5B, box R-3998-3999 Broadcasting and Peace, LoN.

61 Burrows, "Les moyens modernes de diffusion utilisés dans l'intérêt de la paix," 86.

62 "La radio," Série 8793, 2, box 32, file La radio et l'Industrie du disque, IBU.

They not only enabled music exchange, but also created a feeling of international belonging among listeners. Moreover, gramophone recording technology would come to play an important part in creating an artificial sense of live participation in events.

The International Live Program Constructed

Live programming was a complicated affair. An international study on the role of broadcasting in creating a feeling of community proposed:

> We could cite a group of listeners who have requested the transmitting station to always transmit the audience's applause after concerts in a public space: in that way they then felt more «present». It is quite characteristic. There is concern for these delicate problems, but no sufficient evidence to judge whether the efforts to provoke new forms of «community» via radio have already been successful.[63]

It was not easy to create a sense of belonging in broadcasting audiences. The study was dissatisfied with how live broadcasting events up to 1935 had been. There were various options and apparently not every option worked well. Broadcasters also struggled with this issue. The mere idea of having applause after a concert broadcast from a concert hall was not enough. Over the years, the IBU and its members designed unique and serial international events that in their own way tried to create a sense of community. Endless technical innovations changed their options for designing these events and therefore changed their options for envisioning specific kinds of society. The IBU developed a growing awareness of those program formats which it thought best expressed a feeling of international belonging.

The IBU community distinguished two main criteria that determined the quality of society-building via live broadcasting programs. In the first place, a live event would have to give listeners the feeling of participating in the event together with many other people. In the eyes of the IBU such a mutual experience could only happen if listeners were aware that they were hearing exactly the same program as someone else on the European continent listening in. Secondly, it considered that

63 "On pourrait citer le cas d'un groupe d'auditeurs qui ont prié la station émettrice de transmettre toujours les applaudissements du public après la transmission des concerts donnés dans une salle publique: ils se sentaient ainsi plus « présents ». C'est tout à fait characteristique. On se préoccupe certes de ces problems délicats, mais nous ne possédons pas les éléments suffisants pour pouvoir juger si les tentatives de provoquer par la radiodiffusion de nouvelles formes de « communauté » ont déjà été couronnées de succès." Beidler-Wagner et al., *Le rôle intellectuel et éducatif de la radiodiffusion*, 218-219.

the ultimate expression of community depended on an as inclusive as possible participation of all broadcasting stations in Europe and beyond, the latter being significant for production. Not only would high participation in an event create a wider diffusion of the experience, it would ultimately express a sense of communal production, of unity in society.

Initially broadcasters lacked the technology to create high quality live programs that would propagate a social experience among listeners. The quality of wireless was too low to ensure reception of the event across the entire European continent. The quality of a received wireless signal, if received at all, would be too low to ensure a pleasing experience of the live event for all listeners. Furthermore, in the 1920s there was no international relay network that connected all broadcasting stations to therefore secure a Europe-wide diffusion of international live events. Consequently, when the IBU discussed international live events for the first time it had to find alternatives that would at least give the impression that their programs were broadcast live internationally. At the suggestion of the Spanish Radio SA., the IBU began the serial broadcasting of *Nuits Nationales*, National Nights in 1926.[64]

The idea of the Nights was that each European IBU member organization in turn would organize a Night showcasing their country's culture. With a lack of cross-border relaying options, the showcased country created a program script as well as additional promotional material which it sent to the IBU. The IBU would translate the program script into English, German, and French then redistribute it to all IBU members participating in the event. The program script gave everyone the main outline of the Night. If there was space in the script for music and literature, broadcasters everywhere in Europe could choose from a list of pre-selected repertoire. Now imagine a Night featuring the life and culture of Hungary. The Dutch broadcasting organization might have chosen the Hungarian music of Smetana while the BBC chose to broadcast Bartók. Consequently, each Night was broadcast on the same evening and at the same time in every participating European country.[65] Nevertheless, it only approached the idea of communal listening to one unique broadcast. Each station broadcasted individually a Night based on its own national adaptation of the program script.

The IBU considered such nationally appropriated program scripts an inferior quality standard when it came to achieving a community effect. Listeners did not participate in one unique live event and consequently could not optimally experience a shared sense of belonging with listeners in other countries. Although the

64 "Letter from Secretary General Arthur Burrows to IBU members," Série 203 and also see: 204, 18 May 1926, box 85, file Relais, Programmes Nationaux (Nuits Nationales), Général, 1926, IBU.

65 IBU, P.V. Conseil Administrative 1926, 58; IBU, P.V. Conseil Administrative 1927, 97, 109-110.

Nights were broadcast live in each nation independently, they did give an impression of wider participation because the presenters announced all the participating countries. One could say that the Nights were an evening of nationally adapted international programs featuring a Europe of nation states. They did not live up to the idea of high quality international programs.

After the first Night, the IBU considered the series a temporary solution until technology would allow for a true live experience.[66] Already in 1927 when the Nights were still molded in their final format, Burrows dreamt of new program formats, the so-called *Semaines européennes* European weeks, or *Jours européenns* European days.[67] Such live programs would have the advantage that "all European broadcasting organizations could transmit the programs of all countries, which from a psychological point of view would have a marvelous effect on the listeners."[68] In the eyes of Burrows, the international relay network under construction would increase the effect of a communal experience since it would allow the live relay of a local event to all corners of Europe. The "best way to let foreign countries appreciate the music and literature of your countries probably consists of relaying from your studios or another hall, by telephonic circuit, a typical national program performed by your best artists."[69] By 1929 when parts of such a relay network were in place, several countries already began broadcasting the National Nights live. The new technological developments required a different type of program, veritable live events.

The first concerts the IBU designed after the National Nights were the *concerts européens,* the European concerts. Nationally adapted program scripts now belonged to the past.[70] According to the various European program directors involved in the process, the European concerts would be highly representative and were "an artistic event of European importance."[71] Whereas the preparations were

66 "L'avenir des nuits nationales, rapporteur H. Giesecke," Série 1592, 4 nov. 1929, box 85, file Relais, Programmes Nationaux (Nuits Nationales), Général, 1926-1931, IBU.

67 "Commission de Rapprochement, Intellectuel, Social & Artistique: Procès-verbal de séances tenue à Vienne," Série 502, 27 avril 1927, 2, file P.V. Commission des Programmes [anciennement commissions de Rapprochement et Relais], 1926-1939, IBU.

68 "...toutes les Sociétés radiophoniques européennes pourraient transmettre les programmes de tous les pays ce qui, au point de vue psychologique aurait un effet merveilleux sur les écouteurs." "Commission de Rapprochement," Série 502, 2, file P.V. Commissions des Programmes, 1926-1939, IBU.

69 "La meilleure façon de faire apprécier à l'étranger la musique et la littérature de votre pays consiste probablement à relayer de votre studio ou d'une autre salle, par circuit téléphonique, un programme national typique exécuté par vos meilleurs artistes." "Nuits Nationales," Série 1331, annexe à Série 1330, 29 avril 1929, box 85, file Relais, Programmes Nationaux (Nuits Nationals), Général, 1926-1931, IBU.

70 In the course of the 1930s more and more stations would become linked to the international relay network. Consequently, less and less stations would have to use a program script.

71 "un évènement artistique d'importance européenne." "Groupe des échanges de programmes de l'Europe Centrale et de l'Europe Occidentale, oct. 1930, Budapest, 1; Quote in the added document: "Deci-

similar to the Nights series, the showcased country now organized a live concert itself either in a studio or in a concert hall. With the international relay network in place, the entire broadcast was exactly the same for all listeners in Europe.

To explain how this worked: At the prearranged time and date the showcased IBU member, Poland for example, linked the venue where the Polish performance was taking place to the international relay network. The Polish PTT, in collaboration with the other European PTTs, then relayed the concert all across Europe, including countries like Denmark. Once the transmission arrived, the Danish PTT coupled the relay network to the national broadcasting network. The Danish broadcasting organization then transmitted the European Concert over the Danish home channel right into the homes of the Danish listeners. In their national radio schedules the Danes thus found a European program featuring Poland. First the Danes would hear the announcement in Danish, the language of the relaying country. Then they heard the same announcement in Polish, the language of the showcased country, and then in French, the main international language of the time.[72] The presenters mentioned the names of all *countries* participating in the event in each language. Thus, in multiple languages, the presenters created the geographical society of European listeners for the Danish audience. The same procedure happened in all participating European countries. Apart from the first announcement in their own language, each listener in Europe heard the same entire program.

Along similar lines the IBU would also coordinate the unique broadcast *La jeunesse chante au-delà des frontières*, Youth sings across the Frontiers, on October 27, 1935, as well as a new series from 1935 onwards, the *Concerts mondiaux*, the World Concerts. For European program directors, such live events were an expression of the highest quality in cultural standards for broadcasting. The success of these series depended on the inclusive participation of broadcasting organizations in relaying the event. If participation was high, listeners would think they were taking part in a large European society of listeners. Vice versa, if participation was low, the European society effect would disappear. Participation was therefore of the utmost importance for achieving a sense of belonging.

With the development of recording technology and the exchange of recorded music, European program directors could create international programs of higher complexity.[73] The relay of Youth Sings in 1935 had revealed a technical problem.

sions taken at a meeting in Varsovie 25 to 26 Sept. 1930," 2, box 85, file Relais, Concerts Internationaux, 1931 à 1939, Général, IBU.

72 A little later the organization decided to change announcements and skip the introduction in the language of the showcased country. IBU, P.V. Conseil Administrative 1935, 251, 260.

73 The archives do not specify which recording technologies were used or when and where. Innova-

This concerned society-building programs including contributions from as many countries as possible. Unlike a European concert which each evening celebrated the culture of one European country, Youth Sings had aimed to broadcast from as many as thirty-one countries in one evening. Consequently the broadcast lasted for hours. Although successful from a production point of view, listeners lost interest during the evening. The contribution per country was simply too long. Short contributions were technologically impossible. It took the operators of the international relay network considerably longer than one minute to switch and establish a new connection between the contributing country and the rest of the network. In order to have many countries contributing to the content of a live program, the solution lay in recording technology.

Shortly before, and especially after the Youth Sings broadcast, the IBU designed a number of unique events that made use of recording technology. In 1928 already the IBU had been receiving complaints from Swedish broadcasters about the large preparation time and costs for the Nights. Each broadcasting station had to supply the music for the Night in question. It had to choose musical contributions from the list provided and find appropriate musical scores. It then had to find an orchestra and musicians of the highest quality who could practice and then perform the music during the Night. With the sound archives not yet in place, Burrows expressed his regret that there was not yet a follow-up on IBU ideas to facilitate the exchange of recordings that reproduced special forms of national music. Such use and exchange could reduce preparation time and costs substantially.[74] From the mid-1930s onwards, the IBU started using recordings to resolve these organizational and technical issues.

Initially, recording technologies served national live programs, but after the impressive relay of Youth Sings in 1935, the IBU used recordings to produce programs involving many countries. In 1934 and 1935 the Rapprochement Committee created the Christmas Hymns. These were unique events for which organizations produced their own broadcast consisting of hymns, carols and other forms of Christmas music from various countries in Europe. They sent the recordings to the IBU who then re-distributed them to all IBU members who could use the recordings in their broadcast. The recordings allowed broadcasters to demonstrate live broadcasting from foreign countries in their own broadcast.

As follow-up to Youth Sings, the IBU immediately decided to coordinate a new program *Les chansons des Peuples*, that finally became known as *Une heureuse année au monde entier*, A Happy New Year to all of the World. In the eyes of the IBU,

---

tions appeared in gramophone recording as well as for the phonograph.

74 "Nuits Nationales," Série 1152, 1, box 85, file Relais, Programmes Nationaux, Général, IBU.

Figure 6.3 – Live relay of Dvorák's opera Rusulka from the Czech National Theater
For reasons of technical difficulties it had been impossible to transmit the opera from below the stage. As a consequence, the broadcaster had to transmit the music from on-stage.
Source: "National theater, opera Rusulka by Dvorák," box 58 Photographies, IBU. Used by the courtesy of the European Broadcasting Union.

a Happy New Year would have to include contributions from *all* European countries. Contributions could therefore be no longer than one minute. Rather than create many nationally constructed broadcasts as had happened with the Christmas Hymns, the Happy New Year would be broadcast live from Berlin. During the live event the host for the evening, the German RRG, would then integrate live the pre-recorded one minute "pretending to be live" contributions from the various European nations.[75] Gramophone recording technology enabled technically and organizationally impossible live programs. The importance of the inclusiveness of participation in program content outweighed the fact that in reality, the live events were partially recorded.

Besides such complex international live events, the IBU coordinated less demanding live programs as well. It designed for instance the series of *Concerts internationaux*, the International Concerts in 1932. The organization consisted of two countries reserving an evening to exchange programs and have live music from the interested nation. They celebrated the international courtesy between countries.[76] Such bilateral collaboration would increase the number of foreign live music broadcasts over home channels and became an important aspect of diffusing civilization via cultural routes. Though the quality of these broadcasts was high, due to their bilateral character, they did not meet the highest quality standard for international programs.

The best international live program would not only have to reach as large a listening public as possible, but would also integrate as many participating countries in the production, whether in the relay or in the content. In some cases recording technology was a suitably practical solution to complex program formats. Nevertheless, it was genuinely viewed that events should be live at all times, giving the highest quality of communal experience. With changing organizational and technical options over time, the IBU created a wide variety of international programs that expressed a sense of unity in vastly different ways. The society-building effect of these international programs could in no way create a homogenous social experience.

Programs communicated geographical spaces. The standards for suitable feelings of belonging had everything to do with the geographical reach of the programs and the geographical territory communicated via the program content. The

75 IBU, *Broadcasting: An Instrument for the Creation of a Better Understanding Between the Peoples* (Geneva: IBU, 1937); IBU, Conseil Administrative 1936, 51-52.
76 "Commission de Rapprochement," Série 502, 2, file P.V. Commission des Programmes, IBU; "Concerts pouvant faire l'objet de relais internationaux, Réunion des directeurs de programmes, observations de M. K.B. Jirak, Radiojournal, Prague," Série 4763, 11 févr. 1935, 2, box 85, Relais, Concerts européens, Général, 1931 à 1939, IBU.

IBU had always claimed to be an international organization rather than a purely European union of broadcasters. Consequently, its cultural policy, and as such its live programs, did not remain restricted to Europe.

With the further diffusion of the international relay network and the development of the short waves, the IBU managed to bring its series of concerts, as well as entirely new programs, beyond the boundaries of Europe. In the mid-1930s the IBU coordinated a series of global broadcasts. On October 27, 1935 the IBU coordinated the worldwide broadcasting of Youth Sings, and in 1936 it began broadcasting the World Concert series. The Youth Sings program featured "choirs of children and young people in thirty-one countries situated in five continents [who] sang to each other, and to the world at large, the songs they love."[77] Both the content of the event and its geographical reach made the program a unique event of worldwide importance. Preparations took place at IBU headquarters in Geneva, whereas the German RRG organized the relaying of the event on behalf of the IBU on the evening. Berlin not only had a key position in the European relay network, but also possessed a short wave station that could connect Europe with other continents.

Youth Sings had to be organized internationally and was relayed via short wave beyond European borders all the way to Japan, North and South America, as well as some of the Colonies.[78] Moreover, contributions for the event not only came from Europe, but also from Hawaii, Australia, Japan, Uruguay, the South African Union, and the United States. The technical realization of the broadcast was "an intricate piece of work requiring most precise collaboration on the part of the telephonic and radio-telephonic services in each participating country."[79] Two years later Youth Sings would already be termed the "most elaborate and impressive collective manifestation yet organised by the International Broadcasting Union."[80] From the point of view of quality of production, the event created a sense of global union. It reached a worldwide audience and included an impressive number of contributions.

Unlike Youth Sings, the World Concerts followed the tradition of the European Concerts. Each concert showcased a country beyond Europe's borders. The showcased country relayed its concert live via the short waves to a short wave station in Europe. On arrival of the program, the receiving short wave station was then coupled to the international relay network which disseminated the program in

77 IBU, *Broadcasting*, 92.
78 Werner Schwipps, "Deutschland im Weltrundfunkverein 1933-1938," *Rundfunk und Fernsehen* 19, no. 4 (1971): 436.
79 IBU, *Broadcasting*, 92.
80 Ibid.

the same way as the European Concerts. The United States organized the first World Concert on September 20, 1936, followed by the Republic of Argentina in February 1937, and the Dutch East Indies in October 1937.[81] These concerts were not meant to be one-directional from outside to inside Europe. The American concert was broadcast in most European countries, and relayed to Russia, which from the point of view of the material network, also formed part of Europe.[82] In contrast, the third World Concert gained the interest of Australia (ABC), Ceylon, the United States (CBS and NBC), British India, the Dutch East Indies, and Hong Kong. Great divergences in longitude necessitated some of these stations to receive a recording of the World Concert and defer the broadcast to a more appropriate time. The IBU managed to coordinate two-directional broadcasts, with Europe as the relaying center.

Programs like Youth Sings and the World Concerts increasingly corresponded with the IBU's aim to become an international worldwide organization of broadcasters spreading a message of global union (see Chapter 2). The European concerts were the first series to radiate beyond Europe's borders. The American NBC for instance broadcast the 1934 Swedish, 1935 Yugoslavian, 1935 Austrian and 1935 Danish European Concert over its national network.[83] These were followed by Youth Sings and the World Concerts. Live broadcasting consequently moved from a Europe of connected cultural nations to a globe connected by national cultures.

Considering the entirety of international programs designed by the IBU, these programs emphasized the idea of European unity more than global unity. Overall, some twenty-seven European countries organized one or more European Concerts compared to a total of six World Concerts.[84] At all times, Europe remained the center of program coordination. Those World Concerts relayed to other than European countries would diffuse globally from Europe. Perhaps

81 Ibid., 91.

82 The American World Concert found its way to Austria, Czechoslovakia, Denmark, Estonia, Finland, Germany, Italy, Latvia, Lithuania, the Netherlands, Norway, Poland, Rumania, Sweden, Switzerland, Yugoslavia, the UK, and the USSR. NBC, "Survey of international programs arranged by the National Broadcasting Company, 1924-1936," June 15 1938, 77-101, registry file 5B, box R-3999, file 32583/1658, LoN.

83 Though there is no evidence that the European concerts were broadcast elsewhere outside Europe's boundaries, the frequent relaying of these concerts by NBC suggests that they might have been relayed elsewhere as well. NBC, "Survey of international programs," 48-49, 55, registry file 5B, box R-3999, file 32583/1658, LoN.

84 Algeria, Austria, Belgium, Bulgaria, Czechoslovakia, Danzig, Denmark, Egypt, Estonia, Finland, France, Germany, Great Britain, Hungary, Ireland, Latvia, Lithuania, the Netherlands, Norway, Palestine, Poland, Portugal, Rumania, Spain, Sweden, Switzerland, and Yugoslavia. Turkey had also been an IBU member, but was suspended in exactly those years when the IBU broadcast its European Concerts, because of problems with the payment of membership fees.

surprisingly, the European colonies were considered non-European territory. Europe was clearly distinguished from the rest of the world. With a vast increase and diversification of European live events relayed beyond Europe's borders, the number of non-European events would remain far lower. Moreover, broadcasts like the International Concerts and A Happy New Year suggested global events, but were in fact European. Regarding geography, European unity thus dominated international program construction.

Translating Music for Radio

Music conveyed a sense of unification. In 1927, head of the IBU Rapprochement Committee, Oskar Czeija, asked: "Which programs do we have to relay abroad?"[85] The variety of expressions that could shape a culturally desirable society was numerous. One could think of a wide range of musical genres, of radio plays, of concerts, literature, poetry, and discussions related to events. It was not easy to predict program content. Moreover, the medium of broadcasting had different characteristics to all previous forms of listening to music. Kate Lacey argues that reality "was adjusted to the masses, and the masses to reality, with all that that implied for new modes of perception and for new ways of thinking."[86] Consequently, broadcasters tried to adapt music to their medium while upholding their cultural vision of a society that matched their preferred image of civilization. But broadcasters would have a hard time specifying which cultural characteristics created a suitable vision of society.

Broadcasters favored the idea of attentive listening. By the end of the nineteenth century a tradition of attentive listening to music had developed in theaters.[87] Both the IBU community and the cultural elites thought this was valuable. Attentive listening envisioned the idea of education, uplifting the mind to become a culturally and civilized individual. Broadcasters in particular thought up program formats that emphasized this notion.

85 "Quels programmes devons-nous relayer à l'étranger?" "Programmes nationaux, par Oscar Czeija," Série 576, 22 juin 1927, 1, box 85, file Relais, Programmes Nationaux (Nuits Nationals), Général, 1926-1931, IBU.

86 Kate Lacey, "Towards a Periodization of Listening: Radio and Modern Life," *International Journal of Cultural Studies* 3, no. 2 (2000): 285.

87 James Johnson, *Listening in Paris: A Cultural History* (Berkeley: University of California Press, 1995), 228-236; Ther, "The Genre of National Opera in a European Comparative Perspective," 200.

Not all kinds of music that the members usually broadcast in their own national programs met the standards for international series like the National Nights. Essential to educating the masses, was the premise that a musical culture for broadcasting should consist of high art and high-culture music, rather than popular entertainment music. When discussing culture at their first meetings, broadcasters immediately noticed that there was a lack of consistency among members on the definition of music genres. In order to fill programs with high-culture music, they made sure that everyone understood exactly what this was. In the course of three years, the Rapprochement Committee therefore standardized the music genres (Table 6.3). These terms of reference facilitated international discussions and also ensured that the musical standards of the programs coordinated by the IBU would be entirely understood by the various IBU members' program makers. By 1929 the committee had created consistency for musical discussions internationally, nationally, and locally. They had invented a categorized scale to measure which genres suited high culture music.

Program formats for international events like the Nights and the European Concerts would have to be adapted substantially to keep audiences listening attentively to high-culture music. There was a number of recurring problems every time the Rapprochement Committee proposed a new international event. One was integrating high-culture genres like (light) opera and serious music in relation to the wide diversity of languages in Europe. In the view of most members, the repertoire should consist of high-culture music without too much singing. Not being able to understand the lyrics, listeners would lose attention.[88] Given these language complications, the Rapprochement Committee preferred to use instrumental rather than vocal music. Not all international programs would have to adhere to this quality standard. Members were free to design the International Concerts as they pleased. Youth Sings concentrated on choirs. The majority of IBU's international programs, nevertheless, envisioned for its listeners a musical canon of a highly instrumental caliber. Indirectly the IBU thus created the impression that high-culture music was mostly instrumental.

Another problem was bringing high-culture music live to listeners. Many musical scores were destined to be performed by large orchestras with a wide range of instruments. Broadcasting organizations adapted these scores for small chamber orchestras. They then hired a few artists to perform these adaptations live in their broadcasting studios (Figure 6.4). This reduced costs and solved the problem of space. It was much cheaper to organize live events with a limited number of artists.

88 "Programmes nationaux," Série 576, 1-2, box 85, file Relais, Programmes Nationaux, Général, IBU.

Table 6.3 – Standards for genres of music related to broadcasting repertoires, 1929

Title	Description
Music	
Opera	
Light opera	or musical comedy
Serious music	excluding opera or children's music; including religious music, symphonic music, chamber music, instrumental solos, vocal solos, choirs etc.
Light music	excluding light opera or children's music; including orchestras, choirs, solos, cabaret, revue, burlesque etc.
Dance music	music especially designed for dancing
Other categories	
Literature	
Conferences	humanities, sciences, social sciences, foreign language education
General critiques	news or opinions about politics, artistic life, literature, scientific life, inventions, economics, social and domestic life or sports
Religious services	performed for artistic reasons only, not for religious reasons
Children's hour	
Gramophone records	
Exceptional transmissions	
Photodiffusion	experiments with the transmission of pictures

Source: "Procès-verbaux de la Commission de Rapprochement," Série 1274, 28 février 1929, 2-3; Annexe à Série 1271, "Notice explicative ayant trait au tableau statistique série 1272," 1-2, file P.V. Commissions des Programmes [anciennement commissions de Rapprochement et Relais], 1926-1939, IBU.

Moreover, broadcasting studios were usually not large enough to accommodate large orchestras. Equally important were the technological problems that hindered the broadcasting of music played by large groups of instruments of a sufficiently high quality. Symphony orchestras have instruments that can reach very high as well as very low frequencies at once. Initially, the existing technology could not relay these very high and low tones.[89] Transmissions lost their quality. Small chamber orchestras in which one could deliberately choose specific instruments resolved part of this technical shortcoming.[90] Consequently, the IBU transmitted a much

89 "Au Comité Consultatif International des Communications Téléphoniques à Grande Distance, Paris," 4 déc. 1925, 2-3, box 85, file Relais Général I, 1927 et précédents, IBU.

90 Over the years, with the development of the international relay network and the improvement of the quality of the cables which were used, the problems with the high and low tones gradually became things of the past.

Figure 6.4 – German studio for the weekly Sunday Concerts at Köningswusterhausen
Source: "Studio for the weekly Sunday Concerts established by the Deutsche Reichspost at radio station Köningswusterhausen in 1923," box 58 Photographies, IBU. Used by the courtesy of the European Broadcasting Union.

adapted high-culture repertoire to its audiences. It changed original musical formats into simpler versions, perhaps even with a specific choice of instrument.

Other than the complexity of the scores, the length of most high-culture compositions such as operas or symphonies also formed an obstacle when it came to holding the listeners' attention. In general, a National Night or European Concert lasted for one and a half to three hours. Besides, the individual items during such evenings were often too long. Audiences lost interest. The IBU regularly urged its members to shorten the length of the entire event as well as the musical contributions. Four-hour symphonies as performed in theaters outlasted every broadcasting program by far. Broadcasters circumvented this problem by deciding to transmit specific sections such as an opera scene, or the concert overture. These along with the broadcasting programs gave a relatively fragmented and random representation of original musical compositions.

In practice there were many disruptions. In spite of regular warnings, discussions in the IBU Council, and several recommendations, showcased countries often continued to make their contributions too extensive.[91] In 1928 and 1935 the IBU officially shortened the program format of National Nights and European Concerts respectively to a maximum of one hour. World Concerts should even last no longer than thirty minutes.[92] Organizers nevertheless continued to struggle with the length of programs. In 1935 for instance, Youth Sings was widely criticized for its length, even though it was rewarded from the point of view of rapprochement.[93] It was not that easy to change existing musical cultures to suitable program formats that held the listeners' attention and uplifted their spirit.

In the meantime, broadcasters struggled to decide what high-culture music met their quality standards for educative and uplifting programs. National Nights was the Rapprochement Committee's first attempt at fulfilling its cultural policy along the lines of high-culture and high-quality broadcasts of music and literature.[94] The series consisted of classical music and short opera pieces as well as national folk music. Genres like light operas were to be avoided as they would counteract the aim of creating understanding and rapprochement. Light operas often caricatured specifics and habits of nations, and therefore could possibly create disharmony and segregation.[95] In 1929 the Rapprochement Committee discussed reformatting the program at the start of a new series of Nights. The committee decided not to change the musical format. For the time being, the committee had a good idea of what kind of music constituted high culture: classical music, short opera pieces, and national folklore.

At the start of the European Concerts, the IBU community shifted its focus to what genres best suited high-culture music. Unlike the serious touch in most of the genres used for the Nights, the European Concerts favored a lighter yet highly creative genre. Light but artistic music would serve unifying aims better than symphonic oeuvres which lasted too long and needed to be avoided.[96] Since the light music would be of a highly artistic standard, it would still correspond with the IBU's idea of high-culture music. In 1935 the program directors of the IBU member organizations went one step further. At a meeting organized by the IBU Rapprochement Committee, they decided to allow both classical and modern – contemporary classical – works of *all* genres. Massive symphonies, concerts or choral oeuvres still should

91 "Soirées Nationales, Liste des dates de la première série des Nuits Nationales," Série 875, 7 mars 1928, box 85, file Relais, Programmes Nationaux (Nuits Nationales), Général, 1926-1931, IBU.

92 IBU, *Broadcasting*, 91.

93 Ibid., 92.

94 "Procès-verbaux de la Commission de Rapprochement," Série 1274, 28 février 1929, 2-3; "Notice explicative ayant trait au tableau statistique série 1272," Annexe à Série 1271, 1-2, file P.V. Commissions des Programmes [anciennement commissions de Rapprochement et Relais], 1926-1939, IBU.

95 "Programmes nationaux," Série 576, 1-2, box 85, file Relais, Programmes Nationaux, Général, IBU; Fickers and Lommers, "Eventing Europe," 232.

96 "Concerts européens," box 85, Relais, Concerts européens, Général, 1931 à 1939, IBU.

be avoided.[97] The program directors gave the green light for a series of live events showcasing a musical heritage of widely diverse genres. They considered all genres suitable as long as they were of the highest quality and performed by artists of the highest standing.

This clear breach with the initially serious and traditional approach to the high-culture music of the Nights was not appreciated by the international intellectual elite. Especially in the 1930s when collaboration between the OCI, artistic affiliations, and broadcasting organizations increased, it became more difficult to estimate the kind of "high culture" music that could be used in broadcasting. Adjustments made to the music by the broadcasters met with fierce resistance. On the one hand, broadcasters were confronted with dissatisfied listeners who requested more popular and entertaining music. Their international programs attracted fewer and fewer listeners. On the other hand, around the same time the OCI and the international musical elite criticized broadcasting because the cultural level of the programs was too low and too popular. Around the mid-1930s, the IBU found itself in an uncomfortable position.

The OCI and the international music elites objected strongly. Čapek and Bartók in their 1931 resolutions had explicitly stressed the importance of including popular and dance music in an international musical repertoire. Together dance and folk music formed "the true popular music."[98] Their idea, however, was that these popular influences would have to be raised to the level of art song. Fragments of popular music should be reworked into high art music rather than remain popular. This was a modernist approach, but still highlighted the importance of high art music. In 1935 the OCI conducted a study in collaboration with external musical experts. The study concluded that broadcasting programs in general failed to appropriately meet their definition of high art music and educational broadcasting. The musical experts feared that broadcasting would turn on "an instant flow of music, not unlike a faucet or a light switch."[99] Radio broadcasts contained severe stylistic errors, and degraded art music to the level of inferior and (commercial) popular music.

The objections stemmed from habit, from traditional cultures of listening to music. The thought that opera and operetta music were adapted for small orchestras

97 IBU, P.V. Conseil Administrative 1935, 259-261.

98 "La vraie music populaire." "Projet d'un questionnaire établi en collaboration avec L'I.I.C.I. devant être communiqué aux membres de l'Union Internationale de Radiodiffusion," Série 2796, Annexe A, 25 sept. 1931, 3, box 92, file Société des Nations, Section de Coopération Intellectuelle, Général, 1927-1932, IBU.

99 "le robinet à musique, comme le robinet à eau ou le commutateur de la lumière électrique." Beidler-Wagner et al., *Le rôle intellectuel et éducatif de la radiodiffusion*, 218.

or even for two to three performers, according to Beidler-Wagner "often abuses the word 'fantasy.'"[100] It was unimaginable that such performances would uphold the original artistic value of a piece. The place of the performance and the choice of instruments further downgraded the music. Religious music, for instance, should be performed in churches, not in concert halls. Moreover, performing music with homogenous sets of instruments like brass bands or guitars by nature highlighted musicians from the popular (meaning commercial) scene. Music experts recognized that broadcasting could not relay music in exactly the same way as people were accustomed. The music experts felt there was far too much adaptation of the music, making it of an inferior quality that did not express high culture. Short musical excerpts should be broadcast, giving people an impression of what a live experience of music in a theater or a concert hall was like.[101]

By the 1930s the discussion on formats for international programs was gaining more ground in the OCI. With the increasing fear of a new world war, it became ever more urgent to stimulate international minds of peace, collaboration, and understanding. The route the OCI suggested did not correspond with the desires of listeners across Europe. In order to continue capturing the attention of listeners when broadcasting an international program, the IBU aimed to maintain the quality of performance, while making the music more accessible to listeners. Whereas the IBU thought it could achieve an internationalist mind with such adjustments, the OCI did not. Amid quickly deteriorating international relations, the OCI considered it more and more urgent to create quality high-culture music formatted in the best possible manner.

Cultural expressions were an extremely important component of civilization, and became a frequent subject of debate before World War II. Civilization was in a deplorable state, Murray, the head of the OCI said at a meeting in 1935. "It seems as if civilization were caught in a net from which it was trying in vain to escape…civilization…seemed unable to save itself."[102] Discussions on culture and broadcasting programs in the second half of the 1930s were thus increasingly surrounded by an aura of fear and defeat. The OCI's plans for programming became a desperate attempt to give its view of civilization and culture a new and hopeful impetus.

The OCI took matters in its own hands. Having expressed the importance of suitable international broadcasting programs at the International Convention for

100 "abusent si souvent du mot "fantaisie."" Ibid., 211-212.

101 Ibid., 209, 215.

102 LoN, *International Committee on Intellectual Co-Operation, Minutes of the Seventeenth Session, held at Geneva, July 15th, 1935,* LoN doc. ser., C.I.C.I./17th session/P.V.1, 7-8.

the Use of Broadcasting in the Cause of Peace, the OCI organized a meeting to discuss program content in 1938.[103] Although the IBU participated, these discussions would not influence program formats further in the remaining interwar years. International tensions mounted rapidly, and broadcasters experienced increasing difficulty keeping their programs independent of government influence. The IBU had a hard enough time satisfying its own members and audiences, let alone the international cultural elite.

By the time war broke out, broadcasters and cultural elites still had different opinions about what high-culture music was and how it should be formatted. In the end the IBU considered the interests of its members and audience more important than those of the international elite. Its program formats included light though highly artistic music from a wide range of genres and that were short, largely instrumental, and relatively simple impressions. According to the IBU, such an adapted version of high-culture music could equally well propagate the education and uplifting of the mind. These changes were justified. Their efforts would fail if people at home preferred to tune into national programs or other frequencies.

Suitable music was a must. International live programs would achieve feelings of belonging that crossed national boundaries. Paradoxically, in order to serve international and European unity, the IBU stressed the national aspect of its international program series like the Nights, and the European or World Concerts. In 1926, the IBU designed National Nights in such a way that each Night "the spirit of one country passes through all of Europe."[104] By the mid-1930s the World Concerts were still "...to include preference, performances by orchestras of a character peculiar to the country providing the concert, typically national music..."[105] Oddly enough, the IBU community envisioned a musical repertoire for international broadcasting programs that focused on national cultures.

The national focus of these international programs was decided without deep discussions. In 1927 head of the Rapprochement Committee, Oskar Czeija, noted:

> It is highly likely that, as far as it concerns the scope and use of the relays abroad of national programs, the whole world is of the same opinion. Transmissions of this nature serve the cultural and practical scope of

103 See amongst others: LoN, *International Committee on Intellectual Co-Operation, Minutes of the Twentieth Session, held at Geneva, July 14th, 1938,* LoN doc. ser., C.I.C.I./20th session/P.V.7, 57-67.

104 "l'âme d'un même pays passe sur toute l'Europe." "Letter from Secretary General Arthur Burrows," Série 203, box 85, file Relais, Programmes Nationaux, Général, IBU.

105 IBU, *Broadcasting*, 91.

> broadcasting best. By acquainting foreign listeners with national creations of a country – musical works, literature and science – they bring together the countries of foreign listeners and revive mutual sympathy. Such procedures could awaken countries and peoples – encourage travels and develop tourism – so nothing else than the circumstances can contribute to the rapprochement between nations and it is here that the great value of these transmissions lies.[106]

Broadcasters intimately linked their idea of centralizing the national within their international programs to their plans for international collaboration, peace, and interconnectedness. International programs based on national cultures would make people understand one another, have an interest in one another's culture, stimulating cross-border travelling and the like. The national nature of the programs was a prerequisite for people to really get to know one another. Only then could international fellowship truly evolve. Much in line with their other activities, broadcasters therefore in a natural way interconnected the national and the international in their programs. In their opinion, the emphasis on the national was an important quality standard for international programs.

The lack of discussion might give the false impression of a lack of tension between the national and international dimension. Broadcasters experienced difficulties defining the exact commonalities in a shared European or international culture that transcended boundaries. More surprisingly perhaps, they would find it equally difficult to determine the typical features of their national cultures in such international programs.

Although broadcasters had been gaining experience with national programming, it appeared to be more difficult to broadcast national cultures as part of international programs. With the implementation of the Nights for instance, it was still impossible to relay a live broadcast across Europe. You could not get a real impression of other countries' cultures. In the eyes of the IBU, artists would always appropriate foreign scores, because they made interpretations based on their own national cultures. The Nights featured national appropriations of the cultures of other European countries, rather than the original music. The IBU tried to avoid

---

106 "Il est probable que, pour ce qui concerne le but et l'utilité du relais à l'étranger des programmes nationaux, tout le monde est du même avis. Ce sont des transmissions de cette nature qui sont le plus aptes à servir les buts culturels et pratiques de la radiodiffusion. En faisant connaître aux écouteurs étrangers les créations nationales d'un pays, – œuvres musicales, littéraires et scientifiques – elles rapprochent ce pays des écouteurs étrangers et réveillent en eux l'intérêt et la sympathie pour lui. Ce procédé peut faire naître les pays et les peuples – d'où encouragement aux voyages et développement du tourisme – donc rien que des circonstances pouvant contribuer au rapprochement des nations, et c'est en cela que réside la grande valeur de telles transmissions." "Programmes nationaux," Série 576, 1, box 85, file Relais, Programmes Nationaux, Général, IBU.

this by stressing the need to invite artists of the highest national standing to perform. Only the finest artists were capable of approaching the original intention of the musical compositions.[107] With the international relay network in place, national appropriations became a matter of the past. People in Europe and beyond could now experience the original and authentic national music of their neighboring countries.

Reality turned out to be less ideal than these program formats suggested. Whereas series like the European and World Concerts could now feature authentic national music, National Nights and International Concerts featured a combination of authentic and nationally appropriated music. The IBU repeatedly warned its members to focus on their own national music and avoid foreign compositions in their country's contributions. Apparently the IBU considered it likely that its members would choose foreign music, and with reason. Broadcasting organizations made foreign compositions part of their national contributions.[108] In 1934 the French national broadcasting federation offered an International Concert relayed from the *Théâtre National de l'Opéra-Comique.* French artists performed the Italian work of Puccini, *La Bohème*, and Leoncavallo, *Paillasse.*[109] Although broadcasts increasingly featured authentic national music, thus raising the quality of the international programs, they did not entirely eradicate performances that in the tradition of nineteenth century opera interpreted foreign works. Consequently, the musical broadcasting repertoire would never be completely representative of national cultures.

The visions of national cultures were also difficult to determine. The IBU envisioned that national cultures should consist of the diversity of musical styles within their boundaries. Already with the design of the National Nights in 1926 the Rapprochement Committee encouraged IBU members to integrate explicitly folk and dance music into their program scripts. Over the years, the IBU valued highly the creation of programs that accounted for the ethnic diversity in Europe. Everybody would have to feel included in the IBU's international programs. The IBU followed composers like Bartók, who by taking a modernist approach negated the existence of single style national canons.[110] The IBU explicitly chose to include

107 "Nuits Nationales," Série 1331, annexe à Série 1330, box 85, file Relais, Programmes Nationaux, Général, IBU; "Concerts pouvant faire l'objet de relais internationaux," Série 4763, 1-2, box 85, file Relais, Concerts européens, Général, IBU.

108 The question remains of what the scale of such appropriated foreign music has been. The International Concerts have been one of the most popular and thriving series of the IBU, but its filing has been limited. Box 85, file Relais, Concerts internationaux 1931 à 1939, Général, IBU.

109 Fédération Nationale de Radiodiffusion, "Retransmission depuis le Théâtre National de l'OPERA-COMIQUE," Paris le 28 avril 1934, box 85, file Relais, Concerts internationaux 1931 à 1939, Général, IBU.

110 Ther, "The Genre of National Opera in a European Comparative Perspective," 197.

the local. In a way, it thus envisioned a transnational kind of brotherhood of national cultures.

In practice, IBU members had a hard time living up to these music standards. Although the IBU encouraged the integration of folk and dance music, broadcasting organizations often selected high culture, albeit relatively accessible, classical music like the overture to a concert. Instead of popular chansons they chose internationally renowned composers like Dvorák, Mozart, and Beethoven. They clung to the well-known, to the national musical canons. In 1928 IBU Secretary-General Burrows stressed that the Nights had included very few popular chansons and dance music. He made it clear that these genres were allowed and should be encouraged once more.[111]

The live broadcast of the European Concerts moved IBU members further from their classical national musical canons. Broadcasting organizations increasingly created their contributions in accordance with IBU repertoire standards, while giving their concerts an individual flavor. Germany mostly filled concerts with its greatest classical composers. The Netherlands and Norway broadcast more progressive modern classical oeuvres. And a country like Austria positioned itself with light classical music.[112] Especially with the series of European concerts the broadcasters were able to live up to the musical repertoire standards set by the IBU. They communicated a more diversified repertoire to their audiences.

Over the years, folk music would also slowly find a place in the musical broadcasting repertoire. Albeit relatively late, Czechoslovakia would include the work of Leoš Janaček in its third European concert in March 1936. Until then it preferred well-established and more classical composers like Dvorák and Smetana.[113] Poland on the contrary integrated the work of Karol Szymanowski in each of its European concerts. It portrayed him as a modern and western European composer with Polish and Slavic influences in his compositions.[114] In Hungary, the work of Bartók

111 Polish and Czech section in: Box 85, file Concerts Européens [subdivisé par pays], IBU; "Nuits Nationales," Série 1152, 1, box 85, file Relais, Programmes Nationaux, Général, IBU.

112 My warm thanks go to Karin Bijsterveld for her comments on the musical content of the European concerts. Of course the responsibility for the use and interpretation of the findings is mine and mine alone. Further combined musicological and historical music research could give more detailed understanding of the factual visions that have been broadcast and the extent to which the musical content of the European concerts lived up to the IBU's desired inclusion of folk and dance, and thus of the music of the minorities and different races inside the various countries. Attention should be paid to the extent to which western European concerts inserted a popularized and diverse vision of their cultures. See: Box 85, file Concerts Européens [subdivisé par pays], IBU.

113 Série 2912, 13 nov. 1931, IBU; Série 3971, 22 déc. 1933, IBU; Série 5613, 5 fév. 1936, Czech section in box 85, file Concerts Européens [subdivisé par pays], IBU.

114 Série 5438, 31 oct. 1935 on Karol Szymanowski; all documents in the Hungarian section, box 85, Concerts Européens [subdivisé par pays], IBU.

featured alongside that of Franz Liszt who was part of the Hungarian musical canon. Hungary introduced Bartók as a composer whose style was "the development, and artistic burgeoning of the popular chanson."[115] The Rhapsody which the Hungarian broadcasters chose to broadcast was introduced as Bartók's first work of romantic modern Hungary. Bartók described the piece as his first work in his new style and as such, a preparation for his later compositions. Folk music slowly found its way into international programs. Nevertheless, it continued to be outweighed by music chosen from the established national canon. Programs highlighted the national more than the local.

With such choice for national music, broadcasters tended to prefer music of "their" best and most renowned "national" composers. Selecting "national" composers turned out to be more difficult than anticipated and caused disagreements between broadcasters. In 1932 the IBU decided not to be party to these discussions, because of the clear "sensibilities between countries in this domain" and its lack of moral authority on the issue.[116] Could one really speak of national composers? Like artists, composers had a tradition of travelling across Europe and sometimes even farther. Bartók for instance traveled from Czechoslovakia, to the Netherlands, Italy, and Switzerland in 1923 and 1925. In 1929 he would make a grand tour of three weeks in Russia, followed by a series of concerts in Switzerland, Denmark, Great Britain, the Netherlands, Germany, France, Italy, Austria, and Hungary. During his travels he was receptive to a wide variety of western musical influences, notably the German school. He would become inspired by Debussy, Stravinsky, Schoenberg, Beethoven, and Bach.[117] His own work in turn gained widespread recognition internationally as well. Composers in Turkey for instance started to seek ways of using folk material in their music.[118] To what extent could one say that Bartók was representative of Hungarian culture alone?

Bartók was just one of many internationally known composers who traveled and sometimes lived abroad. When broadcasters made their international programs on the basis of their national culture they in fact tried to iconize "their

---

115 "Le style de Bartok est le développement, l'épanouissement artistique de la chanson populaire." "Notice sur le programme du concert européen hongrois," Série 3871, 5-6, box 85, Concerts Européens, IBU.

116 "tenant compte des sensibilités entre pays dans ce domaine." "Procés-verbaux des réunions des commissions de rapprochement et de relais réunis," Série 3246, 10 juin 1932, 4, file P.V. Commissions des Programmes [anciennement commissions de Rapprochement et Relais], 1926-1939, IBU.

117 Gillies, "Bartók, Béla"; Gillies, *Bartók Remembered*, 102-103, 225.

118 Merih Erol, "Modernist and Nationalist Discourses on Music in Southeastern Europe (1900-1945): The Cases of Greece and Turkey." *Prepared for the CAS (Centre for Advanced Study, Sofia) Fellowship program "Regimes of Historicity and Discourses of Modernity and Identity, 1900-1945, in East-Central, Southeastern and Northern Europe (2008-2010)"* (forthcoming as a book chapter in an edited volume after completion of the program, 2009), 12-13.

national" composers. Philipp Ther explains how a German composer like Richard Wagner became universalized during the nineteenth century rather than remaining purely German as Wagner liked to stress himself.[119] One could also ask how contemporaries viewed a composer like George Fridrich Handel who featured prominently in the European Concert organized by Germany in 1935. Handel was German by birth, but had lived his entire life in Great Britain.[120] Was he then a German or a British composer? Positioning Ludwig von Beethoven as an example of one national culture was problematic as well. He was born a German and had lived there part of his life. However he spent thirty-five years in Vienna, Austria, where he had taken lessons with great Austrian composers like Joseph Haydn and Johan Schenk.[121] Beethoven featured in the Austrian European Concert of 1935. Was he then a German or an Austrian composer? And to trigger the discussion: if a contemporary could not say with certainty whether Handel was German or British, or Beethoven was German or Austrian, would one then not have responded instead: Well in any case he is European? The problem remained unresolved, and broadcasters continued to have a free hand in iconizing "their" national composer.

There was some kind of universality attached to culture which flourished in the national diversity of the European Concerts. Composers could not always easily be "iconized" as the composer of one specific nation. Moreover, the national appropriations of foreign music continued, giving a touch of universality to the musical repertoire. Although the example of nationally appropriated music shows the desire to undermine other than authentic national music, the IBU also sought a certain degree of universality. Too much internationality, uniformity or generalization in programs would be the death of the idea that people could begin to understand each other by showing each other their art and life. In turn, too much diversity would also alienate listeners from participating in the broadcasts. They would then be unable to relate to the program content anymore.[122]As a result, the IBU intentionally aimed to create a sense of unity in the diversity of soundscapes, music, and languages which the broadcasts offered together and individually.

An important prerequisite for the success of the serial broadcasts was a certain level of homogeneity. A series like the Traveling Microphone provided what we could term "soundscapes," a sound experience of a particular national landscape integrating local as well as national festivals, ceremonies, or events. Although the

119 Ther, "The Genre of National Opera in a European Comparative Perspective," 198.

120 Anthony Hicks, "Handel, George Frideric," *Grove Music Online. Oxford Music Online*, n.d., http://www.oxfordmusiconline.com/subscriber/article/grove/music/40060. Accessed April 29, 2009.

121 Joseph Kerman et al., "Beethoven, Ludwig van," *Grove Music Online. Oxford Music Online*, n.d., http://www.oxfordmusiconline.com/subscriber/article/grove/music/40026. Accessed April 29, 2009.

122 "Commission de Rapprochement, " Série 502, 3, file P.V. Commissions des Programmes, IBU.

experience centered on national and local cultural impressions from one country, all Travelling Microphone broadcasts provided a soundscape of similar events from many European cultures. Another example was the introductory talks to a series like National Nights. Each Night began by introducing the habits and characteristics of European countries. The showcased countries not only highlighted their national characteristics, but also explicitly positioned themselves in the greater context of European society.

Germany and Italy form a telling example. Germany stressed its modernity, youth, technology, architecture, and dramatic arts like theater and music. However it began its introduction with: "Germany, situated at the center of Europe." Moreover, in its talk, it indirectly referred to its role during the Great War by stressing the recent redefining of its borders and the difficult situation in the country.[123] With such a negative recent past, Germany tried to put itself in a positive light with a hopeful future in an interconnected Europe. Italy, on the contrary, gave a "romantic" impression of its country, geography, and life. It also emphasized its connection with Europe, technologically, culturally, as well as via tourism. People from all over Europe visited Italy and internationally well-known Europeans like Goethe, Nietzsche, Shelley, and Keats liked to write about Italy. Other than social and artistic connections, Italy also stressed its material link with Europe, referring to the "great express, the international trains," that passed through its stations.[124] Both Germany and Italy had their specific characteristics, but they made it clear that they were part of European society, of European history, European relations, and European culture.

The homogeneity or universality in the serial broadcasts should not be boring. The IBU advised its members to contact one another when it was their turn to organize a European Concert. Countries should contact those who had organized a concert the year before. They could then exchange information on issues like the genres which had been used, or the most appropriate forms to give to a concert.[125] The IBU thus encouraged the exchange of knowledge and expertise in organizing an international event like the European Concerts. Members could harmonize and diversify their contributions, creating unity in the diversity of their programming.

Recent and ongoing research on the history of a European culture of music and musical theater supports these findings.[126] Attempts to create a musical culture

123 "Conseil," Série 1530, 14-15, box 85, file Nuits Nationales, IBU.

124 "Causerie devant servir d'introduction à la nuit nationale italienne du 15 octobre 1930," Série 2060, 16 sept. 1930, 1-3, box 85, file Nuits Nationales [subdivisé par pays], 1926-1931, IBU.

125 IBU, P.V. Conseil Administrative 1935, 261.

126 I base these findings mostly on the ongoing research program "Europe and Beyond: Transfers, Networks and Markets for Musical Theatre in Modern Europe, 1740-1960," based at the European University

for broadcasting followed nineteenth century developments in European musical culture. The emphasis on national culture was nothing new. The universality that emerged in these national cultures in Europe was not new either. In the course of the nineteenth century the opera in Europe developed specific national identities, turning the Italian opera into one of many national variations. The creation of national operas depended on "the context of national movements, as well as their reception of music in general and opera in particular."[127] This process of differentiation was accompanied by a process of convergence. The stages of explicitly nationalistic operas were often interchangeable, used for example for French as well as Czech nationalist operas. Moreover, "even composers who had a clear nationalist mission still used a musical language that was compatible with European stylistic conventions of opera."[128] Opera consisted of different traditions, but "was perceived as a European art and thereby essentialized as European."[129]

A similar development took place with musical broadcasting repertoire. Broadcasters had difficulty creating international and mostly European programs with a focus on national cultures. They designed their nationally diverse cultures integrating what was generally perceived as typical European musical culture. People were used to appropriating the music of foreign composers, or the literature and ideas of men like Nietzsche. The Italian broadcasters could therefore easily refer to the connection between the philosopher and their own beautiful country. Composers and compositions traveled around, creating a universal idea of European culture. The IBU has continued in this tradition, integrating contested visions of the locality, nationality, and internationality of the music. The organization integrated unity and diversity in the music.

Success and Failure

The question remains whether the broadcasters succeeded. The aims of the IBU community and the OCI were to educate and uplift the minds of people to create international peace and understanding. They wanted to increase the amount of foreign programs and foreign music over listeners' home channels by means of program exchange and international live events. The success of these programs

Institute in Florence, Italy. See: Haupt and Ther, *Europe and Beyond: Transfers, Networks and Markets for Musical Theatre in Modern Europe, 1740-1960*; Ther, "The Genre of National Opera in a European Comparative Perspective"; Erol, "Modernist and Nationalist Discourses on Music in Southeastern Europe."

127 Ther, "The Genre of National Opera in a European Comparative Perspective," 197.

128 Ibid., 198-199.

129 Ibid., 200.

should not be overestimated nor underestimated.

Organizing international serial broadcasts proved to be a process of trial and error. Several program formats failed to capture the attention of international audiences. They were either improved or replaced by another series. Although IBU members had agreed to the program formats of the Nights and European Concerts, in practice they showed little interest in participating. In 1928 Burrows presented a depressing report. No more than ten out of twenty members had participated in the last Night, which equaled some thirty out of 130 stations. Moreover, those countries that were highlighted also showed a lack of interest. One country had mistakenly provided advertizing material for another participant, resulting in a well-received journal publishing "a photo of Mont Blanc (which is in France) as one of the characteristics of Switzerland."[130] Members considered the organization of the Nights too complex and time consuming, having to collect scores, search for, and organize rehearsals with artists and orchestra in advance.[131] There was a clear lack of interest among IBU members and listeners alike. Nonetheless, the IBU continued to relay the Nights until technical innovations allowed for live broadcasting across borders. It had no alternative. The IBU could in any case offer "to the masses a simple example and easy understanding of the work currently undertaken in various fields to create a better understanding between the peoples."[132]

The subsequent series of European Concerts also attracted limited interest particularly in 1933-1934. Countries confirmed their participation too late and were therefore unable to join in the broadcasts. In contrast to the Nights, the IBU decided to stop the series, and replace it with new ones.[133] Nonetheless, one successor, the National Concerts, would soon prove a complete failure. They were supposed to be half-hour broadcasts that could be relayed up to eight days after the original broadcast. But there were too many offers with a quality too poor to meet IBU standards. The IBU Council stopped the National Concerts and reintroduced the European Concerts with some adaptations to the program format. The Council also introduced obligatory participation for its member organizations. Members were strongly urged with an emphasis on moral behavior to participate in the European Concerts. This approach helped. It increased the success of the program substantially and created a

130 "...la photographie du Mont-Blanc (qui est en France) comme une des particularités de la Suisse." "Nuits Nationales," Série 1152, 1, box 85, file Relais, Programmes Nationaux, Général, IBU.

131 "Avenir des Nuits Nationales, Rapporteur A.R. Burrows, Conseil no. 10 de l'Ordre du Jour," Série 1842, 25 avril 1930, box 85, file Relais, Programmes Nationaux (Nuits Nationales), Général, 1926-1931, IBU.

132 "...donnent aux masses un exemple simple et facile à comprendre des efforts faits actuellement dans plusieurs domaines pour créer une compréhension meilleure entre les peuples." IBU, P.V. Conseil Administrative 1930, 254.

133 IBU, P.V. Conseil Administrative 1934, 213.

larger diffusion of the series across the European continent.[134]

Developing these international series regularly interfered with efforts to create national listening cultures. Similar to their internationally oriented activities, broadcasters were trying to create national broadcasting traditions. In 1936 the Norwegian delegate at the IBU complained on behalf of all Scandinavian countries about the time the European Concerts were scheduled. The concerts clashed with a long tradition of Scandinavian weather reports for many small fishermen. The Norwegian delegate proposed rescheduling the concerts. The IBU Council recognized the urgency of the weather reports but decided that the Scandinavian countries would have to record the live events, and broadcast them at a later time or date.[135] The life-saving character of the weather reports and the habit of broadcasting these at a specific time and date outweighed the immediacy of sharing the European Concerts at the same time as other parts of Europe.

IBU's cultural policy had always aimed to respect "the point of view that no transmission station should be hindered in developing itself nationally and culturally."[136] The National Nights and European Concerts drew the attention of a limited number of IBU member organizations. Although broadcasters attached considerable value to international programs, these interfered with the most important night of the week, Sunday evening. Since Sunday was the one evening that most people in Europe spent at home, the IBU thought it a suitable evening to broadcast international programs. These would then have the widest possible psychological community-building impact on listeners. Sundays, however, had soon become evenings of national listening. As a result, several broadcasting organizations took the matter to the IBU.

In January 1928 Czeija discussed the matter in the Rapprochement Committee and outlined complaints from German and Austrian IBU members. Czeija, who was both head of the committee and director general of Austrian Ravag, explained the complaints made by Austrian listeners about scheduling the Nights on Sunday evenings.

> Us in Vienna, we estimate that it is inopportune to transmit a national program [National Night] on Sundays, as we have the custom of choosing an entertaining program for Sunday evening and our subscribers regret it

134 IBU, P.V. Conseil Administrative 1935, 259-261. For an extensive discussion of the European concerts as serial event in the phase of European unification see: Fickers and Lommers, "Eventing Europe."

135 IBU, P.V. Conseil Administrative 1936, 24, 52.

136 "le point de vue qu'aucune société d'émission ne doit être empêchée de se développer nationalement et culturellement." "Commission de Rapprochement," Série 502, 3, file P.V. Commissions des Programmes, IBU.

> if such a program is lacking. It is not at all certain, that our audience, accustomed to listening to an entertaining program on Sundays, is ready to receive a national program [National Night] which demands the listener's sustained attention. It is rather to expect, that the enthusiasts do not listen to these programs at all, so that the aim of these programs will not be achieved.[137]

Sunday evenings in Austria had become national evenings of relaxation and entertainment. The enduring character of a National Night that required attentive listening did not suit these national developments. Germany expressed similar complaints. In Germany, Sunday evenings had a different program schedule from week days. This would change only a little over the years and acquired the status of a German tradition.[138]

In the following years the IBU would respect these national developments in listening traditions, albeit with one small though important exception. Czeija had proposed to change the broadcasting date to the first day of the week. When the IBU communicated its list with the dates for the next series of Nights, they avoided Sunday evening.[139] The following Nights as well as the entire series of European and International Concerts were broadcast during varying weekdays. The IBU had given up the idea of one fixed evening a week for international scheduling. In agreement with the IBU and its member organizations, the unique character of the World Concerts as well as important unique events like Youth Sings would be celebrated by broadcasting them on a Sunday evening. The IBU had to give in to developing national listening traditions even though this reduced the impact of its international programs on the largest possible audience. The most impressive achievements nonetheless continued to reach all of Europe on the most important listening evening of the week. With a modest success rate and a less receptive audience than anticipated, the failure of the international live events was probably guaranteed.

137 "Nous autres à Vienne, nous estimons qu'il est inopportune d'émettre un programme national le dimanche, car nous avons coutume de choisir un programme divertissant pour les soirées de dimanche et nos abonnés regretteraient si un tel programme venait à manquer. Il n'est pas du tout sûr que notre public, habitué d'écouter un programme divertissant le dimanche, soit disposé à recevoir un programme national qui exige l'attention soutenue de l'écouteur. Il est plutôt à supposer, que les amateurs n'écoutent pas du tout ces programmes, de sorte que le but serait pas atteint." "Fixation des dates pour les programmes nationaux, no. 11 de l'ordre du jour de la Commission de Rapprochement (Série 817), Rapporteur M. O. Czeija," Série 843, 24 janv. 1928, box 85, file Relais, Programmes Nationaux (Nuits Nationales), Général, 1926-1931, IBU.

138 For an exposé on the development of Sunday as a national evening of listening in Germany see: Alexander Badenoch, "Making Sunday What it Actually Should Be: Sunday Radio Programming and the Re-Invention of Tradition in Occupied Germany 1945-1949," *Historical Journal of Film, Radio and Television* 25, no. 4 (2005): 577-598; Alexander Badenoch, *Voices in Ruins: West German Radio across the 1945 Divide* (Basingstoke: Palgrave Macmillan, 2008), 190-195.

139 "Soirées Nationales," Série 875, box 85, file Relais, Programmes Nationaux, Général, IBU.

A brief exploratory study of two Dutch broadcasting organizations' program guides suggests otherwise.[140] Their look and feel, their categorization, as well as the music choices corresponded with international cultural policy efforts promoting the education of the mind. The organizations divided the schedules into a national and an international section. The latter had foreign stations organized alphabetically rather than by country. The international clearly meant European stations only.[141] Furthermore, the scheduled music was live programs alternated with time slots called *Grammofoonmuziek*, gramophone music. The IBU members and the record industry agreed that the guides should refer explicitly to music as recorded contributions. However they were inconsistent when it came to numerical references to records. In 1931 the organization *Vereeniging van Arbeiders Radio Amateurs* (VARA) for instance referred to these numbers in its radio guide *De Radiogids*, but by 1935 it did not anymore.[142] With some inconsistencies, the Dutch guides were presented in a similar language and with similar referencing to both the national, and the international that was interchangeable with Europe.

The music in the guides gives a similar impression. About 90 percent of broadcasting in the Netherlands consisted of foreign music.[143] Articles discussed the life and work of well-known composers who were part of the European music tradition. The guides also promoted programs broadcast over foreign channels. The VARA integrated a feature in its foreign section entitled *Europa zendt*, Europe broadcasts, that summarized interesting foreign broadcasts every day. The *Algemeene Vereeniging Radio Omroep* (AVRO) regularly advertised live programs of high-culture music in its guide *De Radiobode*. The music varied from Corsican folk music broadcast by the Paris PTT to Richard Wagner's *Der Ring der Nibelungen* broadcast via the station in Leipzig.[144] The guides promoted foreign live music, both over their national channels as well as those of their foreign colleagues. The international live programs coordinated by the IBU found a special place in the guides. They regularly promoted international events in *Europa zendt* or with an explicit advertisement for a European Concert. The VARA devoted its

140 I realize that there might be a theoretical discussion about ways to interpret cultural sources like program guides, but I will not enter such discussions. The examples only serve to show a certain consistency between transnationally negotiated standards and nationally issued program guides that found their way into the homes of the actual listener.

141 VARA, *De Radiogids*, (1931, 1932, 1935); AVRO, *De Radiobode* (1931, 1932, 1935). American music was relayed on occasion via British stations. The AVRO for instance regularly advertized the American program "Five hours back," relayed via the British station in Droitwich. AVRO, *De Radiobode* 18, no. 46 (16 nov. 1935), 106.

142 For an example see: VARA, *De Radiogids* 5, no. 47 (14 oct. 1931), 35; VARA, *De Radiogids* 9, no. 45 (1935), 142.

143 IBU, *Broadcasting*, 16-17.

144 AVRO, *De Radiobode* 18, no. 45 (13 nov. 1935), 72; AVRO, *De Radiobode* 18, no. 46 (1935).

Figure 6.5 – Europa zendt – IBU standards in Dutch program guides
Source: AVRO, De Radiobode 18, no. 45 (13 Nov. 1935), 72; AVRO, De Radiobode 18, no. 46 (1935); VARA, De Radiogids 9 no. 50 (19 Oct. 1935), 37, 43; VARA, De Radiogids 9, no. 52 (26 Oct. 1935), cover; Used by the courtesy of the Netherlands Institute for Sound and Vision.

front page to the unique Youth Sings, and integrated a two-page description of the event (Figure 6.5). Furthermore, the technical complexity of such broadcasts was occasionally featured.[145] Overall, the guides were full of references to foreign music, to European musical repertoire and the IBU's international programming

145 For an example see: VARA, *De Radiogids* 5, no. 53 (31 Oct. 1931), 12.

activities. These international standards were visible in the lay-out, the feel, pictures, and advertisements. The language and choice of topic were consistent with the IBU and its members' aims to educate listeners with an international mind.

Can we conclude that these transnational activities were fruitless attempts to educate and influence the European audiences and their listening cultures? According to Wintle, agreeing on ways to communicate group identity beyond borders is relatively easy. It is much more complex, however, to "concur on our representation of the group towards ourselves, our self-representation."[146] With their standardization activities, the IBU and its members were attempting to reach consensus about their cultural self-representation. Many transnational standardization activities have found their way into the homes of people via the program guides and schedules. Broadcasters envisioned and shaped Europe's musical cultural heritage, seeking a basis for new listening cultures. They used different methods to determine the whereabouts of these cultural "rules," Nonetheless, "used in its symbolically charged form," Peter Odermatt argues, "the term culture [itself already] implies that there is such a thing as a homogeneous group, even though there need be no consensus throughout that group as to the actual content of that culture."[147]

Conclusion

Programming was perhaps the most obvious and problematic locus for European unification. Intellectuals generally thought of culture as an important aspect of civilization. Consequently, the question of international live programs, in particular the discussed music programs, as well as a cross-border exchange, touched on the sensitive topic of shaping feelings of belonging. Creating a cultural policy for broadcasting and above all deciding what programs would best express a civilized society, proved to be complicated.

The standardization of a cultural policy for broadcasting remained a much more random affair. Many questions, such as copyrights, were firmly established in national legislation. In order to allow for cross-border flows of music, these national laws required comprehensive fine-tuning efforts. Furthermore, a successful

146 Michael Wintle, "Introduction: Cultural Diversity and Identity in Europe," in *Culture and Identity in Europe: Perceptions of Divergence and Unity in Past and Present*, ed. Michael Wintle (Aldershot; Brookfield [Vt.] USA: Avebury, 1996), 6.

147 Wintle, "Introduction," 6; Quoted from Peter Odermatt, "The Use of Symbols in the Drive for European Integration," in *Yearbook of European Studies: 4: National Identity*, ed. Joep Leerssen and Menno Spiering (Amsterdam; Atlanta, GA: Rodopi, 1991), 219.

policy had to facilitate the practical coordination of programs and their formats as well as standards for a repertoire for broadcasting music. With the wide variety of issues came a wide variety of stakeholders who operated locally, nationally, and/or internationally. They had different opinions and different interdependencies.

By the early 1930s, these stakeholders generally agreed that music united and that broadcasting could create unity among listeners. Music was a universal language that could overcome the large differences in speech in Europe. In the early 1930s they thought that speech could never achieve the same effect.[148] Stakeholders did not necessarily agree on how to realize the unifying force of music organizationally. For instance they had different opinions on program content and the responsibility for program construction. Besides, nobody had experience with the organization and coordination of programs that crossed borders. Defining a cultural policy was therefore a case of learning by doing. Even when stakeholders agreed, the question remained whether the status quo of the technology allowed for the practical realization of their visions.

Over the years, the IBU, and thus the European broadcasting organizations, managed to maintain control over the construction of programs for broadcasting. In particular, when the League's Committee of Letters & Arts envisioned creating an international program committee, the IBU fiercely opposed the idea. The IBU community related its programming activities much more directly to a vision of Europe than its illicit propaganda activities. Moreover, the development of worldwide live broadcasting events only began some five years before the outbreak of war. Consequently, Europe-oriented live programs by far outweighed these global broadcasts in number. Programs featured mostly European artists and European traditions of listening to music.

Over the years stakeholders came up with numerous visions of Europe when negotiating their cultural policy standards. These differed from central clearing houses for sound broadcasting to many visions fulfilling a European culture consisting of a brotherhood of national cultures. The international broadcasting community consisting of IBU experts, OCI members and external musical elites constructed a European culture that reflected a universalized idea of national diversity. The community contributed a relatively safe, generally accepted, high quality, and high-art image of European culture to Europe's cultural heritage in the interwar years. Their musical repertoire had a slightly lower culture profile

148 After ten years of experience with the construction of international programs the IBU concluded that not everything could be said with music. Sometimes speech was needed, because "music, although it has an important mission in social life, cannot express the innermost thoughts of one people to another." As a consequence, the language issue remained a problem with respect to the negotiation of programs in relation to the unification of listeners. IBU, *Broadcasting*, 93.

than pre-broadcasting era European canons of music. They increasingly directed their attention towards accessible high-art music consisting of fragmented pieces that were shorter and more instrumental. Even though these adaptations formed a break with the past, the nature of negotiations of musical culture in general continued in established traditions. Creating a European musical culture was a question of balancing universality and diversity.

Many programming issues remained unresolved at the time of the outbreak of World War II. In particular the question of what makes a good European program was still unclear. The OCI, the IBU, and international cultural elites had not resolved their disagreements on defining high-culture or high-quality broadcasting, nor were they able to conclude a binding international agreement on copyrights or artists' rights. Nonetheless, the European community of broadcasting experts did arrange a considerable number of international live programs, a dynamic program exchange, and their standards based on European ideals of peace and understanding clearly found their way into Dutch program guides. These repeated and unique efforts were all geared to making the international and the European a routine part of listeners' everyday lives. To what extent did they actually succeed?

Chapter 7 Conclusion
Internationalism in Practice

Count Richard Coudenhove-Kalergi had clear views on the role technology should play in the reconstruction of Europe after the Great War. He was the leader of the Pan European Movement, and a well-known lobbyist for European unification. Within a year of the establishment of the IBU, he remarked that technology had basically made Europe get lost in a universalizing world as a result of ever increasing connections.[1] No more than ten years later, the Austrian Oskar Czeija, in his role as director of the IBU Rapprochement Committee, would speak at one of the conferences of the Pan European Movement held in Vienna in 1935. Apparently Coudenhove-Kalergi had now developed an eye for the possibilities that broadcasting offered to promote his European activities.

Czeija's speech made it clear that the IBU ultimately aimed to create a European union, but that it above all pursued "the *tendency* to contribute to the rapprochement of nations..."[italics – s.l.].[2] For the time being, it was more fruitful to practically relate to ongoing rapprochement activities of organizations like the League of Nations. Although Czeija did not undermine the importance of the idea of a European union, he was working on an initiative with a global vision. In his speech he thus gently drew attention away from the European towards the international, an entity that he considered of more interest. *Europe – On Air* has queried precisely this connection between broadcasting construction and the unification of Europe, questioning processes in which actors negotiated specific projects which they articulated in European terms. *Europe – On Air* asks the question how and why actors involved in transnational broadcasting contributed to European unification in the interwar years.

1 Count Richard Coudenhoven-Kalergi, *Paneuropa* (Wien: Paneuropa-Verlag, 1926), 16.

2 "la *tendence* de contribuer au rapprochement des nations..." "L'idée de Pan-Europe et la Radio-diffusion," 1, box 74, file Propagande de l'UIR et des membres, II, Général, 1932-1936, IBU. The speech in abbreviated form has also been referred to by Oskar Czeija, "Paneuropa – Idee und Rundfunk," *Paneuropa* 11, no. 6/8 (1935): 197-199.

Dreams of Feasibility

People who developed broadcasting above all dreamt of feasibility. In the early 1920s, wireless experts and enthusiasts had only just started to develop broadcasting. They tried to build and promote a new tradition of listening to music. Soon they encountered problems which they could not resolve at the national level. They decided to join hands internationally. Together they managed to combine a striking variety of sensitive topics in all sorts of standards. Among others, they tried to resolve definition issues. What exactly was "good," "high-culture," "commercial" or "political" broadcasting? As the above example of Czeija shows, these experts were well aware of the fact that their choices contributed to specific European projects and to the desire not only to build strong national societies, but also forge ties across borders. The need to solve practical problems and the desire to shape a transnational sense of unity among listeners are the two major factors which fuelled the activities of those who were building broadcasting networks beyond national boundaries.

Creating transnational broadcasting was a hard and far from linear process. The internationally active broadcasting experts never took a blue print approach. They organized ad hoc processes of trial and error agreeing that it was always better to pursue the achievable. They understood that contestations and errors were just part of the game of designing a system and a tradition that was absent.[3] In their view, building broadcasting across borders was above all a matter of learning by doing, a matter of "practical internationalism."[4]

The process was heavily influenced by new technological options. One of these was short waves. This development redefined the entire transnational broadcasting system that existed up to the early 1930s. It opened up the possibility of worldwide broadcasting, and therefore challenged the European standards that had only recently been developed. Moreover, many new stations based their systems on this new and unregulated option, challenging the existing ones in all their facets from technical to programming standards. Innovations like the short waves created new problems and opportunities for the actors engaged in the process of building broadcasting networks across borders.

In order to deal with their challenges, the broadcasting experts organized themselves transnationally into a flexible expert community with an open organizational structure. This community consisted of experts who worked within and outside

3 "Détails d'organisation des concerts pouvant faire l'objet d'une diffusion internationale, Réunion des directeurs de programmes, observationos de M. K.B. Jirak, Radiojournal, Prague," Série 4762, 11 fév. 1935, 2, box 85, file Relais Concerts européens, Général, 1931 à 1939, IBU.

4 IBU, *Union Internationale de Radiophonie* (Geneva: IBU, 1926), 15.

government, in business, and with cultural affiliations at local, national, and international level. Open organization structures like those of the IBU, the ITU, and the League of Nations technical committees, were a way of dealing with high-speed, diverse, and interrelated developments. They allowed the expert community to create different forms of collaborations for each of the problems it tackled, acquiring precisely the right kind of expertise required for the problem at hand. These collaborations could be ad hoc, short, long, informal, formal, permanent, or official, involving large or small numbers of experts as the occasion required. However, there were always a small number of individual experts working in different organizations who formed a core community, tying together the various collaborations.

It is interesting to note that the core community dominating these collaborations did not only include leading experts from the big European powers of Germany, Great Britain, and France. The group also consisted of experts from smaller countries and countries located more towards the east such as Austria, Norway, Poland, and Czechoslovakia. These countries in particular embraced a very rich musical culture, which they defined as an important tool for their domestic unification efforts.[5] Consequently, these north, central and east European experts frequently took the lead in solving the most complicated problems. They took the lead in creating an international relay network for music, standards for propaganda broadcasting, as well as international programming and exchange, to name but a few. With the exception of southern Europe, the core community of experts thus reflected Europe in a much broader sense than media historians imply.

Over the years, the IBU in particular developed as a mediating arena where these experts could collaborate. Initially the organization was established to fill a regulatory gap left by other organizations who for the moment decided not to deal with the newly evolving medium. The IBU created a transnational practice and a culture, that by appealing to gentlemanly and civilized behavior tried to make its members and other interested stakeholders adhere to its non-binding recommendations. Over the years it established collaboration with the League of Nations and the International Telegraph Union who would recognize the Union as European expert in broadcasting by the end of the 1920s.

Relations between these organizations changed throughout the years, depending on which issue was being discussed. The IBU managed to employ these organizations as informal executors of its own plans. The League of Nations, as well

5 Philipp Ther, Philipp Ther, "The Genre of National Opera in a European Comparative Perspective," in *The Oxford Handbook to the New Cultural History of Music*, ed. Jane Fulcher (Oxford: Oxford University Press, 2011), 182-208; Philipp Ther, *In der Mitte der Gesellschaft: Operntheater in Zentraleuropa 1915-1914* (Wien; München: Oldenbourg, 2006).

as the International Telegraph Union, would fine-tune into their own international agreements, appropriately adapted IBU recommendations. A statute change in 1929 meant that the IBU from then on also stood for the ITU and PTT administrations in relation to technical affairs. Consequently, its technical standards acquired a more official and powerful status. On occasions the relations between the IBU, ITU, and League of Nations shifted. From the mid-1930s onwards, the League of Nations for instance continued and then completed the international regulation of propaganda broadcasting, while the IBU had to drop this issue. As a result of internal political and ideological litigation, the IBU reluctantly chose to withdraw. The Union did not always manage to take a stand.

The transnational expert community thus discovered that the basis for collaboration was fragile. Changing international relations influenced the development of transnational broadcasting in all its facets. Progress in relations signified progress in broadcasting construction. The signing of the Kellogg-Briand Pact in 1928 for instance brought hope and improved relations. In its wake, the IBU obtained collaboration from the League of Nations to take up all kinds of problems from propaganda to copyrights on cross-border flows of music. In addition, the USSR took action to develop closer and collaborative initiatives with the other European countries, whereas the League of Nations created its own station, Radio Nations. This flow of new broadcasting initiatives slowed down when international relations worsened again in the course of the 1930s with the rise of harsh nationalist sentiments. Fear of war and international quarrels made it increasingly difficult to accomplish progress in every aspect of transnational broadcasting construction.

As a way of handling these fragile international relations, a number of IBU experts developed specific techno-political diplomatic skills. They emphasized the technological side of their work. The Union also claimed it did not interfere in political problems and international relations. Yet these experts managed to deal with all kinds of politically sensitive issues including fine-tuning systems with different politico-ideological purposes such as the IBU and Radio Moscow. When international relations deteriorated during the 1930s, the increasing international tensions forced the IBU to withdraw from and reduce its overtly political work. The IBU techno-political diplomats managed to continue their politically neutral attitude in processes highlighting the technological side of their task. Inevitably, the changing international relations jeopardized and weakened IBU's role as mediator in transnational broadcasting construction efforts. Other experts and organizations in the transnational community took over the responsibility for solving political problems. The underlying collaborative relationships nevertheless remained intact. Everything the broadcasting community undertook happened with an eye to achievability.

Envisioning Europe on Air

Shaping relationships between people and creating identities always had top priority. This book has been an inquiry into the question of whether some sort of European agenda was part of the process. It is clear that the efforts described did not relate to a European project in a univocal way nor did they create a single, straightforward European broadcasting network. The one exception was Radio Luxembourg. Broadcasting experts in Europe mostly had global ambitions. They wanted to create international identities, not European ones per se. Europe, however, hovered somewhere in the background as a geographical focus or as a practical in-between solution.

The geographical focus of Europe, for instance, often remained implicit. Experts only defined Europe if their activities required clarification on boundary work. In the 1920s, the definition focused on the natural fringes of Europe with the North Sea in the west, the Ural Mountains in the east, and the Sahara in the south. By the time the short waves opened up the possibility of worldwide broadcasting, an established system was already in place in Europe. A further development of this system beyond the European region needed to be fine-tuned with others. These fine-tuning processes renegotiated Europe in the light of the wider world, heading towards a transnational Eurocentric global space for broadcasting. In the entire process of building a broadcasting system beyond national borders, most of the time the geographical notion of Europe remained very vague.

This was all happening as part of what the IBU called a practical internationalist approach. The term flourished widely at the time, mostly referring to technical organizations like the International Telegraph Union or an organization such as the IBU.[6] By using "internationalism" the IBU related to the ideals of collaboration, rapprochement, and peace as pursued by the transnational elites and the aims of international organizations like the League of Nations. "Practical" referred to commerce, to making profit, as well as to efficiency and innovation instead of a political approach. Practical internationalism seems to relate closely to the analytic term technocratic internationalism as defined by Johan Schot and Vincent Lagendijk, which they see as the context philosophy for infrastructure construction in Europe in areas like electricity and transport.[7]

6 Francis Paul Walters, *A History of the League of Nations* (London: Oxford University Press, 1952), 7.
7 Johan Schot and Vincent Lagendijk, "Technocratic Internationalism in the Interwar Years: Building Europe on Motorways and Electricity Networks," *Journal of Modern European History* 6, no. 2 (2008): 196-217.

Schot and Lagendijk do not explicitly link the idea of technocratic internationalism to traditions of European thought and culture. In contrast, the IBU community on occasion consciously specified this relationship. In particular with respect to its program-related activities, Europe entered as a historically defined culture and set of behavioral manners which the IBU community viewed as deeply rooted in the ideology of internationalism. They felt that their program efforts should contribute to the shaping of Europe's civilization and culture, in short to building European cultural unification. The IBU for instance explicitly sought to identify that one program which could give the best possible expression of European civilization and culture. At times at the forefront, but mostly hidden, Europe hovered somewhere within the philosophy or newly developing tradition of practical internationalism.

Other actors outside the IBU designed inherently different kinds of broadcasting systems, negotiating from an entirely or partially different viewpoint. The initial idea of Radio Luxembourg for instance adhered to this tradition, albeit from the point of view of European economic unity. In contrast, Radio Moscow was right from the start an explicit political tool for building a Socialist new world. It did not aim for understanding and collaboration between nations, but focused on one international socialist society. Consequently, not all promoters of broadcasting systems between the wars can be said to fit the technocratic internationalist philosophy in its entirety or even partially. Similarly, most of these viewpoints did focus on the world as a whole, rather than specifically on Europe.

Opinions might differ about the success or failure of these interwar transnational broadcasting communities. On the one hand, the transnational experts built a network, broadcast "international" programs, made arrangements for the use of broadcasting in the context of propaganda, war and peace, created an institutional network, and managed to have their ideas trickle down into national program guides. This was no mean performance during such turbulent interwar years. On the other hand, the international tensions of the 1930s, as well as the outbreak of World War II, prevented the final realization and effectiveness of the transnational broadcasting systems they envisioned. As a result, several initiatives were less effective than planned or met with less enthusiasm to participate than its designers had hoped. By the end of the 1930s many questions remained unresolved.

One should not conclude on the basis of these developments that these experts' efforts were in any way marginal or meaningless. They succeeded in creating a European space, a transnational broadcasting community, and some kind of European regulatory culture for broadcasting. After the war, these found a new start in the European Broadcasting Union as well as the East European

Organization for Radio and Television, continuing up to present-day television programs. In addition, the International Convention for the Use of Broadcasting in the Cause of Peace was taken up after the war by UNESCO, even though the effectiveness of the Convention remained dubious given the role of transnational broadcasting during the Cold War years.[8] Interwar projects have thus influenced the direction of the agenda for broadcasting right up to the present day.

8 In 2005 UNESCO adopted a new Convention that directly built on the interwar year efforts to expand the idea of an all inclusive European culture of diversity. The convention affirms that "freedom of thought, expression and information, as well as diversity of the media, enable cultural expressions to flourish within societies." See: UNESCO, *Convention on the Protection and Promotion of the Diversity of Cultural Expressions* (Paris: UNESCO, 2005), 2, http://unesdoc.unesco.org/images/0014/001429/142919e.pdf. Accessed September 23, 2010. To date the EBU stays in close contact with UNESCO and seriously participates in its ideological work. The EBU has established a Cultural Diversity group and provides a Cultural Diversity Toolkit to all its members and interested parties. Its policy finds its way into all television and radio program schedules of its member public broadcasters. Interview with Pierre Duret, Project manager EBU Eurovision TV, conducted in Geneva on February 12, 2008; http://www.ebu.ch/en/legal/topical/leg_t_gats_top.php. Accessed September 23, 2010.

Appendix

MEMBERS OF THE INTERNATIONAL BROADCASTING UNION 1925-1944[1]

ACTIVE MEMBERS

European zone

		Date of entry
Albania	Enti Shqiptare Audicijoni Radiofonike I Tirana	25.VI.1941
Algeria	Gouvernement général de l'Algérie, Service central des PTT, Alger Joined as an associate member: Joined as an active member:	 6.VI.1932 13.V.1933
Austria	*Oesterreichische Radio-Verkehrs- A.G. (RAVAG), Vienna Incorporated in the RRG, Berlin in 1938.	5.IV.1925
Belgium	* Radio Belgique S.A. Since 1.II.1931: Institut national belge de Radiodiffusion N.V. Radio-Louvain, Louvain Since 1.II.1931: Institut national belge de Radiodiffusion Institut national belge de Radiodiffusion (INR) , Brussels	18.III.1925 21.XI.1929 11.II.1931
Bulgaria	Radio-Sofia, Sofia	13.III.1937
Croatia	Hrvatski Krugoval, Zagreb	4.VI.1942
Czechoslovakia	* Radiojournal, Prague Since 10.VI.1939: Cesky Rozhlas, Prague	18.III.1925
Danzig	Danziger Rundfunk – Sendebetrieb der Post- und Telegraphen-Verwaltung, Danzig Since 15.X.1935: Landespostdirektion der Freien Stadt Danzig, Danzig Incorporated in the RRG, Berlin in 1939.	11.II.1931
Denmark	Radioraadet, Copenhagen Since 15.II.1941: Statsradiofonien, Copenhagen	23.III.1926
Egypt	Egyptian State Telegraphs and Telephones, Cairo Collaboration suspended 23.V.1942.	4.X.1933

1 IBU, *Twenty Years of Activity of the International Broadcasting Union* (Geneva: IBU, 1945), 65-76.

Estonia	S.A. Raadio Ringhääling, Tallinn Since 1.VII.1934: Riigi Ringhääling, Tallinn	6.VI.1932
Finland	Osakeyhtiö Suomen Yleisradio, Helsinki Since 15.V.1934: O.Y. Suomen Yleisradio A.B., Helsinki Since 31.V.1944: Oy. Yleisradio Ab., Helsinki	12.IX.1927
France	* Le Petit Parisien, Paris Afterwards: Compagnie générale d'Energie radioélectrique, Poste Parisien, Paris Collaboration suspended June 1940. * Compagnie française de Radiophonie, Radio-Paris, Paris Taken up again on 12.IX.1933 by Service de la Radiodiffusion de l'Administration française des Postes et des Télégraphes, Paris Became on 29.VII.1939: Radiodiffusion Nationale, Paris (afterwards Vichy) Became in October 1944: Radiodiffusion Française, Paris Radio-Toulouse, La Radiophonie du Midi, Toulouse Collaboration suspended from 1929 onwards. Rejoined the Union under its new name: S.A. La Radiophonie du Midi, Toulouse Incorporated in the Radiodiffusion Française in October 1944. Syndicat de la Radiophonie lyonnaise, Lyon Afterwards became: Radio-Lyon Emissions S.A., Lyon Collaboration ceased on 17.VI.1932. Rejoined the IBU under its new name: S.A. Radio-Lyon, Lyon Incorporated in the Radiodiffusion Française in October 1944. S.A. Radio-Méditerranée. Antibes Incorporated in the Radiodifusion Française in October 1944. Société méridionale de Radiodiffusion, Nîmes Incorporated in the Radiodiffusion Française in October 1944.	18.III.1925 5.IV.1925 23.III.1926 24.VI.1943 23.III.1926 24.VI.1943 24.VI.1943 24.VI.1943
Germany	* Reichs-Rundfunk-G.m.b.H., Berlin	18.III.1925
Great Britain	British Broadcasting Company Ltd., London Since 1.I.1927: British Broadcasting Corporation, London Collaboration suspended 18.III.1941.	18.III.1925

Greece	Greek Broadcasting Service, Athens Since December 1938: Broadcasting Department of the State Under-Secretariat for Press and Touring, Athens Since 2.VI.1941: Greek Broadcasting Society A.A. (AERE). Athens	l.VII.1938 26.XI.1941
Hungary	Magyar Telefonhirmondo es Radio R.T., Budapest	24.III.1926
Ireland	Broadcasting Service, Department of Posts and Telegraphs, Dublin Since 29.XII.1937: Direction of Broadcasting, General Post Office, Dublin	3.IX.1928
Italy	Unione Radiofonica Italiana, Milan Since 17.XI.1927: Ente Iteliano per Ie Audizioni Radiofoniche (EIAR), Rome Since the autumn 1944: Radio Audizioni Italia (RAI), Rome	23.III.1926
Latvia	Department of Posts and Telegraphs, Riga Since 1.IV.1937: Latvijas Radiofons, Riga	10.II.1933
Lithuania	General Direction of Posts and Telegraphs, Kaunas	20.II.1935
Morocco	Office chérifien des Postes, Télégraphes et Téléphones, Rabat Joined as associate member: Joined as active member:	 11.II.1931 25.VI.1941
Netherlands	* Nederlandsche Seintoestellen Fabriek (NSF), Hilversum Afterwards acquired by another group. N.V. Nederlandsche Draadlooze Omroep, Hilversum Withdrawn from the IBU on 28.V.1929 because it did not fulfil the required conditions anymore. Algemeene Vereeniging Radio-Omroep (AVRO), Amsterdam Katholieke Radio-Omroep (KRO), Amsterdam Nederlandsche Christelijke RadioVereeniging (NCRV), Hilversum Vereeniging van Arbeiders Radio Amateurs (VARA), Hilversum Vrijzinnig Protestantsche Radio Omroep (VPRO), Amsterdam Membership resigned 21.VI.1935. All these societies dissolved 27.XII.1940 and replaced by the State organisation: Rijksradio-Omroep de Nederlandsche Omroep, Hilversum	18.III.1925 30.V.1929 30.V.1929 30.V.1929 30.V.1929 30.V.1929
Norway	* Kringkastingselskapet A.S., Oslo Since 1. VII.1933: Norsk Rikskringkasting, Oslo	8.VI.1925
Palestine	Palestine Broadcasting Service, Jerusalem Collaboration suspended 8.VI.1941.	20.II.1936

Poland	Polskie Radjo S.A., Warsaw	11.V.1927
Portugal	Administração geral dos Correios, Telegrafos e Telefones, Lisbon Since 14.IX.1940: Emissora nacional de Radiodifusão, Lisbon	23.V.1930
Rumania	Societatea de Difuziune Radiotelefonica, Bucharest Since 9.VII.1937: Societatea romana de Radiodifuziune, Bucharest	1.VI.1928
Slovakia	Slovensky Rozhlas, Bratislava	25.IV.1940
Spain	* Asociación Nacional de Radiodifusión, Barcelona On 1.II.1927 incorporated in the: Union Radio S.A., Madrid Became in 1942: Societad Española de Radiodifusión, Madrid Collaboration suspended Since the Spanish civil war. Servicio de Radiodifusión de la Republica Española, Barcelona (before Madrid and Valencia) Closed down as a result of changes brought about by political events. Sección Radiocomunicación, Dirección de Telegrafos, Comisión de Obras Públicas y Comunicaciones, Burgos (of the Franco Government) At first as special member; afterwards: Departamento Radio del Servicio Nacional de Propaganda, Burgos Admitted as active member: Afterwards: Departamento Radio, Servicio de Telecomunicación – Ingenieria – Ministerio de la Gobernación, Barcelona (afterwards Madrid)	25.II.1925 23.III.1926 29.VI.1938 27.II.1939
Sweden	Aktiebolaget Radiotjänst, Stockholm	23.III.1926
Switzerland	* Société des Emissions “Radio-Genève, Geneva Radiogenossenschaft Bern, Berne Société romande de Radiophonie (Radio-Lausanne), Lausanne On 24.II.1931 all the Swiss broadcastings societies were grouped under the name of: Société suisse de Radiodiffusion (SSR), Berne Since 29.VIII.1939: Service de la Radiodiffusion suisse (SRS), Berne	18.III.1925 23.III.1926 28.IX.1927 29.VI.1931
Tunisia	Office des Postes, Télégraphes et Téléphones, Tunis	25.IV.1940
Turkey	Turkish Wireless Telephony Company, Stambul Collaboration suspended 30.VI.1931. Broadcasting Service of the Department of Posts and Telegraphs, Ataturk Since 1.VI.1940: Radio -Ankara, Press Department, Ankara	28.IX.1927 27.II.1939

Yugoslavia	Wireless Telegraphy Company, Belgrade Ceased functioning on the entry into service of: Radio A.D. Beograd, Belgrade Radiostanica Zagreb, Zagreb Radio Oddajna, Ljubljana	23.III.1926 30.V.1929 1.VI.1928 21.V.1930

**Foundation member having taken part in the constituent assemblies of the IBU (London and Geneva, 1925).*

SPECIAL MEMBER

		Date of entry
Vatican City	Vatican Broadcasting Service, Vatican City	15.II.1936

ASSOCIATE MEMBERS

Africa

		Date of entry
Kenya	British East African Broadcasting Company, Nairobi Membership resigned 26.IX.1930.	1.VI.1928
South Africa	African Broadcasting Company, Ltd., Johannesburg Membership resigned 30.VII.1930. Rejoined the IBU on 18.VI.1934. Became on 1.VIII.1936: South African Broadcasting Corporation, Johannesburg	30.V.1929

North and South America

		Date of entry
Argentina	Estación de Radiodifusión del Estado, Dirección general de Correos y Telegrafos (Servicio de Radio-comunicaciones), Buenos Aires Asociación de Broadcasters Argentinos (ADEBA), Buenos Aires Closed down, 1.III.1940.	29.VI.1938 1.VII.1936
Canada	James Richardson and Sons, Ltd, Winnipeg Membership resigned 30.III.1932 as a result of a projected re-organisation of the Canadian Radio. Canadian Broadcasting Corporation, Ottawa Collaboration suspended 29.VIII.1940.	1.XII.1931 13.III.1937
Cuba	Telephone Company, Havana Dissolved 20.II.1935.	12.V.1927

Haiti	Radio-Haïti, Port-au-Prince Not in the IBU since 3.VI.1942.	27.II.1939
Mexico	Departamento Autónomo de Prensa y Publicidad, Mexico Membership resigned 12.II.1944.	29.VI.1938
Peru	Administration of Posts and Telegraphs of Peru, Lima Not in the IBU since 12.V.1933.	12.V.1927
Porto Rico	Radio Corporation of Porto Rico, San Juan Membership resigned 17.VII.1940.	20.II.1935
United States	Columbia Broadcasting System, Inc., New York Membership suspended 16.XII.1941. Mutual Broadcasting System, New York National Association of Broadcasters, Washington Membership resigned 6.III.1940. National Broadcasting Company, New York The Tribune Company, Chicago Incorporated in the Mutual Broadcasting System World Wide Broadcasting Foundation, New York Membership resigned 12.VII.1940. Northwestern Broadcasting, Inc., Minnesota Acquired by the Columbia Broadcasting System in 1932. Head of the Lakes Broadcasting Company, Superior Membership resigned 28.VII.1932.	21.V.1930 29.VI.1938 13.III.1937 12.V.1927 29.VI.1931 27.II.1939 29.VI.1931 29.VI.1931
Uruguay	Servicio Oficial de Difusión Radio Eléctrica (SODRE), Montevideo	21.VI.1939

Asia

		Date of entry
Burma	Burma State Broadcasting Service, General Direction of Posts and Telegraphs, Rangoon Membership resigned 14.XII.1942.	21.VI.1939
China	The Central Broadcasting Station (XGOA), Nanking (afterwards Chungking)	30.VI.1936
India (British)	The Indian Broadcasting Service, Bombay Membership resigned 30.VI.1932. Rejoined under the name of: All-India Radio, New Delhi Collaboration suspended 7.VI.1941.	12.V.1927 13.V.1937
Indies (Dutch)	N.V. Nederlandsch-Indische Radio Omroep Maatschappij (NIROM), Amsterdam	20.II.1939
Iran	Wireless Telegraphy Department, Pahlevi, at the Ministry of Posts, Telegraphs and Telephones, Teheran	1.VII.1939
Japan	Nihon Hoso Kyokai (JOCK Broadcasting Station), Nagoya Membership resigned 12.VI.1934 on account of its incorporation in: Nippon Hoso Kyokai (Broadcasting Corporation of Japan), Tokio	12.V.1927 28.IX.1927
Manchukuo	Mandshu Denshin Denwa Kabushiki Kaisha, Hsinking	24.VI.1943

Europe

		Date of entry
Netherlands	N.V. Philip's Omroep Holland-Indie (PHOHI), Eindhoven Not in the IBU since May 1st, 1940, when its services were suspended.	28.XI.1934

Oceania

		Date of entry
Australia	Amalgamated Wireless (Australasia), Ltd., Sydney	12.V.1927
	Central Broadcasting, Ltd., Adelaide Membership resigned 13.IX.1928.	12.V.1927
	National Broadcasting Service, Melbourne Collaboration suspended 30.IV.1941.	21.XI.1929
	Australian Broadcasting Company, Ltd., Sydney Closed down 30.VI.1932.	5.XII.1929
	Australian Broadcasting Commission, Sydney Collaboration suspended 10.IV.1941.	10.II.1933
New Zealand	New Zealand Broadcasting Board, Zealand Wellington Since 1.VII.1936: New Zealand National Broadcasting Service, Wellington Collaboration suspended 10.X.1941.	12.VI.1934

Bibliography

Archival Sources

European Broadcasting Union Archives, Geneva (EBU)
P1 Programme, "Notre Monde" Our World, 1967

International Broadcasting Union Archives, Geneva (IBU)
Box 02 Adhésions et démissions
Box 04 Amateurs, Association d'...
Box 13 Commission Technique 1937-1945
Box 14 Conférences et conventions internationales
Box 15 Contrôle de la matière radiodiffusée
Box 17 Désarmement moral et propagande inadmissible
Box 31 Graphiques
Box 32 Gramophones
Box 33 Histoire de l'UIR
Box 53 Statut International de la radio, 1930-1941
Box 58 Photos
Box 59 Politique et Radio
Box 74 Propagande de l'Union des membres
Box 80ter Commission de Rapprochement, 1927-1950
Box 85 Relais 1926-1950
Box 86 Relations Internationales
Box 92 Société des Nations 1925-1940
Box 94 Stations à ondes longues – moyennes – intermédiaires
Box 94bis Stations d'amateurs, privées et clandestines
Box 95bis Programmes
Conseil Administrative, Documentes "Séries", 1927-1950
Commission des Programmes [anciennement Commissions de Rapprochement et Relais], 1926-1939
Procès-verbaux, Assemblée Générale 1926-1945
Procès-verbaux, Conseil Administrative 1925-1945
Procès-verbaux, Commission Mixte, de Relais, Spéciale
Procès-verbaux, Commission Technique 1925-1940
Questionnaires, no. 1-869; no. 877-2278
85, Concerts européens [subdivisé par pays], 1931-1939
85, Concerts internationaux [mondiaux] [subdivisé par pays], crée en 1936
85, Nuits nationales [subdivisé par pays], 1926-1931

International Telecommunication Union Library and Archives, Geneva (ITU)

Series 41 Z 11, Circulaires télégraphiques no. 849-1004 (1928-1933)
Series 41 Z 11, Circulaires télégraphiques no. 1171, 1196, 1257, 1275, 1278 (1947-1953)
Fonds de Berne, UIR docs, liasses no. 0907-0911, 1929-1938, 1945-1946
Fonds de Berne, OIR, liasses sans numéro, 1946-1947, 1951-1952

League of Nations Archives, Geneva (LoN)

Registry file 5B Intellectual Co-Operation
Registry file 9G Communications and Transit Section: Posts and Telegraphs
Registry file 14 Transit
Broadcasting and Peace 1938 Expert Group, minutes of meetings
Commission of Enquiry for European Union, minutes of meetings
Committee of Letters & Arts, minutes of meetings
Committee for Moral Disarmament files
Communications and Transit Committee, minutes of meetings
International Committee on Intellectual Co-Operation, minutes of meetings
Personal Files Gijsbert Frans van Dissel

The Netherlands Institute for Sound and Vision, Hilversum, the Netherlands

De Omroepgids, Nederlandsche Christelijke Radio-Vereeniging (NCRV), 1925-1940
De Katholieke Radio Gids, Katholieke Radio Omroep (KRO), 1925-1940
De Radiogids, Vereeniging van Arbeiders Radio Amateurs (VARA), 1927-1940
Vrije Geluiden, Vrijzinnig Protestantse Radio Omroep (VPRO), 1927-1974
De Radiobode, Algemeene Vereeniging Radio Omroep (AVRO), 1929-1958

Interviews

Formal

Raina Konstantinova, Director of EBU Radio Department, June 20, 2006
Werner Rumphorst, Director of EBU Legal Department, June 22, 2006
Pierre-Yves Tribolet, Head of Euroradio / Classics of the EBU Radio Department, June 21, 2006
Pierre Duret, Project manager EBU Eurovision TV of the EBU Television Department, February 12, 2008

Informal

Jean-Réveillon, Secretary-General of EBU, June 22, 2006
Raina Konstantinova, Director of EBU Radio Department, February 13, 2008
Avril Mahon-Roberts, Director of EBU Communications Department, November 23, 2006. (Many occasions since)
David Wood, Head of New Media, EBU Technical Department, June 20, 2006
Georges Béry, Senior engineer Eurovision Technical Service / Eurovision Operations, EBU, January 27, 2006
Frans de Jong, Senior engineer Technical Department, EBU, November 23, 2006
Matthias Coinchon, Senior engineer Technical Department, EBU, November 23, 2006

Newspaper and Magazine Articles

"Monthly Commentary: The Industry's Plan for Broadcasting." *Wireless World* 51, no. 9 (1945): 257.
"Mr. P.P. Eckersley: First Chief Engineer of the B.B.C." *The Times*, March 15, 1963.
"Plan for Europe: Broadcasting Allocations on Engineering Principles." *Wireless World* 51, no. 9 (1945): 258.
"Real International Broadcasting: Proposals for a European Alliance." *Wireless World* 50, no. 8 (1944): 234-235.
Czeija, Oskar. "Paneuropa – Idee und Rundfunk." *Paneuropa* 11, no. 6/8 (1935): 197-199.
Valensi, Georges. "La téléphonie européenne." *L'Europe Nouvelle* 417 (1926): 204.

Published Documentation

European Broadcasting Union Documents

Brack, Hans. "Aspects of the Legal Committee's Work within the Framework of the EBU, 1965-1972." *EBU-Review* 24, no. 1 (1973): 37-43.
——. *The Evolution of the EBU through its Statutes from 1950 to 1976*. Geneva: EBU, 1976.
EBU. *25 Years European Broadcasting Union*. Geneva: EBU, 1974.
——. "25 Years of the European Broadcasting Union: A retrospect." *EBU-Review* 26, no. 1 (1975): 10-27.
——. "Six pères fondateurs de la TSF." *Revue Technique*, no. 263 (Spring 1995): 82-96.
"Merger: 1 January 1993: A New Era for Eurovision." *EBU Dossiers: 50 Years of Eurovision*, no. 1 (2004): 30-33.
Sherman, Charles E. "The International Broadcasting Union: A Study in Practical Internationalism." *EBU-Review* 25, no. 3 (1974): 32-36.
Sherman, Charles E., and John Ruby. "The Eurovision News Exchange." *Journalism Quarterly* 51, no. 3 (1974): 478-485.
Wallenborn, Léo. "From IBU to EBU: The Great European Broadcasting Crisis (Part I)." *EBU-Review* 29, no. 1 (January 1978): 25-34.
——. "From IBU to EBU: The Great European Broadcasting Crisis (Part II)." *EBU-Review* 29, no. 2 (March 1978): 22-30.

International Broadcasting Union Documents

Braillard, Raymond. "L'avenir de la radiodiffusion sur ondes courtes et la conférence du Caire." *Radiodiffusion*, no. 5 (November 1937): 13-20.
Burrows, Arthur R. "La Convention Internationale sur l'Emploi de la Radiodiffusion dans l'Intérêt de la Paix." *Radiodiffusion*, no. 3 (October 1936): 26-33.
——. "Les moyens modernes de diffusion utilisés dans l'intérêt de la paix." *Radiodiffusion*, no. 5 (November 1937): 83-91.
Dubois, Antoine. "L'Activité de la radiodiffusion internationale dans le domaine du rapprochement." *Radiodiffusion*, no. 1 (1935): 41-46.
IBU. *Broadcasting: An Instrument for the Creation of a Better Understanding between the Peoples*. Geneva: IBU, 1937.
—— *Extrait des documents de la Conférence Radiotélégraphique Internationale de Madrid (1932): COMMISSION TECHNIQUE*. Geneva: Selection of all formal Madrid proceedings on broadcasting, 1932.
——. *First Intercontinental Meeting of Broadcasting Organisations, Paris, February 27 – March 6 1936: Preparatory Documents and Minutes of the Meeting*. Geneva: IBU Office, 1936.

——. *International Union of Broadcasting Organisations – Union Internationale de Radiophonie.* Geneva: IBU, 1926.
——. *L'Union Internationale de Radiodiffusion: Son histoire – son activité.* Genève: IBU, 1944.
——. *L'activité de l'Union Internationale de Radiodiffusion de 1939 à 1946.* Genève: IBU, 1946.
——. "La radiodiffusion: Instrument de rapprochement international, Numéro spécial." *Radiodiffusion*, no. 4 (January 1937).
——. "Le folklore et la radiodiffusion dans les États Baltes." *Radiodiffusion*, no. 5 (November 1937): 78-82.
——. *Maurice Rambert, 1866-1941.* Geneva: IBU, 1941.
——. *Memorandum on the present situation of European broadcasting from the point of view of interference, and the best means by which this may be remedied.* Presented to the Telegraphic Administrations by the Technical Committee of the Union Internationale de Radiodiffusion, Technical Publications of the Union Internationale de Radiodiffusion 1. Brussels: IBU, 1932.
——. *The Importance of Broadcasting.* Geneva: IBU, 1932.
——. *The Problems of Broadcasting.* Geneva: IBU, 1930.
——. *Twenty Years of Activity of the International Broadcasting Union.* Geneva: IBU, 1945.
——. *Union Internationale de Radiophonie.* Genève: IBU, 1926.
Reichs-Rundfunk G.m.b.H. "La radiodiffusion Allemande et l'entente entre les peuples." *Radiodiffusion*, no. 5 (November 1937): 49-55.

International Telecommunication Union Documents

Carte des circuits Internationaux d'Europe spécialement établis ou aménagés pour transmettre la musique. Berne: BUIT, 1932, 1933, 1935, 1937, 1939, 1948.
Documents de la Conférence Européenne de Radiodiffusion, Montreux, Mars/Avril 1939. Berne: BUIT, 1939.
Documents de la Conférence Européenne des Radiocommunications Lucerne, MAI/JUIN 1933. Berne: BUIT, 1933.
Documents de la Conférence Internationale des Radiocommunications du Caire (1938), Tome I: Propositions transmises au Bureau de l'Union pour être soumises à la Conférence: Propositions, notes, déclarations, études, remarques soumises pendant la Conférence. Berne: BUIT, 1938.
Documents de la Conférence Internationale des Radiocommunications du Caire (1938), Tome II: Travaux de la Conférence (Délibérations, Actes définitifs). Berne: BUIT, 1938.
Documents de la Conférence Radioélectrique européenne de Prague 1929. Berne: BUIT, 1929.
Documents de la Conférence Radiotélégraphique Internationale de Madrid (1932): Tome I: Propositions transmises au Bureau de l'Union pour être soumises à la Conférence: Propositions, notes, déclarations, études, remarques soumises pendant la Conférence. Berne: BUIT, 1932.
Documents de la Conférence Radiotélégraphique Internationale de Madrid (1932): Tome II: Travaux de la Conférence (Délibérations, Actes définitifs). Berne: BUIT, 1932.
Documents de la Conférence Radiotélégraphique Internationale de Washington (1927): Tome I: Propositions soumises à la conférence. Berne: BUIT, 1928.
Documents de la Conférence Radiotélégraphique Internationale de Washington (1927): Tome II: Travaux de la Conférence (Délibérations, Actes définitifs). Berne: BUIT, 1928.
Règlement général des radiocommunications (Révision du Caire, 1938); Protocole final (Révision du Caire, 1938); Règlement additionnel des radiocommunications (Révision du Caire); Protocol additionnel aux actes de la Conférence internationale des radiocommunications du Caire, 1938, signé par les délégués des pays de la région européenne annexés à la Convention

internationale des télécommunications (Madrid, 1932). Berne: BUIT, 1938.
CCIR: 50th Anniversary. ITU, 1979.
ITU: A hundred years of international cooperation 1865-1965. Division of Journal, Library and Documentation, ITU General Secretariat, 1965.
"Un réseau international de radiodiffusion en Europe." *Journal des Télécommunications* 12, no. 5 (1945): 62-63.

League of Nations Documents

Advisory and Technical Committee for Communications and Transit, Procès-Verbal of the Fourth Session, held at Geneva, April 23rd-30th, 1923. LoN doc. ser., C.486.M.202.1923.VIII. Geneva: LoN, 1923.
Advisory and Technical Committee for Communications and Transit, Procès-Verbal of the Fifth Session, held at Geneva, Aug. 29th-Sept. 1st, 1923. LoN doc. ser., C.667.M.267.1923.VIII. Geneva: LoN, 1923.
Advisory and Technical Committee for Communications and Transit, Procès-Verbal of the Sixth Session, held at Geneva, March 12th-14th, 1924. LoN doc. ser., C.196.M.61.1924.VIII. Geneva: LoN, 1924.
Advisory and Technical Committee for Communications and Transit, Procès-Verbal of the Seventh Session, held at Geneva, Nov. 26th-29th, 1924. LoN doc. ser., C.72.M.36.1925.VIII. Geneva: LoN, 1925.
Advisory and Technical Committee for Communications and Transit, Procès-Verbal of the Eigth Session, held at Geneva, July 24th-30th, 1925. LoN doc. ser., C.621.M.203.1925.VIII. Geneva: LoN, 1925.
Advisory and Technical Committee for Communications and Transit, Minutes of the Sixteenth Session, held at Geneva, May 28th-June 2nd, 1931. LoN doc. ser., C.417.M.173.1931.VIII. Geneva: LoN, 1931.
Broadcasting and Peace: Studies and Projects in the Matter of International Agreements. Paris: International Institute of Intellectual Co-Operation, 1933.
Comment faire connaître la Société des Nations et développer l'esprit de Coopération internationale: Recommendations du Sous-Comité d'Experts de la Commission internationale de Coopération intellectuelle de la Société des Nations. Geneva, 1927.
Commission Consultative et Technique des Communications et du Transit, Procès-Verbal de la 9ième session, tenue à Genève, 12 au 17 juillet 1926. LoN doc. ser., C.595.M.231.1926.VIII. Geneva: LoN, 1926.
Comité d'Experts chargé d'élaborer des suggestions concernant l'utilisation de la radiodiffusion dans l'intérêt de la paix, Procès-verbal provisoire de la deuxième séance, tenue à Genève, 17 juin 1938. LoN doc. ser., C.I.C.I./C.E.E.R./P.V.1-C.I.C.I./C.E.E.R./P.V.2. Geneva: LoN, 1938.
Conference for the Conclusion of an International Convention for the Use of Broadcasting in the Cause of Peace, Action to be taken as a result of the Conference, Report by the French Representative, Geneva, October 9th, 1936. LoN doc. ser., C.440.1936.XII. Geneva: LoN, 1936.
Conference for the Reduction and Limitation of Armaments, Moral Disarmament, held at Geneva, July 25th, 1932. LoN doc. ser. Conf. D. 138. Geneva: LoN, 1932.
Conference for the Reduction and Limitation of Armaments, Committee for Moral Disarmament, Text Adopted by the Committee. LoN doc. ser., Conf. D/C.D.M./36. Geneva: LoN, Nov. 21st 1933.
Establishment of a League of Nations Radio-Telegraphic Station. LoN doc. ser., A.22.1928.VIII. Geneva: LoN, March 1928.

Fleury, A. "Comment le siège de la Société des Nations est devenue l'Office européens des Nations Unies." In *The League of Nations 1920-1946: organization and accomplishments: a retrospective of the First International Organization for the Establishment of World Peace*, 81-85. Geneva; New York: UN, 1966.

—. "La Suisse et Radio Nations." In *The League of Nations in Retrospect: Proceedings of the Symposium*, edited by UN Office at Geneva. Library. League of Nations Archives and Historical Collections Unit, 196-220. Berlin; New York: Walter de Gruyter, 1983.

International Committee on Intellectual Co-Operation, Minutes of the Eighth Session, held at Geneva, July 26th-29th, 1926. LoN doc. ser., C.462.M.181.1926.XII (C.I.C.I/8th Session/P.V.). Geneva: LoN, 1926.

International Committee on Intellectual Co-Operation. Minutes of the Tenth Session, held at Geneva, July 25th-30th, 1928. LoN doc. ser., C.533.M.160.1928.XII. Geneva: LoN, Oct. 1928.

International Committee on Intellectual Co-Operation, Minutes of the Eleventh Session, held at Geneva, July 22nd-26th, 1929. LoN doc. ser., C.342.M.121.1929.XII. Geneva: LoN, 1929.

International Committee on Intellectual Co-Operation, Minutes of the Thirteenth Session, held at Geneva, July 20th-25th, 1931. LoN doc. ser., C.471.M.201.1931.XII (C.I.C.I./13th session/ P.V.). Geneva: LoN, 1931.

International Committee on Intellectual Co-Operation, Minutes of the Fourteenth Session, Minutes of the First Meeting, held at Geneva, July 18th, 1932. LoN doc. ser., C.I.C.I./14th Session/P.V.1-2. 7-8. Geneva: LoN, 1932.

International Committee on Intellectual Co-Operation, Minutes of the Fifteenth Session, held at Geneva, November 21st, 1933. LoN doc. ser., C.I.C.I./15th Session/P.V.9. Geneva: LoN, 1933.

International Committee on Intellectual Co-Operation, Minutes of the Sixteenth Session, held at Geneva, July 17th, 1934. LoN doc. ser., C.I.C.I./16th session/P.V.4, 11. Geneva: LoN, 1934.

International Committee on Intellectual Co-Operation, Minutes of the Seventeenth Session, held at Geneva, July 15th, 1935. LoN doc. ser., C.I.C.I./17th session/P.V.1. Geneva: LoN, 1935.

International Committee on Intellectual Co-Operation, Minutes of the Nineteenth Session, held at Geneva, July 15th-17th, 1937. LoN doc. ser., C.I.C.I./19th session/P.V.1-8. Geneva: LoN, 1937.

International Committee on Intellectual Co-Operation, Minutes of the Twentieth Session, held at Geneva, July 11th-15th, 1938. LoN doc. ser., C.I.C.I./20th session/P.V1-10. Geneva: LoN, 1938.

International Committee on Intellectual Co-Operation, Minutes of the Twenty-first Session, held at Geneva, July 20th, 1939. LoN doc. ser., C.I.C.I./21st session/P.V.7. Geneva: LoN, 1939.

International Committee on Intellectual Co-Operation, Report by the Committee on the Work of its Twelfth Session. LoN doc. ser., A.21.1930.II. Geneva: LoN, 1930.

International Committee on Intellectual Co-Operation, Report of the Committee on the Work of its Seventeenth Plenary Session. LoN doc. ser., C.290.M.154.1935.XII. Geneva: LoN, 1935.

Intellectual Co-Operation Organisation, Modern Means of Spreading Information Utilised in the Cause of Peace. LoN doc. ser., A.19.1938.XII. Geneva: LoN, 1938.

Kimmich, Ch.M. "Germany and the League of Nations." In The League of Nations in Retrospect: Proceedings of the Symposium, 118-127. Berlin; New York: Walter de Gruyter, 1983.

Moral Disarmament, Memorandum from the Polish Government. LoN doc. ser., C.602.M.240.1931.IX [Conf. D. 16]. Geneva: LoN, Sept. 23rd, 1931.

Organisation for Communications and Transit, Report on the work of the organisation between the Eight and Ninth ordinary Session of the Assembly, September 20th 1928. LoN doc. ser., A.73.1928.VIII. Geneva: LoN, 1928.

Radio-Telegraphic Station for the Purpose of ensuring Independent Communications for the League of Nations at Times of Emergency, Supplementary Report of the Advisory and Technical Committee for Communications and Transit. LoN doc. ser., A.23.1928.VIII. Geneva: LoN, 1928.

Radio-Telegraphic Station for the Purpose of Ensuring Independent Communications for the League of Nations at Times of Emergency, Note and Memorandum from the Swiss Federal Government. LoN doc. ser., A.31.1928.VIII. Geneva: LoN 1928.
Second Committee, 1928 Assembly, Minutes of the Eight meeting, held at Geneva, September 19th 1928. LoN doc. ser., A.II/P.V.8. Geneva: LoN, 1928.
Ten Years of World Co-Operation. London: Secretariat of the League of Nations, 1930.
Travaux de la Commission et de l'Institut International de la Coopération Intellectuelle, Rapport de la sixième commission à l'Assemblée, Rapporteur: M. G. Gallavresi (délégué de l'Italie). LoN doc. ser., A.74.1928.XII. Genève: LoN, 1928.
Verbatim Record of the 10th Ordinary Session of the Assembly of the League of Nations, 6th Plenary Meeting. LoN doc. ser., A.10.1929. Geneva: LoN, 1929.

UNESCO Documents

Convention on the Protection and Promotion of the Diversity of Cultural Expressions. Paris: UNESCO, 2005. http://unesdoc.unesco.org/images/0014/001429/142919e.pdf.

Scholarly Books, Articles, and Dissertations

Abshire, David M. *International Broadcasting: A New Dimension of Western Diplomacy*. The Washington Papers. Beverly Hills: Sage Publications, 1976.
Aitken, Hugh G.J. *Syntony and Spark: The Origins of Radio*. New York: John Wiley & Sons, 1976.
Anastasiadou, Irene. *Constructing Iron Europe: Transnationalism and Railways in the Interbellum*. Amsterdam: 2012.
Applegate, Celia, and Pamela Potter. "Germans as the 'People of Music': Genealogy of an Identity." In *Music and German National Identity*, edited by Celia Applegate and Pamela Potter, 1-35. Chicago; London: University of Chicago Press, 2002.
Applegate, Celia, and Pamela Potter, eds. *Music and German National Identity*. Chicago: University of Chicago Press, 2002.
Badel, Laurence. *Un milieu libéral et européen: Le grand commerce français, 1925-1948*. Paris: Comité pour l'histoire économique et financière, 1999.
Badenoch, Alexander. "Making Sunday What it Actually Should Be: Sunday Radio Programming and the Re-Invention of Tradition in Occupied Germany 1945-1949." *Historical Journal of Film, Radio and Television* 25, no. 4 (2005): 577-598.
——. "Touring between War and Peace: Imagining the 'Transcontinental Motorway,' 1930-1950." *Journal of Transport History* 28, no. 2 (2007).
——. *Voices in Ruins: West German Radio across the 1945 Divide*. Basingstoke: Palgrave Macmillan, 2008.
——. "Myths of the European Network: Constructions of Cohesion in Infrastructure Maps." In *Materializing Europe: Transnational Infrastructures and the Project of Europe*, edited by Alexander Badenoch and Andreas Fickers, 47-77. London: Palgrave, 2010.
Badenoch, Alexander, and Andreas Fickers. "Introduction: Europe Materializing? Toward a Transnational History of European Infrastructures." In *Materializing Europe: Transnational Infrastructures and the Project of Europe*, edited by Alexander Badenoch and Andreas Fickers, 1-25. London: Palgrave, 2010.
Barnouw, Erik. *A History of Broadcasting in the United States: Vol. I-III*. New York: Oxford University Press, 1970.
Bayly, Christopher Allan, Sven Beckert, Matthew Connelly, Isabel Hofmeyer, Wendy Kozol, and Patricia Seed. "AHR Conversation: On Transnational History." *The American History Review* 111, no. 5 (2006): 1441-1464.

Beidler-Wagner, F.W., H. Bonnet, K. Bonnevie, H. Matheson, A.R. Burrows, W. Davies, H. Frisch, et al. *Le rôle intellectuel et éducatif de la radiodiffusion*. Cahiers, 2. Paris: Institut International de Coopération Intellectuelle, 1935.

Bilby, Kenneth. *The General: David Sarnoff and the Rise of the Communications Industry*. 1st ed. New York: Harper & Row, 1986.

Boeckmann, Kurt von. *Von der Arbeit des Deutschen Kurzwellen-senders*. S.l.: S.n., 1939.

Boer, Pim den. "Europe to 1914: The Making of an Idea." In *The History of the Idea of Europe*, edited by Kevin Wilson and Jan van der Dussen, 13-82. London: Routledge, 1993.

Borchmeyer, Dieter. *Drama and the World of Richard Wagner*. Princeton: Princeton University Press, 2003.

Borneman, John, and Nick Fowler. "Europeanization." *Annual Review of Anthropology* 26 (1997): 487-514.

Botstein, Leon. "Modernism." *Grove Music Online. Oxford Music Online*, n.d. http://www.oxfordmusiconline.com:80/subscriber/article/grove/music/40625.

Bourdon, Jérôme. "Comment écrire une histoire transnationale des médias? L'exemple de la télévision en Europe." *Les Temps des Médias. Dossier: Espaces européens et transferts culturels* 11, no. 2 (2008): 164-181.

Briggs, Asa. "Peter Pendleton Eckersley." *The History of Broadcasting in the United Kingdom, Volume I: The Birth of Broadcasting*. Oxford [u.a.]: Oxford University Press, 1961.

—. *The History of Broadcasting in the United Kingdom, Volume II: The Golden Age of Wireless*. Oxford [u.a.]: Oxford University Press, 1995.

—. *The History of Broadcasting in the United Kingdom, Volume III: The War of Words*. Oxford: Oxford University Press, 1995.

—. *The History of Broadcasting in the United Kingdom, Volume I-V*. Oxford: Oxford University Press, 1961.

—. *Oxford Dictionary of National Biographies*, 2008. http://www.oxforddnb.com/view/article/37385.

Brinkmann, Sören. "Bilder eines Krieges: Europa und der Bürgerkrieg in Spanien." In *Europäische Öffentlichkeit: Transnationale Kommunikation seit dem 18. Jahrhundert*, edited by Jörg Requate and Martin Schulze Wessel, 250-272. Frankfurt am Main; New York: Campus, 2002.

Brochand, Christian. *Histoire générale de la radio et de la télévision en France*. Paris: La Documentation française, 1994.

Browne, Donald R. *International Radio Broadcasting*. New York, N.Y.: Praeger, 1982.

Bugge, Peter. "The Nation Supreme: The Idea of Europe 1914-1945." In *The History of the Idea of Europe*, edited by Kevin Wilson and Jan van der Dussen, 83-149. London: Routledge, 1993.

—. "'Shatter Zones': The Creation and Re-Creation of Europe's East." In *Ideas of Europe Since 1914: The Legacy of the First World War*, edited by Menno Spiering and Michael Wintle, 47-68. New York: Palgrave, 2002.

Burrows, Arthur R. *The History of Broadcasting*. London: Cassell, 1924.

—. "Broadcasting outside the United States." *Annals of the American Academy of Political and Social Science* 177 (1935): 29-41.

Bussière, Eric. *La France, la Belgique et l'organisation économique de l'Europe, 1918-1935*. Paris: Comité pour l'histoire économique et financière de la France, 1992.

Carreras, Albert, Andrea Giuntini, and Michèle Merger, eds. *European Networks, 19th – 20th Centuries: New Approaches to the Formation of a Transnational Transport and Communications System*. B8 Proceedings Eleventh International Economic History Congres Milan, September 1994. Milan: Università Bocconi, 1994.

CBS. *Spanning the Seven Seas: A Résumé of Inter-Continental Broadcasting in 1935*. New York; London: Columbia Broadcasting Organization, 1935.

Chalmers, Kenneth. *Béla Bartók*. London: Phaidon Press Ltd, 1995.
Clavin, Patricia. *The Great Depression in Europe, 1929-1939*. New York, N.Y.: St. Martin's Press, 2000.
——. "Europe and the League of Nations." In *Twisted Paths: Europe 1914-1945*, edited by Robert Gerwarth, 325-354. Oxford: Oxford University Press, 2007.
Clavin, Patricia, and Jens-Wilhelm Wessel. "Transnationalism and the League of Nations: Understanding the Work of its Economic and Financial Organisation." *Contemporary European History* 14 (2005): 465-492.
Codding, George A. *The International Telecommunication Union: An Experiment in International Cooperation*. Leiden: Brill, 1952.
Cordier, Andrew W. "European Union and the League of Nations." *Geneva Special Studies* 2, no. 6 (1931): 3-27.
Coudenhoven-Kalergi, Count Richard. *Paneuropa*. Wien: Paneuropa-Verlag, 1926.
Couldry, Nick. *Media Rituals: A Critical Approach*. London; New York: Routledge, 2003.
Crisell, Andrew. *An Introductory History of British Broadcasting*. 2nd ed. London; New York: Routledge, 2002.
Currid, Brian. *A National Acoustics: Music and Mass Publicity in Weimar and Nazi Germany*. Minneapolis: University of Minnesota Press, 2006.
Davies, Alan. "The First Radio War: Broadcasting in the Spanish Civil War, 1936-1939." *Historical Journal of Film, Radio and Television* 19, no. 4 (1999): 473-513.
Davies, Norman. *Europe: A History*. London: Pimlico, 1997.
Dayan, Daniel, and Elihu Katz. *Media Events: The Live Broadcasting of History*. Cambridge, Mass.: Harvard University Press, 1992.
Degenhardt, Wolfgang. *Die Entstehung und Entwicklung der europäischen Partnerschaft im Fernsehbereich 1950-1970: Zur historischen Betrachtung eines komplexen Sensemaking-Prozesses*. Siegen: Universität Siegen, 2002.
Degenhardt, Wolfgang, and Elisabeth Strautz. *Auf der Suche nach dem europäischen Programm: Die Eurovision 1954-1970*. Baden-Baden: Nomos Verlagsgesellschaft, 1999.
Delanty, Gerard. *Inventing Europe: Idea, Identity, Reality*. New York: St. Martin's Press, 1995.
Demény, János, ed. "Bartók to Octavian Beu, January 10, 1931." In *Béla Bartók: Letters*. London; Budapest: Faber and Faber Ltd.; Corvina Press, 1971.
Deschamps, Etienne. "L'Européen (1929-1940): A Cultural Review at the Heart of the Debate on European Identity." *European Review of History* 9, no. 1 (2002): 85-98.
Dissel, Gijsbert Frans van. *De internationale organisatie van het radio-electrische verkeer: Voordrachten gehouden door comm. Brenot voor de afdeling der radio-telegrafie van de école supérieure d'électricité gedurende het semester 1920-1921*. Leiden: A.W. Sijthoff, n.d.
Douglas, Susan. *Inventing American Broadcasting, 1899-1922*. Baltimore: Johns Hopkins University Press, 1988.
——. *Listening in: Radio and the American Imagination*. Minneapolis, Minn.; London: University of Minnesota Press, 2004.
Druick, Zoë. "The International Educational Cinematograhp Insitute, Reactionary Modernism, and the Formation of Film Studies." *Canadian Journal of Film Studies* 16, no. 1 (2007): 80-97.
Duggan, Stephen. "Radio: An Agency for International Understanding or Friction, In 'Radio: The Fifth Estate', Special Issue." *Annals of the American Academy of Political and Social Science* 177 (1935): 176-178.
Dukes, Paul, ed. *Frontiers of European culture*. Lewiston: E. Mellen Press, 1996.
Dunlap, Orrin E. *Marconi, the Man and his Wireless*. New York: Arno Press, 1971.
Dycke, Vernon van. "The Responsibility of States for International Propaganda." *The American Journal of International Law* 34, no. 1 (1940): 473-513.

Eagleton, Terry. *The Idea of Culture*. Oxford: Blackwell Publishing, 2000.
Eckersley, Peter P. *The Power Behind the Microphone*. London: J. Cape, 1941.
Ellwood, David W. *Rebuilding Europe: Western Europe, America and Postwar Reconstruction*. London; New York: Longman, 1992.
Erol, Merih. "Modernist and Nationalist Discourses on Music in Southeastern Europe (1900-1945): The Cases of Greece and Turkey." *Prepared for the CAS (Centre for Advanced Study, Sofia) Fellowship program "Regimes of Historicity and Discourses of Modernity and Identity, 1900-1945, in East-Central, Southeastern and Northern Europe (2008-2010)"*. Forthcoming as a book chapter in an edited volume after completion of the program, 2009.
Eugster, Ernest. *Television Programming Across National Boundaries: The EBU and OIRT Experience*. Dedham, MA: Artech House, 1983.
Fickers, Andreas "Sichtbar hörbar! Radioapparat und Stadt: Knoten im vernetzten Kommunikationsraum." In *Zentralität und Raumgefüge der Grossstadte im 20. Jahrhundert*, edited by Clemens Zimmermann, 83-103. Stuttgart: Franz Steiner Verlang Stuttgart, 2006.
—. "National Barriers for an Imag(e)ined European Community: The Technopolitical Frames of Postwar Television Development in Europe." In *Northern Lights: Film and Media Studies Yearbook 2005* 4 (2006): 15-35.
—. "Broadcasting as Critical Infrastructure: A Story of European Fine-Tuning and Techno-Political References". Stockholm, 2008.
—. "Broadcasting." In *The Palgrave Dictionary of Transnational History*, edited by Pierre-Yves Saunier and Akira Iriye, 106-109. Basingstoke; New York: Palgrave Macmillan, 2009.
—. "The Techno-politics of Colour: Britain and the European Struggle for a Colour Television Standard." *Journal of British Cinema and Television* 7, no. 1 (2010): 95-114.
—. "In Search of the 'Greenwich of the Air': Techno-Political Diplomacy in European Frequency Allocation and Control (1925-1952)." *History and Technology* (forthcoming).
Fickers, Andreas, and Pascal Griset, eds. *Eventing Europe: Electronic Information and Communication Spaces in Europe*. Forthcoming with Palgrave Macmillan, 2013.
Fickers, Andreas, and Catherine Johnson. "Transnational Television History: A Comparative Approach." *Media History* 16, no. 1 (2010): 1-11.
Fickers, Andreas, and Suzanne Lommers. "Eventing Europe: Broadcasting and the Mediated Performances of Europe." In *Materializing Europe: Transnational Infrastructures and the Project of Europe*, edited by Andreas Fickers and Alexander Badenoch, 225-251. London: Palgrave, 2010.
Figes, Orlando. *Natasha's Dance: A Cultural History of Russia*. New York: The Penguin Press, 2002.
Finch, George A. "A Pact of Non-Aggression." *The American Journal of International Law* 27, no. 4 (1933): 725-732.
Fisher, Harold A. *The EBU: Model for Regional Cooperation in Broadcasting*. 68. Lexington, Ky.: Association for Education in Journalism, 1980.
Fleury, Antoine. "La Suisse et Radio Nations." In *The League of Nations in Retrospect: Proceedings of the Symposium*, edited by UN Office at Geneva. Library. League of Nations Archives and Historical Collections Unit, 196-220. Berlin; New York: Walter de Gruyter, 1983.
Fulbrook, Mary, ed. *Europe Since 1945*. Oxford; New York: Oxford University Press, 2001.
Geddes, Keith, and Gordon Bussey. *The Setmakers: A History of the Radio and Television Industry*. London: BREMA, 1991.
Giesecke, Heinrich. "Der Weltrundfunkverein: Seine Enststehung und sein Wirken bis zur Luzerner Konferenz." In *Weltfunkverein: Studien zum Weltrundfunk und Fernsehrundfunk*. Heidelberg; Berlin; Magdeburg: Kurt Vowinckel Verlag, 1941.
Gillies, Malcolm. *Bartók Remembered*. London: Faber and Faber Ltd, 1990.

——. "Bartók, Béla." *Grove Music Online. Oxford Music Online*, n.d. http://www.oxfordmusi-conline.com/subscriber/article/grove/music/40686pg3.
Goebbels, Joseph. *Der Kongress zur Nüremberg 1934*. Munich: Zentralverlag der NSDAP; Franz Eher Nachfolger GmbH, 1934.
——. "Der Rundfunk als achte Groβmacht." In *Signale der neuen Zeit. 25 ausgewählte Reden von Dr. Joseph Goebbels*, 197-207. Munich: Zentralverlag der NSDAP, 1938.
Gosden, Peter. "Sir (Henry) Frank Heath." *Oxford Dictionary of National Biographies*, 2008. http://www.oxforddnb.com/view/article/33791?docPos=14.
Graham, Helen. *The Spanish Republic at War, 1936-1939*. Cambridge: Cambridge University Press, 2002.
Greaves, Harold R.G. *The League Committees and World Order: A Study of the Permanent Expert Committees of the League of Nations as an Instrument of International Government*. London: Oxford University Press; H. Milford, 1931.
Griset, Pascal. "Innovation and Radio Industry in Europe during the Interwar Period." In *Innovations in the European Economy between the Wars*, edited by François Caron, Paul Erker, and Wolfram Fischer, 37-63. Berlin; New York: Walter de Gruyter, 1995.
Haas, Peter M. "Introduction: Epistemic Communities and International Policy Coordination." *International Organization* 46, no. 1 (1992): 1-35.
Haupt, Heinz-Gerhard, and Philipp Ther. *Europe and Beyond: Transfers, Networks and Markets for Musical Theatre in Modern Europe, 1740-1960*. Project proposal. Florence: European University Institute, 2008.
Hay, Denys. *Europe: The Emergence of an Idea*. New York: Harper & Row, 1966.
Headrick, Daniel. *The Tentacles of Progress: Technology Transfer in the Age of Imperialism, 1850-1940*. New York: Oxford University Press, 1988.
——. *The Invisible Weapon: Telecommunications and International Politics, 1851-1945*. New York: Oxford University Press, 1991.
Heater, Derek. *The Idea of European Unity*. Leicester: Leicester University Press, 1992.
Heister, H.S. von. "Heinrich Giesecke und Kurt Magnus." In *Funkköpfe: 46 Literarische Porträts*, edited by Karl Wilczynski, 8-10. Berlin: Verl. Funk-Dienst, 1927.
Henrich-Franke, Christian. "Die «EBU Screening Sessions»: Wandlungen des europäischen Markts für Fernsehprogramme 1963-1985." *Rundfunk und Geschichte* 31, no. 1 (2005): 17-25.
——. *Globale Regulierungsproblematiken in historischer Perspektive: Der Fall des Funkfrequenzspektrums 1945-1988*. Baden-Baden: Nomos Verlagsgesellschaft, 2006.
——. "Cookies for ITU: The Role of Social Networks in Standardization Processes." In *Bargaining Norms – Arguing Standards: Negotiating Technical Standards*, edited by Judith Schueler, Andreas Fickers, and Anique Hommels, 86-97. The Hague: Stichting Toekomstbeeld der Techniek, 2008.
——. "Creating Transnationality through an International Organization?" *Media History* 16, no. 1 (2010): 67-81.
Hermon, Elly. "«Le désarmement moral en tant que facteur dans les relations internationales pendant l'entre-deux-guerres»." *Historical Papers/Communications historiques* 22, no. 1 (1987): 198-211.
Heuvel, J.H.J. van den. "Dubois, Antoine (1887-1956)." *Biografisch Woordenboek van Nederland. Deel 1*. 's Gravenhage, 1979. URL:http://www.inghist.nl/Onderzoek/Projecten/BWN/lemmata/bwn1/duboisa.
Hicks, Anthony. "Handel, George Frideric." *Grove Music Online. Oxford Music Online*, n.d. http://www.oxfordmusiconline.com/subscriber/article/grove/music/40060.
Hilmes, Michele. *Only Connect: A Cultural History of Broadcasting in the United States*. Belmont, CA: Wadsworth/Thomson Learning, 2002.

——. *Network Nations: A Transnational History of British and American Broadcasting*. New York: Routledge, 2011.

Hobsbawm, Eric. *The Age of Extremes: A History of the World, 1914-1991*. London: Abacus, 1994.

Hughes, Thomas. *Networks of Power: Electrification in Western Society, 1880-1930*. Baltimore: Johns Hopkins University Press, 1983.

——. "The Evolution of Large Technical Systems." In *The Social Construction of Technological Systems: New Directions in the Sociology and History of Technology*, edited by Wiebe Bijker, Thomas Hughes, and Trevor Pinch, 51-82. Cambridge: MIT Press, 1987.

Huth, Arno. *La radiodiffusion: Puissance mondiale*. Paris: Librairie Gallimard, 1937.

Ifversen, Jan. "The Crisis of European Civilization after 1918." In *Ideas of Europe Since 1914: The Legacy of the First World War*, edited by Menno Spiering and Michael Wintle, 14-46. New York: Palgrave, 2002.

International Conciliation. *Washington Conference on the Limitation of Armaments*. Vol. 169. Greenwich: American Association for International Concilliation, 1921.

"International Convention Concerning the Use of Broadcasting in the Cause of Peace." *The American Journal of International Law* 32, no. 3 (1938): 113-120.

Iriye, Akira. *Cultural Internationalism and World Order*. Baltimore: Johns Hopkins University Press, 1997.

——. *Global Community: The Role of International Organizations in the Making of the Contemporary World*. Berkeley: University of California Press, 2002.

Jacot, Bernard Louis, and D.M.B. Collier. *Marconi – Master of Space: An Authorised Biography of the Marchese Marconi*. London: Hutchinson & Co, 1935.

Jirgens, Eckhard. *Der deutsche Rundfunk der 1. Tschechoslowakischen Republik: Eine Bestandsaufnahme*. Frankfurt am Main: Lang, 2005.

Johnson, James. *Listening in Paris: A Cultural History*. Berkeley: University of California Press, 1995.

Kaftanov, S.V., and Soviet Union. "Gosudarstvennyi Komitet po radioveshchan i iu i televiden i iu." *Radio and Television in the USSR*. Washington: U.S. Joint Publications Research Service, 1961.

Kerman, Joseph, Alan Tyson, Scott G. Burnham, Douglas Johnson, and William Drabkin. "Beethoven, Ludwig van." *Grove Music Online. Oxford Music Online*, n.d. http://www.oxfordmusiconline.com/subscriber/article/grove/music/40026.

Krupicka, Miroslav. *Czech Radio History*. Radio Prague, 1996. http://www.radio.cz/en/article/49702.

Laak, Dirk van. "Der Begriff „Infrastruktur" und was er vor seiner Erfindung besagte." *Archiv für Begriffsgeschichte* 41 (1999): 280-294.

——. "Infra-Strukturgeschichte." *Geschichte und Gesellschaft* 27 (2001): 367-393.

——. "Technological Infrastructure, Concepts and Consequences." *ICON Journal of the International Committee for the History of Technology* 10 (2004): 53-64.

Laborie, Léonard. "La France, L'Europe et l'ordre international des communications (1865-1959)". Ph.D. diss., Université Paris IV – Sorbonne, 2006.

Lacey, Kate. "Towards a Periodization of Listening: Radio and Modern Life." *International Journal of Cultural Studies* 3, no. 2 (2000): 279-288.

Lagendijk, Vincent. *Electrifying Europe: The Power of Europe in the Construction of Electricity Networks*. Amsterdam: Aksant, 2008.

——. "'To Consolidate Peace'? The International Electro-Technical Community and the Grid for a United States of Europe, 1929-1937." *Journal of Contemporary History* (forthcoming 2012).

Laqua, Daniel. "Internationalisme ou affirmation de la nation? La coopération intellectuelle transnationale dans l'entre-deux-guerres." *Critique – Internationale, Une autre approche de*

la globalisation: socio-histoire des organisations internationales (1900-1940) 52 (2011): 51-67.
——. "Transnational Intellectual Cooperation, the League of Nations and the Problem of Order." *Journal of Global History* 6, no. 2 (2011): 223-247.
Lenin, Vladimir I. "On the Slogan for a United States of Europe." In *Lenin Collected Works. Originally in Sotsial Demokrat, No. 4, 1915*, edited by Marxist Internet Archive, 21, 339-343. Moscow: Progress Publishers, 1974.
Lerg, Winfried B. *Rundfunkpolitik in der Weimarer Republik, Band 1*. München: Deutscher Taschenbuch Verlag, 1980.
Luchaire, Julien. *Le désarmement moral*. Paris: Librairie Valois, 1932.
Mackenzie, Alexander J. *Propaganda Boom*. London: J. Gifford, 1938.
MacMillan, Margaret. *Peacemakers: The Paris Conference of 1919 and its Attempt to End War*. London: John Murray, 2001.
Maisonneuve, Sophie. "Le disque et la musique classique en Europe, 1877-1949: L'invention d'un médium musical, entre mutations d'écoute et formation d'un patrimoine". Ph.D. diss., European University Institute, Department of History & Civilization, 2002.
Malmborg, Mikael af, and Bo Stråth. "Introduction: The National Meanings of Europe." In *The Meaning of Europe: Variety and Contention within and among Nations*, edited by Mikael af Malmborg and Bo Stråth, 1-26. Oxford; New York: Berg, 2002.
Martin, Marc. "Publicité et programmes radiophoniques dans la France d'avant-guerre." In *Histoire des programmes, histoire des jeux à la radio et à la télévision*, edited by Michèle de Bussierre, Caroline Mauriat, and Cécile Méadel, 119-130. Paris: GEHRA, 1986.
Matelski, Marilyn J. *Vatican Radio: Propagation by the Airwaves*. Media and Society Series. Westport, Conn.: Praeger, 1995.
Mattelart, Armand. *Networking the World, 1794-2000*. Minneapolis: University of Minnesota Press, 2000.
Mayntz, Renate, and Thomas Hughes. *The Development of Large Technical Systems*. Frankfurt am Main: Campus, 1988.
McChesney, Robert W. *Telecommunications, Mass Media, and Democracy: The Battle for the Control of U.S. Broadcasting, 1928-1935*. New York: Oxford University Press, 1993.
McMahon, Richard. "Cultures of Communication: New Historical Perspectives." *European Review of History: Revue européenne d'histoire* 16, no. 1 (2009): 1-14.
Méadel, Cécile. *Histoire de la radio des années trente: Du sans-filiste à l'auditeur*. Paris: Anthropos; Diffusion Economica, 1994.
Miazek, Ryszard. *Przeminelo z radiem: opowiesc o Zygmuncie Chamcu – zalozycielu i pierwszym dyrektorze Polskiego Radia*. Rzeszów; Warszawa: Wydawnictwo Wyzszej Szkoly Informatyki i Zarzadzania; Polskie Radio, 2005.
Mikkeli, Heikki. *Europe as an Idea and an Identity*. New York: St. Martin's Press, 1998.
Misa, Thomas, and Johan Schot. "Inventing Europe: Technology and the Hidden Integration of Europe. Introduction." *History and Technology* 21, no. 1 (2005): 1-22.
Moores, Shaun. "'The Box on the Dresser': Memories of Early Radio and Everyday Life." *Media, Culture and Society* 10 (1988): 23-40.
Müller, Michael G., and Cornelius Torp. "Conceptualising Transnational Spaces in History." *European Review of History: Revue européenne d'histoire* 16, no. 5 (2009): 609-617.
Münzberg, Edler von. "Weingartner, (Paul) Felix." *Grove Music Online. Oxford Music Online*, n.d. http://www.oxfordmusiconline.com:80/subscriber/article/grove/music/30042.
Murphy, Craig. *International Organization and Industrial Change: Global Governance Since 1850*. New York: Oxford University Press, 1994.
Nelson, Michael. *War of the Black Heavens: The Battles of Western Broadcasting in the Cold War*. London; Washington: Brassey's, 1997.

Nichols, Richard. *Radio Luxembourg, the Station of the Stars: An Affectionate History of 50 Years of Broadcasting*. London: W.H. Allen, 1983.

Odermatt, Peter. "The Use of Symbols in the Drive for European Integration." In *Yearbook of European Studies: 4: National Identity*, edited by Joep Th. Leerssen and Menno Spiering, 217-240. Amsterdam; Atlanta, GA: Rodopi, 1991.

Pagden, Anthony, ed. *The Idea of Europe from Antiquity to the European Union*. Cambridge: Cambridge University Press, 2001.

Paulu, Burton. *Radio and Television Broadcasting in Eastern Europe*. Minneapolis: University of Minnesota Press, 1974.

Pegg, Carl. *Evolution of the European Idea, 1914-1932*. Chapel Hill: University of North Carolina Press, 1983.

Pemberton, Jo-Anne. "New Worlds for Old: The League of Nations in the Age of Electricity." *Review of International Studies* 28, no. 2 (2002): 311-336.

Pession, Giuseppe. *Guglielmo Marconi con figure nel testo e 16 tavole in rotocalco*. I Grandi italiani: Collana di biografie diretta da Luigi Federzoni 15. Torino: Unione tipografico-editrice torinese, 1941.

Pohle, Heinz. *Der Rundfunk als Instrument der Politik: Zur Geschichte des deutschen Rundfunk von 1923/38*. Hamburg: Verlag Hans Bredow-Institut, 1955.

Pons, Silvio. *Stalin and the Inevitable War, 1936-1941*. London: Frank Cass, 2002.

Ræsted, Arnold. *Europe and the Atlantic World*. Oslo: I kommisjon hos Aschehoug, 1958.

Reith, John. *Broadcast over Britain*. London: Hodder and Stoughton Ltd., 1924.

Renoliet, Jean-Jaques. *L'Unesco oubliée: La Société des nations et la coopération intellectuelle, 1919-1946*. Paris: Publications de la Sorbonne, 1999.

Ripert, George. *Hommage au professeur V.V. Pella: Promoteur de l'unification du droit criminel, artisan insigne de l'organisation d'une justice pénale internationale*. Paris: s.n., 1939.

Roshwald, Aviel, and Richard Stites, eds. *European Culture in the Great War: The Arts, Entertainment, and Propaganda, 1914-1918*. Cambridge: Cambridge University Press, 1999.

Rougemont, Denis de. *The Idea of Europe*. New York: Macmillan, 1966.

Rupp, Leila. *Worlds of Women: The Making of an International Women's Movement*. Princeton, N.J.: Princeton University Press, 1997.

Saunier, Pierre-Yves. "Learning by Doing: Notes about the Making of the Palgrave Dictionary of Transnational History." *Journal of Modern Economic History* 6, no. 2 (2008): 159-179.

Scannell, Paddy, and David Cardiff. "Serving the Nation: Public Service Broadcasting before the War." In *Popular Culture: Past and Present*, edited by Bernard Waites, Tony Bennett, and Graham Martin, 161-188. London: Croom Helm London, 1982.

Schipper, Frank. *Driving Europe: Building Europe on Roads in the Twentieth Century*. Amsterdam: Aksant, 2008.

Schipper, Frank, Vincent Lagendijk, and Irene Anastasiadou. "New Connections for an Old Continent: Rail, Road and Electricity in the League of Nations Organisation for Communications and Transit." In *Materializing Europe: Transnational Infrastructures and the Project of Europe*, edited by Alexander Badenoch and Andreas Fickers, 113-143. London: Palgrave, 2010.

Schlögl, Reinhard. *Oskar Czeija: Radio- und Fernsehpionier, Unternehmer, Abenteurer*. Wien: Böhlau, 2005.

Schot, Johan."Transnational Infrastructures and the Rise of Contemporary Europe". Working document no. 1, www.tie-project.nl, 2003.

—. "Transnational Infrastructures and the Origins of European Integration." In *Materializing Europe: Transnational Infrastructures and the Project of Europe*, edited by Alexander Badenoch and Andreas Fickers, 82-109. London: Palgrave, 2010.

Schot, Johan, and Vincent Lagendijk. "Technocratic Internationalism in the Interwar Years: Building Europe on Motorways and Electricity Networks." *Journal of Modern European History* 6, no. 2 (2008): 196-217.

Schot, Johan, and Frank Schipper. "Experts and European Transport Integration, 1945-1958." *Journal of European Public Policy* 18, no. 2 (March 1, 2011): 274-293.

Schröter, Harm. *Americanization of the European Economy: A Compact Survey of American Economic Influence in Europe Since the 1880s*. Boston, MA: Springer, 2005.

Schueler, Judith. *Materialising Identity: The Co-Construction of the Gotthard Railway and Swiss National Identity*. Amsterdam: Aksant, 2008.

Schwipps, Werner. "Deutschland im Weltrundfunkverein 1933-1938." *Rundfunk und Fernsehen* 19, no. 4 (1971): 429-441.

Schwoch, James. *The American Radio Industry and its Latin American Activities, 1900-1939*. Urbana: University of Illinois Press, 1990.

Searchinger, César. *Hello America! Radio Adventures in Europe*. Cambridge, Mass.: Houghton Mifflin, 1938.

Shore, Chris. *Building Europe: The Cultural Politics of European Integration*. London; New York: Routledge, 2000.

Silva, Ana Paula. "Shaping the Portuguese Empire in the Twentieth Century: The Telegraph and the Radio." *ICON Journal of the International Committee for the History of Technology* 7 (2001): 106-122.

Smith, Anthony D. "National Identity and the Idea of European Unity." *International Affairs* 68, no. 1 (1992): 55-76.

Smulyan, Susan. *Selling Radio: The Commercialization of American Broadcasting, 1920-1934*. Washington: Smithsonian Institution Press, 1994.

Spiering, Menno, and Michael Wintle, eds. *Ideas of Europe Since 1914: The legacy of the First World War*. New York: Palgrave, 2002.

Spohrer, Jennifer. "Ruling the Airwaves: Radio Luxembourg and the Origins of European National Broadcasting, 1929-1950". Ph.D. diss., Columbia University, 2008.

——. "Paradise or Perdition? European Commercial Broadcasters and American Radio, 1920-1960". Amsterdam, 2009.

Steffen, Katrin, and Martin Kohlrausch. "The Limits and Merits of Internationalism: Experts, the State and the International Community in Poland in the First Half of the Twentieth Century." *European Review of History: Revue européenne d'histoire* 16, no. 5 (2009): 715-737.

Steiner, Kilian J.L. "Buchbesprechung – Oskar Czeija: Radio und Fernsehpionier, Unternehmer, Abenteurer." *Zeitschrift für Unternehmensgeschichte* 52, no. 1 (2007): 120.

Steiner, Zara S. *The Lights that Failed: European International History, 1919-1933*. Oxford; New York: Oxford University Press, 2005.

Stråth, Bo. "Introduction: Europe as a Discourse." In *Europe and the Other and Europe as the Other*, edited by Bo Stråth, 7-42. Brussels; New York: P.I.E.-P. Lang, 2000.

Stray, Christopher, ed. *Gilbert Murray Reassessed: Hellenism, Theatre, and International Politics*. Oxford: Oxford University Press, 2007.

Street, Seán. *Crossing the Ether: British Public Service Radio and Commercial Competition, 1922-1945*. Eastleigh: John Libbey Publishing, 2006.

Ther, Philipp. *In der Mitte der Gesellschaft: Operntheater in Zentraleuropa 1915-1914*. Wien; München: Oldenbourg, 2006.

——. "The Genre of National Opera in a European Comparative Perspective." In *The Oxford Handbook to the New Cultural History of Music*, edited by Jane Fulcher, 182-208 Oxford: Oxford University Press, 2011.

Théry, Franck. *Construire l'Europe dans les années vingt: L'action de l'Union paneuropéenne sur la scène franco-allemande, 1924-1932*. Euryopa. Genève: Institut européen de l'Université de Genève, 1998.

Tomeš a kolektiv, J. *Český biografický slovník XX. století*. Vol. 3. Praha: Litomyšl; Paseka; P. Meissner, 1999.
Tomlinson, John D. "The International Control of Radiocommunications". Ph.D. diss., Université de Genève, 1938.
Vleuten, Erik van der. "Infrastructures and Societal Change: A View from the Large Technical System Field." *Technology Analysis & Strategic Management* 16, no. 3 (2004): 395-414.
—. "Understanding Network Societies: Two Decades of Large Technical System Studies." In *Networking Europe: Transnational Infrastructures and the Shaping of Europe, 1850-2000*, edited by Erik van der Vleuten and Arne Kaijser, 279-314. Sagamore Beach: Science History Publications/USA, 2006.
—. "Toward a Transnational History of Technology: Meanings, Promises, Pitfalls." *Technology and Culture* 49, no. 4 (2008): 974-994.
Vleuten, Erik van der, Irene Anastasiadou, Vincent Lagendijk, and Frank Schipper. "Europe's System Builders: The Contested Shaping of Transnational Road, Electricity and Rail Networks." *Contemporary European History* 16, no. 3 (2007): 321-347.
Vleuten, Erik van der, and Arne Kaijser. "Prologue and Introduction: Transnational Networks and the Shaping of Contemporary Europe." In *Networking Europe: Transnational Infrastructures and the Shaping of Europe, 1850-2000*, edited by Erik Van der Vleuten and Arne Kaijser, 1-22. Sagamore Beach: Science History Publications/USA, 2006.
Waddington, Lorna L. "The Anti-Komintern and Nazi Anti-Bolshevik Propaganda in the 1930s." *Journal of Contemporary History* 42, no. 4 (2007): 573-594.
Waever, Ole. "Europe Since 1945: Crisis to Renewal." In *The History of the Idea of Europe*, edited by Jan van der Dussen and Kevin Wilson, 151-214. London: Routledge, 1993.
Walters, Francis Paul. *A History of the League of Nations*. London: Oxford University Press, 1952.
Webster, Andrew. "The Transnational Dream: Politicians, Diplomats and Soldiers in the League of Nations' Pursuit of International Disarmament, 1920-1938." *Contemporary European History* 14, no. 4 (2005): 493-518.
White, Harry, and Michael Murphy. *Musical Constructions of Nationalism: Essays on the History and Ideology of European Musical Culture, 1800-1945*. Cork: Cork University Press, 2001.
Wiener, Antje, and Thomas Diez, eds. *European Integration Theory*. Oxford; New York: Oxford University Press, 2004.
Williams, Trevor I. *A History of Technology: Vol VII The Twentieth Century c. 1900 to c. 1950 Part II*. Oxford: Clarendon Press, 1978.
Wilson, Kevin, and Jan van der Dussen, eds. *The History of the Idea of Europe*. London: Routledge, 1993.
Wintle, Michael. "Introduction: Cultural Diversity and Identity in Europe." In *Culture and Identity in Europe: Perceptions of Divergence and Unity in Past and Present*, edited by Michael Wintle, 1-8. Aldershot; Brookfield [Vt.] USA: Avebury, 1996.
Wit, Onno de. "Radio tussen Verzuiling en Individualisering." In *Techniek in Nederland in de Twintigste Eeuw: Deel V Transport en Communicatie*, edited by Johan Schot and Onno de Wit, 203-229. Eindhoven: Stichting Historie der Techniek, 2002.
Wood, James. *History of International Broadcasting*. London: Peregrinus, 1992.
Wormbs, Nina. "Standardising Early Broadcasting in Europe: A Form of Regulation." In *Bargaining Norms – Arguing Standards: Negotiating Technical Standards*, edited by Judith Schueler, Andreas Fickers, and Anique Hommels, 112-120. The Hague: Stichting Toekomstbeeld der Techniek, 2008.
—. "'A Simple Arithmetic Task': Technopolitics in Early Frequency Planning for Broadcasting in Europe". Karlstad, 2009.

——. "Invisible with Global Reach: Radio Spectrum for Broadcasting in Europe". Sofia, Bulgaria, 2010.

——. "Technology-Dependent Commons: The Example of Frequency Spectrum for Broadcasting in Europe in the 1920s." *International Journal of the Commons* (submitted).

Zaidi, Waqar. "Technology and the Reconstruction of International Relations: Liberal Internationalist Proposals for the Internationalisaton of Aviation and the International Control of Atomic Energy in Britain, USA and France, 1920-1950". Ph.D. diss., University of London (Imperial College), 2008.

Zeller, Rüdiger. *Die EBU – Union Européene de Radio-Télévision (UER) – European Broadcasting Union (EBU): Internationale Rundfunkkooperation im Wandel.* Baden-Baden: Nomos Verlagsgesellschaft, 1999.

Summary

In the eyes of the ruling international elites, the Great War had been driven by a rising nationalism that left Europe's civilization in shambles. The elites argued that a new modernism, combined with a world that was increasingly technologically interconnected, was to blame for the ruins, anguish, and hatred that dominated peoples' minds after 1918. These intellectual elites tried to rebuild Europe's civilization and create awareness beyond national boundaries. In this context, broadcasting experts in Europe built the first radio broadcasting infrastructures inside and beyond their borders. This book examines if, how, and why the promoters of broadcasting linked their activities in the interwar period to projects that aimed to unite Europe.

The book describes five cross-border issues that the emerging transnational community of broadcasting experts worked to resolve. These issues concerned the institutionalization of broadcasting; the construction of networks; the interactions between broadcasting systems such as Radio Moscow, Radio Nations, Vatican Radio, and Radio Luxembourg; the role of broadcasting within a broader international context of warfare and peace building; and international programming efforts. The book is mainly based on research in the archives of international organizations such as the International Broadcasting Union, the League of Nations, and the International Telegraph Union. These organizations could all be seen as European system builders. They functioned as arenas in which various actors simultaneously negotiated the futures of both transnational broadcasting and Europe.

Chapter 2 examines the "birth of an idea." Broadcasting was originally a private activity whereby states usually granted concessions to companies that would then operate (and sometimes construct) the stations. Several of the operational and construction problems mentioned above could not be resolved without a certain amount of international collaboration. However, international organizations such as the League of Nations and the International Telegraph Union (ITU) did not face such problems. They argued that the rapidly developing technology outpaced legislative action. As a result, ten European broadcasting organizations established the International Broadcasting Union (IBU) in 1925. The IBU's institutional structure followed the European tradition of combining a technical approach with the ideals of international peace and rapprochement. A wave of nationalizing broadcasting in Europe challenged this structure. Eventually, shared European traditions, combined with the central role of a small core group of experts in the IBU, ensured minimal state interference. The IBU was able to continue to seek cross-border collaboration and became the key player for interwar broadcasting in Europe.

Chapter 3 deals with the pressing problem of network construction. Radio signals traveled through the air freely, which caused interference with signals across borders and major inefficiencies in the network. The IBU, at first alone, but later in collaboration with national PTT administrations, the League of Nations' Communication and Transit Committee, the ITU and its consultative committee for long-distance telephony, drafted plans to allocate frequencies in Europe. These activities secured efficient national wireless broadcasting. Furthermore, the gradually growing transnational community of broadcasting experts complemented these wireless systems with an interconnected European relay network for broadcasting via wire and cable. This network, coupled to the national wireless networks, facilitated the exchange of music and broadcasting of international programs throughout Europe. The construction processes were contested along the way. Throughout the 1930s, the development of short waves challenged the recently established European frequency allocation standards. Attempts to standardize these short waves resulted in an unplanned global regionalization of broadcasting.

Chapter 4 challenges the IBU and ITU network efforts by focusing on their interaction with other structures, such as Radio Moscow, Radio Nations, Vatican Radio, and Radio Luxembourg. These stations, with the exception of Radio Moscow, were expanding at a time when the IBU and ITU efforts were well on their way to becoming the European standard. The ideas of the new stations' promoters, which were mostly globally-oriented, did not necessarily coincide with the IBU's vision of a Europe with national wireless broadcasting systems. Fine-tuning these standards with the structure of Radio Moscow, for instance, would redefine the eastern boundaries of the European network. Furthermore, the structure of Radio Nations and Radio Luxembourg threatened the very idea of a Europe made up of nation states. These systems favored a pan-national approach to the organization and network construction of broadcasting. International fine-tuning of these different systems in Europe could usually only take place via "*technified*" discussions that bridged ideological and political differences.

Chapter 5 looks more closely at the role of broadcasting in the context of war and peace. Illicit propaganda broadcasting created serious problems on a continent with as many states as Europe and in an era of rapidly changing international relations. Any attempt to solve this problem depended on how well international relations were progressing. In a reluctant international atmosphere, the IBU initially took a gentleman's approach, requesting that its members broadcast in a civilized way without offending people in other countries. When international relations improved in the late 1920s, the League recognized the idea of positive propaganda. The subsequent close collaboration between the League and the IBU endorsed an international mindset based on European values of civilization and Enlightenment. The collaboration came to a halt when the rise in harsh nationalism caused international relations to deteriorate in the 1930s, thereby disrupting a well-oiled European broadcasting system. Nazi Germany forced the IBU to cease any activity related to power politics. The IBU officially dropped out of propaganda regulation, but a core of IBU experts continued to individually facilitate the League on this matter. Solutions came too late – a new world war broke out in 1939.

Chapter 6 explores international programming. Programs and music touched the heart of culture and had a direct impact on feelings of belonging and identity. Any effort to standardize transnational cultural broadcasting policy, build international programs, and compile a musical repertoire suitable for broadcasting became an intricate matter. IBU experts and intellectual elites, both within and outside the League of Nations, disagreed substantially on what constituted a "high quality," "high culture," or "suitable" program. Consequently, many programming issues remained unresolved in 1939. Over the years, the various promoters of broadcasting expressed different concepts of the kind of unity that international programs should convey to listeners. They related programming more directly to the creation of European unity than their other broadcasting activities. Ultimately, their programs communicated a European culture that reflected a universal idea of national diversity. The promoters contributed a relatively safe, generally accepted, high-quality, and high-art image of European culture to Europe's cultural heritage in the interwar years. They balanced the local and national diversity of Europe with the international unity of Europe.

Europe – On Air concludes that interwar promoters of broadcasting did indeed connect their activities to projects for European unification. To most of these promoters, broadcasting was a matter of practical internationalism. In the first instance, this meant resolving practical problems from a technical standpoint. The promoters formed a cross-organizational and flexible transnational expert community that could adapt to the problems at hand. Diffuse personal networks offered a way out when official routes failed. Secondly, practical internationalism meant that the promoters' ultimate goal was to contribute to the internationalist ideals of peace and rapprochement worldwide. With the exception of Radio Luxembourg, systems such as Radio Moscow, Radio Nations and Vatican Radio had global aims, albeit for different reasons. As always, Europe was shimmering somewhere in the background, whether entering into construction efforts and interests for a practical in-between solution, as a geographical space, as a tradition of commerce and organization, or as an explicit goal for cultural unification and civilization.

Opinions vary regarding the success or failure of these interwar efforts. On one hand, the promoters of broadcasting created suitable international institutions, material networks, a great variety of "international" programs, and managed to let their vision trickle down into European program guides. On the other hand, increasing tension during the 1930s and the outbreak of the Second World War hindered the implementation and effective employment of their efforts. Many issues remained unresolved. *Europe – On Air* argues that the efforts of these promoters were not in vain. They have managed to create a European space, a community, and a kind of European regulatory culture for broadcasting. Equally important, their efforts have found a new start in post-Second World War organizations like the European Broadcasting Union and the United Nations Educational, Scientific, and Cultural Organization. In this way, interwar projects have influenced the broadcasting agenda right through to today.

Foundation for the History of Technology & Aksant Academic Publishers
Technology and European History Series
Ruth Oldenziel and Johan Schot
(Eindhoven University of Technology)
Series Editors

The Technology and European History series seeks to present scholarship about the role of technology in European history in the nineteenth and twentieth centuries. The series focuses on how technical communities, nation-states, businesses, social groups, and other actors have contested, projected, performed, and reproduced multiple representations of Europe while constructing and using a range of technologies. The series understands Europe both as an intellectual construct and material practice in relation to spaces inside as well as outside Europe. In particular, the series invites studies focusing on Europe's (former) colonies and on the two new superpowers of the twentieth century: the United States of America and the Soviet Union. Interdisciplinary work is welcomed. The series will offer a platform for scholarly works associated with the Tensions of Europe Network to find their way to a broader audience. For more information on the network and the series see: www.tensionsofeurope.eu

Books in series

1. Judith Schueler, *Materialising identity. The co-construction of the Gotthard Railway and Swiss national identity* (Amsterdam, June 2008)
2. Vincent Lagendijk, *Electrifying Europe. The power of Europe in the construction of electricity networks* (Amsterdam, August 2008)
3. Frank Schipper, *Driving Europe. Building Europe on roads in the twentieth century* (Amsterdam, September 2008)
4. Adri Albert de la Bruhèze and Ruth Oldenziel (editors), *Manufacturing Technology, Manufacturing Consumers. The Making of Dutch Consumer Society* (Amsterdam, January 2009)
5. Irene Anastasiadou, *Constructing Iron Europe. Transnationalism and Railways in the Interbellum* (Amsterdam, March 2012).
6. Milena Veenis, *Material Fantasies. Expectations of the Western Consumer World among East Germans* (Amsterdam, March 2012)
7. Suzanne Lommers, *Europe – On Air, Interwar Projects for Radio Broadcasting* (Amsterdam, April 2012)

Foundation for the History of Technology

The Foundation for the History of Technology (SHT) aims to develop and communicate knowledge that increases our understanding of the critical role of technology in the history of the Western world. Since 1988 the foundation has been supporting scholarly research in the history of technology. This has included large-scale national and international research programs and numerous individual projects, many in collaboration with Eindhoven University of Technology. SHT also coordinates the international research network Tensions of Europe: Technology and the Making of Europe. For more information see: www.histech.nl